This book comes with access to more content online.

Watch videos, take practice tests,
and study with flashcards!

Register your book or ebook at
www.dummies.com/go/getaccess

Select your product, and then follow the prompts
to validate your purchase.

You'll receive an email with your PIN and instructions.

1001

Practice Problems

Statistics

A Wiley Brand

by The Experts
at Dummies

Statistics: 1001 Practice Problems For Dummies®

Published by: **John Wiley & Sons, Inc.**, 111 River Street, Hoboken, NJ 07030-5774, www.wiley.com

Copyright © 2022 by John Wiley & Sons, Inc., Hoboken, New Jersey

Published simultaneously in Canada

For general information on our other products and services, please contact our Customer Care Department within the U.S. at 877-762-2974, outside the U.S. at 317-572-3993, or fax 317-572-4002. For technical support, please visit https://hub.wiley.com/community/support/dummies.

Wiley publishes in a variety of print and electronic formats and by print-on-demand. Some material included with standard print versions of this book may not be included in e-books or in print-on-demand. If this book refers to media such as a CD or DVD that is not included in the version you purchased, you may download this material at http://booksupport.wiley.com. For more information about Wiley products, visit www.wiley.com.

Library of Congress Control Number: 2022934804

ISBN 978-1-119-88359-3 (pbk); ISBN 978-1-119-88360-9 (ebk); ISBN 978-1-119-88361-6 (ebk)

SKY10034115_040722

Contents at a Glance

Table of Contents

Introduction

One thousand and one practice problems for statistics! That's probably more than a professor would assign you in one semester (we hope!). And it's more than you'd ever want to tackle in one sitting (and we don't recommend you try). So why so many practice problems, and why this book?

Many textbooks are pretty thin on exercises, and even those that do contain a fair number of problems can't focus on all aspects of each topic. With so many problems available in this book, you get to choose how many problems you want to work on. And the way these problems are organized helps you quickly find and dig into problems on particular topics you need to study at the time. Whether you're into the normal distribution, hypothesis tests, the slope of a regression line, or histograms, it's all here and easy to find.

Then there's the entertainment factor. What better way to draw a crowd than to invite people over for a statistics practice problems marathon!

What You'll Find

This book contains 1,001 statistics problems divided into 17 chapters, organized by the major statistical topics in a first-semester introductory course. The problems basically take on three levels:

>> **Statistical literacy:** Understanding the basic concepts of the topic, including terms and notation

>> **Reasoning:** Applying the ideas within a context

>> **Thinking:** Putting ideas and concepts together to solve more difficult problems

In addition to providing plenty of problems to work on in each chapter, this book also provides worked-out solutions with detailed explanations, so you aren't left high and dry if you get a wrong answer. So you can rest assured that when you work for 30 minutes on a problem, get an answer of 1.25, and go to the back of the book to see that the correct answer is actually 1,218.31, you'll find a detailed explanation to help you figure out what went wrong with your calculations.

How This Workbook Is Organized

This book is divided into two main parts: the questions and the answers.

Part 1: The Questions

The questions in this book center on the following areas:

>> **Descriptive statistics and graphs:** After you collect and review data, your first job is to make sense of it. The way to do that is two-fold: (1) Organize the data in a visual manner so you can see it and (2) crank out some numbers that describe it in a basic way.

>> **Random variables:** A *random variable* is a characteristic of interest that varies in a random fashion. Each type of random variable has its own pattern in which the data falls (or is expected to fall), along with its own mean and standard deviation for the data. The pattern of a random variable is called its *distribution*.

The random variables in this book include the binomial, the normal (or *Z*), and the *t*. For each random variable, you practice identifying its characteristics, seeing what its pattern (distribution) looks like, determining its mean and standard deviation, and, most commonly, finding probabilities and percentiles for it.

>> **Inference:** This term can seem complex (and word on the street says it is), but inference basically just means taking the information from your data (your sample) and using it to draw conclusions about the group you're interested in (your population).

The two basic types of statistical inferences are confidence intervals and hypothesis testing:

- You use *confidence intervals* when you want to make an estimate regarding the population — for example, "What percentage of all kindergarteners in the United States are obese?"

- You use a *hypothesis test* when someone has a supposed value regarding the population, and you're putting it to the test. For example, a researcher claims that 14 percent of today's kindergarteners are obese, but you question whether it's really that high.

>> The underpinnings needed for both types of inference are margin of error, standard error, sampling distributions, and the central limit theorem. They all play a major role in statistics and can be somewhat complex, so make sure you spend time on these elements as a backdrop for confidence intervals and hypothesis tests.

>> **Relationships:** One of the most important and common uses of statistics is to look for relationships between two random variables. If variables are categorical (such as gender), you explore relationships by using two-way tables containing rows and columns, and you examine relationships by looking at and comparing percentages among and within groups. If both variables are numerical, you explore relationships graphically by using scatter plots, quantify them by using correlation, and use them to make predictions (one variable predicting the other) by using regression. Studying relationships helps you get at the essence of how statistics is applied in the real world.

>> **Surveys:** Before you analyze data in all the ways mentioned in this list, you have to collect the data. Surveys are one of the most common means of data collection; the main ideas of surveys to address with practice are planning a survey, selecting a representative sample of individuals to survey, and carrying out the survey properly. The main goal in all of these areas is to avoid *bias* (systematic favoritism). Many types of bias exist, and in this book, you practice identifying and seeing ways to minimize them.

Part 2: The Answers

This part provides detailed answers to every question in this book. You see how to set up and work through each problem and how to interpret the answer.

Beyond the Book

In addition to what you're reading right now, this book comes with a free, access-anywhere Cheat Sheet that includes tips and other goodies you may want to have at your fingertips. To get this Cheat Sheet, simply go to www.dummies.com and type **Statistics 1001 Dummies Cheat Sheet** into the Search box.

The online practice that comes free with this book offers you the same 1,001 questions and answers that are available here, presented in a multiple-choice format. The beauty of the online problems is that you can customize your online practice to focus on the topic areas that give you trouble. If you're short on time and want to maximize your study, you can specify the quantity of problems you want to practice, pick your topics, and go. You can practice a few hundred problems in one sitting or just a couple dozen, and whether you can focus on a few types of problems or a mix of several types. Regardless of the combination you create, the online program keeps track of the questions you get right and wrong so you can monitor your progress and spend time studying exactly what you need.

To gain access to the online practice, you simply have to register. Just follow these steps:

1. **Register your book or ebook at Dummies.com to get your PIN. Go to** www.dummies.com/go/getaccess.

2. **Select your product from the dropdown list on that page.**

3. **Follow the prompts to validate your product, and then check your email for a confirmation message that includes your PIN and instructions for logging in.**

If you don't receive this email within two hours, please check your spam folder before contacting us through our Technical Support website at http://support.wiley.com or by phone at 877-762-2974.

Now you're ready to go! You can come back to the practice material as often as you want — simply log in with the username and password you created during your initial login. No need to enter the access code a second time.

Your registration is good for one year from the day you activate your PIN.

What you'll find online

The online practice that comes free with this book offers you the same 1,001 questions and answers that are available here, presented in a multiple-choice format. Multiple-choice questions force you to zoom in on the details that can make or break your correct solution to the problem. Sometimes one of the possible wrong answers will catch you in the act of making a certain error. But that's great because after you identify a particular error (often a common error that many others make as well), you'll know not to fall into that trap again.

The beauty of the online problems is that you can customize your online practice — that is, you can select the types of problems and the number of problems you want to work on. The online program keeps track of your performance so you can focus on the areas where you need the biggest boost.

You can access this online tool by using a PIN code, as described in the next section. Keep in mind that you can create only one login with your PIN. Once the PIN is used, it's no longer valid and is nontransferable. So you can't share your PIN with others after you've established your login credentials. In other words, the problems are yours and only yours!

How to register

To gain access to the online practice, all you have to do is register. Just follow these simple steps:

1. **Register your book or ebook at Dummies.com to get your PIN. Go to** www.dummies.com/go/getaccess.

2. **Select your product from the dropdown list on that page.**

3. **Follow the prompts to validate your product, and then check your email for a confirmation message that includes your PIN and instructions for logging in.**

If you do not receive this email within two hours, please check your spam folder before contacting us through our Technical Support website at http://support.wiley.com or by phone at 877-762-2974.

Now you're ready to go! You can come back to the practice material as often as you want — simply log on with the username and password you created during your initial login. There is no need to enter the access code a second time.

Your registration is good for one year from the day you activate your PIN.

Where to Go for Additional Help

The written solutions for the problems in this book are designed to show you what you need to do to get the correct answer to those particular problems. Although a bit of background information is injected at times, the solutions aren't meant to teach the material outright. Solutions to the problems on a given topic contain the normal statistical language, symbols, and formulas that are inherent to the topic, with the assumption that you're familiar with them.

If you're ever confused about why a problem is done a certain way, or you want more info to fill in between the lines, or you just feel like you need to go back and refresh your memory on some of the topics, several *For Dummies* books are available as a reference, including *Statistics For Dummies*, *Statistics Essentials For Dummies*, and *Statistics Workbook For Dummies*, all written by Deborah J. Rumsey, PhD, and published by Wiley.

1

The Questions

Statistics can give anyone problems. Terms, notation, formulas — where do you start? You start by practicing problems that hone the right skills. This book gives you practice — 1,001 problems worth of practice, to be exact. Working problems like these helps you figure out what you do and don't understand about setting up, working out, and interpreting your answers to statistics problems. Here's the breakdown in a nutshell:

Warm up with statistical vocabulary, descriptive statistics, and graphs (Chapters 1 through 3).

Work with random variables, including the binomial, normal, and t-distributions (Chapters 4 through 6).

Decipher sampling distributions and margin of error and build confidence intervals for one- and two-population means and proportions (Chapters 7 through 10).

Master the general concepts of hypothesis testing and perform tests for one- and two-population means and proportions (Chapters 11 through 13).

Get behind the scenes on collecting good data and spotting bad data in surveys (Chapter 14).

Explore relationships between two quantitative variables, using correlation and simple linear regression (Chapters 15 and 16).

Look for relationships between two categorical variables, using two-way tables and independence (Chapter 17).

Chapter **1**

Basic Vocabulary

Everything's got its own lingo, and statistics is no exception. The trick is to get a handle on the lingo right from the get-go, so when it comes time to work problems, you'll pick up on cues from the wording and get going in the right direction. You can also use the terms to search quickly in the table of contents or the index of this book to find the problems you need to dive into in a flash. It's like with anything else: As soon as you understand what the language means, you immediately start feeling more comfortable.

The Problems You'll Work On

In this chapter, you get a bird's-eye view of some of the most common terms used in statistics and, perhaps more importantly, the context in which they're used. Here's an overview:

>> The big four: population, sample, parameter, and statistic

>> The statistics terms you'll calculate, such as the mean, median, standard deviation, z-score, and percentile

>> Types of data, graphs, and distributions

>> Data analysis terms, such as confidence intervals, margin of error, and hypothesis tests

What to Watch Out For

Pay particular attention to the following:

>> Pick out the big four in every situation; they'll follow you wherever you go.

>> Really get the idea of a distribution; it's one of the most confusing ideas in statistics, yet it's used over and over — so nail it now to avoid getting hammered later.

>> Focus not only on the terms for the statistics and analyses you'll calculate but also on their interpretation, especially in the context of a problem.

Picking Out the Population, Sample, Parameter, and Statistic

1–4 You're interested in knowing what percent of all households in a large city have a single woman as the head of the household. To estimate this percentage, you conduct a survey with 200 households and determine how many of these 200 are headed by a single woman.

1. In this example, what is the population?

2. In this example, what is the sample?

3. In this example, what is the parameter?

4. In this example, what is the statistic?

Distinguishing Quantitative and Categorical Variables

5–6 Answer the problems about quantitative and categorical variables.

5. Which of the following is an example of a quantitative variable (also known as a numerical variable)?

(A) the color of an automobile

(B) a person's state of residence

(C) a person's zip code

(D) a person's height, recorded in inches

(E) Choices (C) and (D)

6. Which of the following is an example of a categorical variable (also known as a qualitative variable)?

(A) years of schooling completed

(B) college major

(C) high-school graduate or not

(D) annual income (in dollars)

(E) Choices (B) and (C)

Getting a Handle on Bias, Variables, and the Mean

7–11 You're interested in the percentage of female versus male shoppers at a department store. So one Saturday morning, you place data collectors at each of the store's four entrances for three hours, and you have them record how many men and women enter the store during that time.

7. Why can collecting data at the store on one Saturday morning for three hours cause bias in the data?

(A) It assumes that Saturday shoppers represent the whole population of people who shop at the store during the week.

(B) It assumes that the same percentage of female shoppers shop on Saturday mornings as any other time or day of the week.

(C) Perhaps couples are more likely to shop together on Saturday mornings than during the rest of the week, bringing the percentage of males and females closer than during other times of the week.

(D) The subjects in the study weren't selected at random.

(E) All of these choices are true.

8. Because a variable is a characteristic of each individual on which data is collected, which of the following are variables in this study?

(A) the day you chose to collect data

(B) the store you chose to observe

(C) the gender of each shopper who comes in during the time period

(D) the number of men entering the store during the time period

(E) Choices (C) and (D)

9. In this study, _____ is a categorical variable, and _____ is a quantitative variable.

10. Which chart or graph would be appropriate to display the proportion of males versus females among the shoppers?

(A) a bar graph

(B) a time plot

(C) a pie chart

(D) Choices (A) and (C)

(E) Choices (A), (B), and (C)

11. How would you calculate the mean number of shoppers per hour?

Understanding Different Statistics and Data Analysis Terms

12–17 Answer the problems about different statistics and data analysis terms.

12. Which of the following data sets has a median of 3?

(A) 3, 3, 3, 3, 3

(B) 2, 5, 3, 1, 1

(C) 1, 2, 3, 4, 5

(D) 1, 2, 4, 4, 4

(E) Choices (A) and (C)

13. Susan scores at the 90th percentile on a math exam. What does this mean?

14. You took a survey of 100 people and found that 60% of them like chocolate and 40% don't. Which of the following gives the distribution of the "chocolate versus no chocolate" variable?

(A) a table of the results

(B) a pie chart of the results

(C) a bar graph of the results

(D) a sentence describing the results

(E) all of the above

15. Suppose that the results of an exam tell you your z-score is 0.70. What does this tell you about how well you did on the exam?

16. A national poll reports that 65% of Americans sampled approve of the president, with a margin of error of 6 percentage points. What does this mean?

17. If you want to estimate the percentage of all Americans who plan to vacation for two weeks or more this summer, what statistical technique should you use to find a range of plausible values for the true percentage?

Using Statistical Techniques

18–19 You read a report that 60% of high-school graduates participated in sports during their high-school years.

18. You believe that the percentage of high-school graduates who played sports is higher than what was reported. What type of statistical technique do you use to see whether you're right?

19. You believe that the percentage of high-school graduates who played sports in high school is higher than what's in the report. If you do a hypothesis test to challenge the report, which of these p-values would you be happiest to get?

(A) $p = 0.95$

(B) $p = 0.50$

(C) $p = 1$

(D) $p = 0.05$

(E) $p = 0.001$

Working with the Standard Deviation

20 Solve the problem about standard deviation.

20. Which data set has the highest standard deviation (without doing calculations)?

(A) 1, 2, 3, 4

(B) 1, 1, 1, 4

(C) 1, 1, 4, 4

(D) 4, 4, 4, 4

(E) 1, 2, 2, 4

Chapter 2

Descriptive Statistics

Descriptive statistics are statistics that describe data. You've got the staple ingredients, such as the mean, median, and standard deviation, and then the concepts and graphs that build on them, such as percentiles, the five-number summary, and the box plot. Your first job in analyzing data is to identify, understand, and calculate these descriptive statistics. Then you need to interpret the results, which means to see and describe their importance in the context of the problem.

The Problems You'll Work On

The problems in this chapter focus on the following big ideas:

>> Calculating, interpreting, and comparing basic statistics, such as mean and median, and standard deviation and variance

>> Using the mean and standard deviation to give ranges for bell-shaped data

>> Measuring where a certain value stands in a data set by using percentiles

>> Creating a set of five numbers (using percentiles) that can reveal some aspects of the shape, center, and variation in a data set

What to Watch Out For

Pay particular attention to the following:

>> Be sure you identify which descriptive statistic or set of descriptive statistics is needed for a particular problem.

>> After you understand the terminology and calculations for these descriptive statistics, step back and look at the results — make comparisons, see if they make sense, and find the story they tell.

>> Remember that a percentile isn't a percent, even though they sound the same! When used together, remember that a percentile is a cutoff value in the data set, while a percentage is the amount of data that lies below that cutoff value.

>> Be aware of the units of any descriptive statistic you calculate (for example, dollars, feet, or miles per gallon). Some descriptive statistics are in the same units as the data, and some aren't.

Understanding the Mean and the Median

21–32 Solve the following problems about means and medians.

21. To the nearest tenth, what is the mean of the following data set? 14, 14, 15, 16, 28, 28, 32, 35, 37, 38

22. To the nearest tenth, what is the mean of the following data set? 15, 25, 35, 45, 50, 60, 70, 72, 100

23. To the nearest tenth, what is the mean of the following data set? 0.8, 1.8, 2.3, 4.5, 4.8, 16.1, 22.3

24. To the nearest thousandth, what is the mean of the following data set? 0.003, 0.045, 0.58, 0.687, 1.25, 10.38, 11.252, 12.001

25. To the nearest tenth, what is the median of the following data set? 6, 12, 22, 18, 16, 4, 20, 5, 15

26. To the nearest tenth, what is the median of the following data set? 18, 21, 17, 18, 16, 15.5, 12, 17, 10, 21, 17

27. To the nearest tenth, what is the median of the following data set? 14, 2, 21, 7, 30, 10, 1, 15, 6, 8

28. To the nearest hundredth, what is the median of the following data set? 25.2, 0.25, 8.2, 1.22, 0.001, 0.1, 6.85, 13.2

29. Compare the mean and median of a data set that has a distribution that is skewed right.

30. Compare the mean and the median of a data set that has a distribution that is skewed left.

31. Compare the mean and the median of a data set that has a symmetrical distribution.

32. Which measure of center is most resistant to (or least affected by) outliers?

Surveying Standard Deviation and Variance

33–48 *Solve the following problems about standard deviation and variance.*

33. What does the standard deviation measure?

34. According to the 68-95-99.7 rule, or the empirical rule, if a data set has a normal distribution, approximately what percentage of data will be within one standard deviation of the mean?

35. A realtor tells you that the average cost of houses in a town is $176,000. You want to know how much the prices of the houses may vary from this average. What measurement do you need?

(A) standard deviation

(B) interquartile range

(C) variance

(D) percentile

(E) Choice (A) or (C)

36. What measure(s) of variation is/are sensitive to outliers?

(A) margin of error

(B) interquartile range

(C) standard deviation

(D) Choices (A) and (B)

(E) Choices (A) and (C)

37. You take a random sample of ten car owners and ask them, "To the nearest year, how old is your current car?" Their responses are as follows: 0 years, 1 year, 2 years, 4 years, 8 years, 3 years, 10 years, 17 years, 2 years, 7 years. To the nearest year, what is the standard deviation of this sample?

38. A sample is taken of the ages in years of 12 people who attend a movie. The results are as follows: 12 years, 10 years, 16 years, 22 years, 24 years, 18 years, 30 years, 32 years, 19 years, 20 years, 35 years, 26 years. To the nearest year, what is the standard deviation for this sample?

39. A large math class takes a midterm exam worth a total of 100 points. Following is a random sample of 20 students' scores from the class:

Score of 98 points: 2 students

Score of 95 points: 1 student

Score of 92 points: 3 students

Score of 88 points: 4 students

Score of 87 points: 2 students

Score of 85 points: 2 students

Score of 81 points: 1 student

Score of 78 points: 2 students

Score of 73 points: 1 student

Score of 72 points: 1 student

Score of 65 points: 1 student

To the nearest tenth of a point, what is the standard deviation of the exam scores for the students in this sample?

40. A manufacturer of jet engines measures a turbine part to the nearest 0.001 centimeters. A sample of parts has the following data set: 5.001, 5.002, 5.005, 5.000, 5.010, 5.009, 5.003, 5.002, 5.001, 5.000. What is the standard deviation for this sample?

41. Two companies pay their employees the same average salary of $42,000 per year. The salary data in Ace Corp. has a standard deviation of $10,000, whereas Magna Company salary data has a standard deviation of $30,000. What, if anything, does this mean?

42. In which of the following situations would having a small standard deviation be most important?

(A) determining the variation in the wealth of retired people

(B) measuring the variation in circuitry components when manufacturing computer chips

(C) comparing the population of cities in different areas of the country

(D) comparing the amount of time it takes to complete education courses on the Internet

(E) measuring the variation in the production of different varieties of apple trees

43. Suppose that you're comparing the means and standard deviations for the daily high temperatures for two cities during the months of November through March.

$$\text{Sunshine City: } \mu = 46\degree\text{F}; \sigma = 18\degree\text{F}$$

$$\text{Lake Town: } \mu = 42\degree\text{F}; \sigma = 8\degree\text{F}$$

What's the best analysis for comparing the temperatures in the two cities?

44. Everyone at a company is given a year-end bonus of $2,000. How will this affect the standard deviation of the annual salaries in the company that year?

45. Calculate the sample variance and the standard deviation for the following measurements of weights of apples: 7 oz, 6 oz, 5 oz, 6 oz, 9 oz. Express your answers in the proper units of measurement and round to the nearest tenth.

46. Calculate the sample variance and the standard deviation for the following measurements of assembly time required to build an MP3 player: 15 min, 16 min, 18 min, 10 min, 9 min. Express your answers in the proper units of measurement and round to the nearest whole number.

47. Calculate the standard deviation for these speeds of city traffic: 10 km/hr, 15 km/hr, 35 km/hr, 40 km/hr, 30 km/hr. Express your answers in the proper units of measurement and round to the nearest whole number.

48. Which of the following data sets has the same standard deviation as the data set with the numbers 1, 2, 3, 4, 5? (Do this problem without any calculations!)

(A) Data Set 1: 6, 7, 8, 9, 10

(B) Data Set 2: −2, −1, 0, 1, 2

(C) Data Set 3: 0.1, 0.2, 0.3, 0.4, 0.5

(D) Choices (A) and (B)

(E) None of the data sets gives the same standard deviation as the data set 1, 2, 3, 4, 5.

Employing the Empirical Rule

49–56 Use the empirical rule to solve the following problems.

49. According to the empirical rule (or the 68-95-99.7 rule), if a population has a normal distribution, approximately what percentage of values is within one standard deviation of the mean?

50. According to the empirical rule (or the 68-95-99.7 rule), if a population has a normal distribution, approximately what percentage of values is within two standard deviations of the mean?

51. If the average age of retirement for the entire population in a country is 64 years and the distribution is normal with a standard deviation of 3.5 years, what is the approximate age range in which 95% of people retire?

52. Last year's graduates from an engineering college, who entered jobs as engineers, had a mean first-year income of $48,000 with a standard deviation of $7,000. The distribution of salary levels is normal. What is the approximate percentage of first-year engineers that made more than $55,000?

53. What is a necessary condition for using the empirical rule (or 68-95-99.7 rule)?

54. What measures of data need to be known to use the empirical (68-95-99.7) rule?

55. The quality control specialists of a microscope manufacturing company test the lens for every microscope to make sure the dimensions are correct. In one month, 600 lenses are tested. The mean thickness is 2 millimeters. The standard deviation is 0.000025 millimeters. The distribution is normal. The company rejects any lens that is more than two standard deviations from the mean. Approximately how many lenses out of the 600 would be rejected?

56. Biologists gather data on a sample of fish in a large lake. They capture, measure the length of, and release 1,000 fish. They find that the standard deviation is 5 centimeters, and the mean is 25 centimeters. They also notice that the shape of the distribution (according to a histogram) is very much skewed to the left (which means that some fish are smaller than most of the others). Approximately what percentage of fish in the lake is likely to have a length within one standard deviation of the mean?

Measuring Relative Standing with Percentiles

57–64 Solve the following problems about percentiles.

57. What statistic reports the relative standing of a value in a set of data?

58. What is the statistical name for the 50th percentile?

59. Your score on a test is at the 85th percentile. What does this mean?

60. Suppose that in a class of 60 students, the final exam scores have an approximately normal distribution, with a mean of 70 points and a standard deviation of 5 points. One student's score places them in the 90th percentile among students on this exam. What must be true about this student's score?

61. On a multiple-choice test, your actual score was 82%, which was reported to be at the 70th percentile. What is the meaning of your test results?

62. Seven students got the following exam scores (percent correct) on a science exam: 0%, 40%, 50%, 65%, 75%, 90%, 100%. Which of these exam scores is at the 50th percentile?

63. Students scored the following grades on a statistics test: 80, 80, 82, 84, 85, 86, 88, 90, 91, 92, 92, 94, 96, 98, 100. Calculate the score that represents the 80th percentile.

64. Some of the students in a class are comparing their grades on a recent test. The first student says they almost scored in the 95th percentile. The second student says they scored at the 84th percentile. The third student says they scored at the 88th percentile. The fourth student says they almost scored in the 70th percentile. The fifth student says they scored at the 95th percentile. Rank the five students from highest to lowest in their grades.

Delving into Data Sets and Descriptive Statistics

65–80 Solve the following problems about data sets and descriptive statistics.

65. Which of the following descriptive statistics is least affected by adding an outlier to a data set?

(A) the mean

(B) the median

(C) the range

(D) the standard deviation

(E) all of the above

66. Which of the following statements is incorrect?

(A) The median and the 1st quartile can be the same.

(B) The maximum and minimum value can be the same.

(C) The 1st and 3rd quartiles can be the same.

(D) The range and the IQR can be the same.

(E) None of the above.

67. Test scores for an English class are recorded as follows: 72, 74, 75, 77, 79, 82, 83, 87, 88, 90, 91, 91, 91, 92, 96, 97, 97, 98, 100. Find the 1st quartile, median, and 3rd quartile for the data set.

68. The average annual returns over the past ten years for 20 utility stocks have the following statistics:

1st quartile = 7

Median = 8

3rd quartile = 9

Mean = 8.5

Standard deviation = 2

Range = 5

Give the five numbers that make up the five-number summary for this data set.

69. Bob attempts to calculate the five-number summary for a set of exam scores. His results are as follows:

Minimum = 30

Maximum = 90

1st quartile = 50

3rd quartile = 80

Median = 85

What is wrong with Bob's five-number summary?

70. Which of the following data sets has a mean of 15 and standard deviation of 0?

(A) 0, 15, 30

(B) 15, 15, 15

(C) 0, 0, 0

(D) There is no data set with a standard deviation of 0.

(E) Choices (B) and (C)

71. The starting salaries (in dollars) of a random sample of 125 university graduates were analyzed. The following descriptive statistics were calculated and typed into a report:

Mean: 24,329

Median: 20,461

Variance: 4,683,459

Minimum: 18,958

Q_1: 22,663

Q_3: 29,155

Maximum: 31,123

What is the error in these descriptive statistics?

72. Which of the following statements is true?

(A) Fifty percent of the values in a data set lie between the 1st and 3rd quartiles.

(B) Fifty percent of the values in a data set lie between the median and the maximum value.

(C) Fifty percent of the values in a data set lie between the median and the minimum value.

(D) Fifty percent of the values in a data set lie at or below the median.

(E) All of the above.

73. Which of the following relationships holds true?

(A) The mean is always greater than the median.

(B) The variance is always larger than the standard deviation.

(C) The range is always less than the IQR.

(D) The IQR is always less than the standard deviation.

(E) None of the above.

74. Suppose that a data set contains the weights of a random sample of 100 newborn babies in units of pounds. Which of the following descriptive statistics isn't measured in pounds?

(A) the mean of the weights

(B) the standard deviation of the weights

(C) the variance of the weights

(D) the median of the weights

(E) the range of the weights

75. Which of the following is not a measure of the spread (variability) in a data set?

(A) the range

(B) the standard deviation

(C) the IQR

(D) the variance

(E) none of the above

76. A data set contains five numbers with a mean of 3 and a standard deviation of 1. Which of the following data sets matches those criteria?

(A) 1, 2, 3, 4, 5

(B) 3, 3, 3, 3, 3

(C) 2, 2, 3, 4, 4

(D) 1, 1, 1, 1, 1

(E) 0, 0, 3, 6, 6

77. A supermarket surveyed customers one week to see how often each customer shopped at the store every month. The data is shown in the following graph. What are the best measures of spread and center for this distribution?

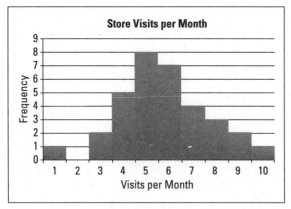

© John Wiley & Sons, Inc.

78. Students took a test that had 20 questions. The following graph shows the distribution of the scores. What are the best measures of spread and center for the data?

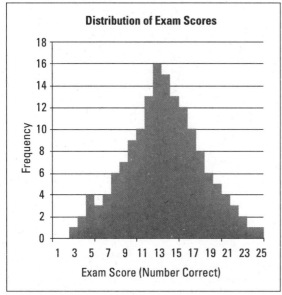

© John Wiley & Sons, Inc.

79. An Internet company sells computer parts and accessories. The annual salaries for all the employees have the following parameters:

Mean: $78,000

Median: $45,000

Standard deviation: $40,800

IQR (interquartile range): $12,000

Range: $24,000 to $2 million

What are the best measures of spread and center for the data?

80. The distribution of scores for a final exam in math had the following parameters:

Mean: 83%

Median: 94%

Standard deviation: 7%

IQR (interquartile range): 9%

Range: 65% to 100%

What are the best measures of spread and center for the data?

Chapter **3**

Graphing

G raphs should be able to stand alone and give all the information needed to identify the main point quickly and easily. The media gives the impression that making and interpreting graphs is no big deal. However, in statistics, you work with more complicated data, consequently taking your graphs up a notch.

The Problems You'll Work On

A good graph displays data in a way that's fair, makes sense, and makes a point. Not all graphs possess these qualities. When working the problems in this chapter, you get experience with the following:

>> Identifying the graph that's needed for the particular situation at hand

>> Graphing both categorical (qualitative) data and numerical (quantitative) data

>> Putting together and correctly interpreting histograms

>> Highlighting data collected over time, using a time plot

>> Spotting and identifying problems with misleading graphs

What to Watch Out For

Some graphs are easy to make and interpret, some are hard to make but easy to interpret, and some graphs are tricky to make and even trickier to interpret. Be ready to handle the latter.

>> Be sure to understand the circumstances under which each type of graph is to be used and how to construct it. (Rarely will someone actually tell you what type of graph to make!)

>> Pay special attention to how a histogram shows the variability in a data set. Flat histograms can have a lot of variability in the data, but flat time plots have none — that's one eye-opener.

>> Box plots are a huge issue. Making a box plot itself is one thing; understanding the do's and (especially) the don'ts of interpreting box plots is a whole other story.

Interpreting Pie Charts

81–86 The following pie chart shows the proportion of students enrolled in different colleges within a university.

Enrollment by College

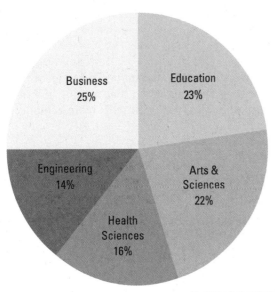

Illustration by Ryan Sneed

81. Which college has the largest enrollment?

82. If some students were enrolled in more than one college, what type of graph would be appropriate to show the percentage in each college?

    (A) the same pie chart

    (B) a separate pie chart for each college showing what percentage are enrolled and what percentage aren't

    (C) a bar graph where each bar represents a college and the height shows what percentage of students are enrolled

    (D) Choices (B) and (C)

    (E) none of the above

83. What percentage of students is enrolled in either the College of Education or the College of Health Sciences?

84. What percentage of students is not enrolled in the College of Engineering?

85. How many students are enrolled in the College of Health Sciences?

86. If 25,000 students are enrolled in the university, how many students are in the College of Arts & Sciences?

Considering Three-Dimensional Pie Charts

87 Answer the following problem about three-dimensional pie charts.

87. What characteristic of three-dimensional pie charts (also known as "exploding" pie charts) makes them misleading?

Interpreting Bar Charts

88–94 *The following bar chart represents the post-graduation plans of the graduating seniors from one high school. Assume that every student chose one of these five options. (**Note:** A gap year means that the student is taking a year off before deciding what to do.)*

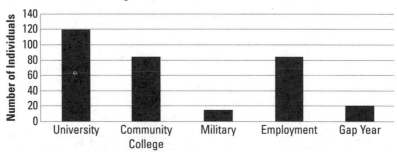

High-School Seniors' Post-Graduation Plans

© John Wiley & Sons, Inc.

88. What is the most common post-graduation plan for these seniors?

89. What is the least common post-graduation plan for these seniors?

90. Assuming that each student has chosen only one of the five possibilities, about how many students plan to either take a gap year or attend a university?

91. How many total students are represented in this chart?

92. What percentage of the graduating class is planning on attending a community college?

93. What percentage of the graduating class is not planning to attend a university?

94. This bar chart displays the same information but is more difficult to interpret. Why is this the case?

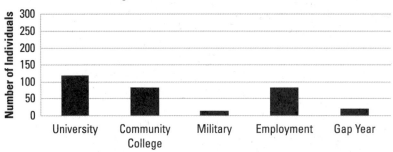

High-School Seniors' Post-Graduation Plans

© John Wiley & Sons, Inc.

Introducing Other Graphs

95–96 Solve the following problems about different types of graphs.

95. What type of graph would be the best choice to display data representing the height in centimeters of 1,000 high-school football players?

96. Is the order of bars significant in a histogram?

Interpreting Histograms

97–105 The following histogram represents the body mass index (BMI) of a sample of 101 U.S. adults.

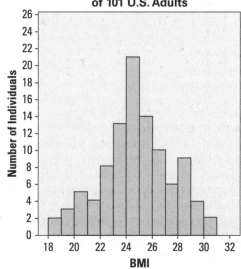

Body Mass Index (BMI) for a Sample of 101 U.S. Adults

© John Wiley & Sons, Inc.

97. Why are there no gaps between the bars of this histogram?

98. What does the *x*-axis of this histogram represent?

99. What do the widths of the bars represent?

100. What does the *y*-axis represent?

101. How would you describe the basic shape of this distribution?

102. What is the range of the data in this histogram?

103. Judging by this histogram, which bar contains the average value for this data (considering "average value" as similar to a balancing point)?

104. How many adults in this sample have a BMI in the range of 22 to 24?

105. What percentage of adults in this sample have a BMI of 28 or higher?

Digging Deeper into Histograms

106–112 *The following histogram represents the reported income from a sample of 110 U.S. adults.*

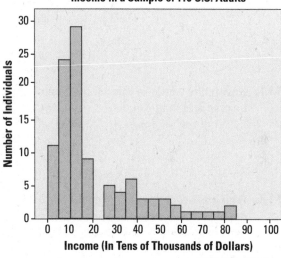

Income in a Sample of 110 U.S. Adults

Income (In Tens of Thousands of Dollars)

© *John Wiley & Sons, Inc.*

106. How would you describe the shape of this distribution?

107. What would be the most appropriate measure of the center for this data?

108. Which value will be higher in this distribution, the mean or the median?

109. What is the lowest possible value in this data?

110. What is the highest possible value in this data?

111. How many adults in this sample reported an income less than $10,000?

112. Which bar contains the median for this data? (Denote the bar by using its left end-point and its right endpoint.)

Comparing Histograms

113–119 *The following histograms represent the grades on a common final exam from two different sections of the same university calculus class.*

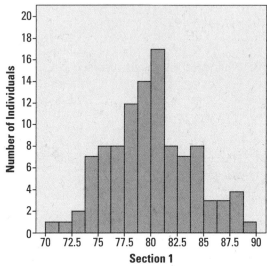

Illustration by Ryan Sneed

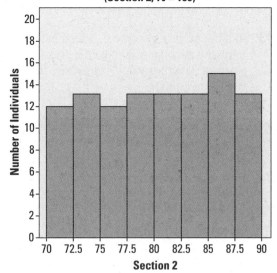

Illustration by Ryan Sneed

113. How would you describe the distributions of grades in these two sections?

114. Which section's grade distribution has the greater range?

115. How do you expect the mean and median of the grades in Section 1 to compare to each other?

116. How do you expect the mean and median of the grades in Section 2 to compare to each other?

117. Judging by the histogram, what is the best estimate for the median of Section 1's grades?

118. Judging by the histogram, which interval most likely contains the median of Section 2's grades?

(A) below 75

(B) 75 to 77.5

(C) 77.5 to 82.5

(D) 85 to 90

(E) above 90

119. Which section's grade distribution do you expect to have a greater standard deviation, and why?

Describing the Center of a Distribution

120 Solve the following problem about the center of a distribution.

120. For the 2013 to 2014 season, salaries for the 450 players in the NBA ranged from slightly less than $1 million to more than $30 million, with 19 players making more than $15 million and about half making $2 million or less. Which would be the best statistic to describe the center of this distribution?

Interpreting Box Plots

121–128 The following box plot represents data on the GPA of 500 students at a high school.

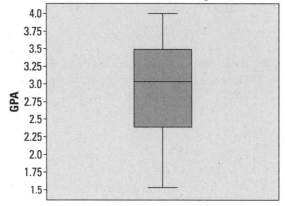

GPA for 500 Students in a High School

Illustration by Ryan Sneed

121. What is the range of GPAs in this data?

122. What is the median of the GPAs?

123. What is the IQR for this data?

124. What does the scale of the numerical axis signify in this box plot?

125. Where is the mean of this data set?

126. What is the approximate shape of the distribution of this data?

127. What percentage of students has a GPA that lies outside the actual box part of the box plot?

128. What percentage of students has a GPA below the median in this data?

Comparing Two Box Plots

129–133 The following box plots represent GPAs of students from two different colleges; call them College 1 and College 2.

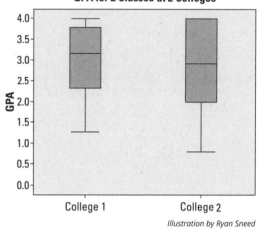

GPA for 2 Classes at 2 Colleges

Illustration by Ryan Sneed

129. What information is missing on this graph and on the box plots?

(A) the total sample size

(B) the number of students in each college

(C) the mean of each data set

(D) Choices (A) and (B)

(E) Choices (A), (B), and (C)

130. Which data set has a greater median, College 1 or College 2?

131. Which data set has the greater IQR, College 1 or College 2?

132. Which data set has a larger sample size?

133. Which data set has a higher percentage of GPAs above its median?

Comparing Three Box Plots

134–139 These side-by-side box plots represent home sale prices (in thousands of dollars) in three cities in 2012.

Home Sale Prices in Three Cities in 2012

Illustration by Ryan Sneed

134. From high to low, what is the order of the cities' median home sale prices?

135. If the number of homes sold in each city is the same, which city has the most homes that sold for more than $72,000?

136. Assuming 100 homes sold in each city in 2012, which city has the most homes that sold for more than $72,000?

137. Which city has the smallest range in home prices?

138. Which of the following statements is true?

(A) More than half of the homes in City 1 sold for more than $50,000.

(B) More than half of the homes in City 2 sold for more than $75,000.

(C) More than half of the homes in City 3 sold for more than $75,000.

(D) Choices (A) and (B).

(E) Choices (B) and (C).

139. Which of the following statements is true?

(A) About 25% of homes in City 1 sold for $75,000 or more.

(B) About 25% of homes in City 2 sold for $75,000 or more.

(C) About 25% of homes in City 2 sold for $98,000 or more.

(D) About 25% of homes in City 3 sold for $75,000 or more.

(E) Choices (A) and (C).

Interpreting Time Charts

140–146 *The data in the following time chart shows the annual high-school dropout rate for a school system for the years 2001 to 2011.*

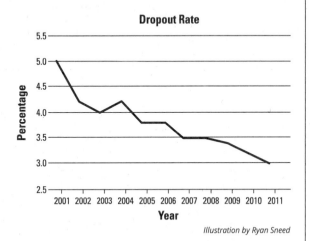

Dropout Rate

Illustration by Ryan Sneed

140. What is the general pattern in the dropout rate from 2001 to 2011?

141. What was the approximate dropout rate in 2005?

142. What was the approximate change in the dropout rate from 2001 to 2011?

143. What was the approximate change in the dropout rate from 2003 to 2004?

144. This time chart displays the same data, but why is it misleading?

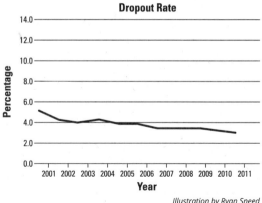

Dropout Rate

Illustration by Ryan Sneed

145. Why do the numbers on this plot represent dropout rates instead of the number of dropouts?

146. This time chart displays the same data, but why is it misleading?

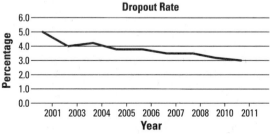

Dropout Rate

Illustration by Ryan Sneed

Getting More Practice with Histograms

147–148 *The following three histograms represent reported annual incomes, in thousands of dollars, from samples of 100 individuals from three professions; call the different incomes Income 1, Income 2, and Income 3.*

147. How would you describe the approximate shape of these distributions (Income 1, Income 2, and Income 3)?

148. All three samples have the same range, from $35,000 to $65,000, but they differ in variability. Put the incomes of these three professions in order in terms of their variability, from largest to smallest, using their graphs.

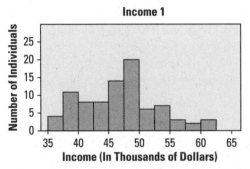

Income 1

Illustration by Ryan Sneed

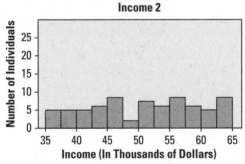

Income 2

Illustration by Ryan Sneed

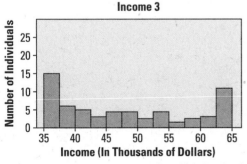

Income 3

Illustration by Ryan Sneed

Chapter 4

Random Variables and the Binomial Distribution

Random variables represent quantities or qualities that randomly change within a population. For example, if you ask random people what their stress level is on a scale from 0 to 10, you don't know what they're going to say. But you do know what the possible values are, and you may have an idea of which numbers are likely to be reported more often (like 9 or 10) and less often (like 0 or 1). In this chapter, you focus on random variables: their types, their possible values and probabilities, their means, their standard deviations, and other characteristics.

The Problems You'll Work On

In this chapter, you see random variables in action and how you can use them to think about a population. Here are some items on the menu:

>> Distinguishing discrete versus continuous random variables

>> Finding probabilities for a random variable

>> Calculating and interpreting the mean, variance, and standard deviation of a random variable

>> Finding probabilities, mean, and standard deviation for a specific random variable, the binomial

What to Watch Out For

The problems in this chapter involve notation, formulas, and calculations. Paying attention to the details will make a difference.

>> Understand the notation really well; several symbols are floating around in this chapter.

>> Be able to interpret your results, not just do the calculations, including the proper use of units.

>> Know the ways to find binomial probabilities; pay special attention to the normal approximation.

Comparing Discrete and Continuous Random Variables

149–154 Solve the following problems about discrete and continuous random variables.

149. Which of the following random variables is discrete?

(A) the length of time a battery lasts

(B) the weight of an adult

(C) the percentage of children in a population who have been vaccinated against measles

(D) the number of books purchased by a student in a year

(E) the distance between a pair of cities

150. Which of the following random variables isn't discrete?

(A) the number of children in a family

(B) the annual rainfall in a city

(C) the attendance at a football game

(D) the number of patients treated at an emergency room in a day

(E) the number of classes taken in one semester by a student

151. Which of the following random variables is discrete?

(A) the proportion of a population that voted in the last election

(B) the height of a college student

(C) the number of cars registered in a state

(D) the weight of flour in a sack advertised as containing ten pounds

(E) the length of a phone call

152. Which of the following random variables is continuous?

(A) the number of heads resulting from flipping a coin 30 times

(B) the number of deaths from plane crashes in a year

(C) the proportion of the American population that believes in ghosts

(D) the number of films produced in Canada in a year

(E) the number of people arrested for auto theft in a year

153. Which of the following random variables is continuous?

(A) the number of seniors in a college

(B) the number of gold medals won at the 2012 Summer Olympics by athletes from Germany

(C) the number of schools in a city

(D) the number of registered physicians in the United States

(E) the amount of gasoline used in the Unites States in 2012

154. Which of the following random variables isn't continuous?

(A) the proportion of adults on probation in a state

(B) the population growth rate for a city

(C) the amount of money spent by a household for food over a year

(D) the number of bird species observed in an area

(E) the length of time it takes to walk ten miles

Understanding the Probability Distribution of a Random Variable

155–157 *The following table represents the probability distribution for X, the employment status of adults in a city.*

X	P(X)
Employed full-time	0.65
Employed part-time	0.10
Unemployed	0.07
Retired	0.18

155. If you select one adult at random from this community, what is the probability that the individual is employed part-time?

156. If you select one adult at random from this community, what is the probability that the individual isn't retired?

157. If you select one adult at random from this community, what is the probability that the individual is working either part-time or full-time?

Determining the Mean of a Discrete Random Variable

158–159 Let X be the number of classes taken by a college student in a semester. Use the formula for the mean of a discrete random variable X to answer the following problems:

$$\mu_x = \sum x_i p_i$$

158. If 40% of all the students are taking four classes, and 60% of all the students are taking three classes, what is the mean (average) number of classes taken for this group of students?

159. If half of the students in a class are age 18, one-quarter are age 19, and one-quarter are age 20, what is the average age of the students in the class?

Digging Deeper into the Mean of a Discrete Random Variable

160–163 In the following table, X represents the number of automobiles owned by families in a neighborhood.

X	P(X)
0	0.25
1	0.60
2	
3	0.05

160. What is the missing value in this table (representing the number of automobiles owned by two families in a neighborhood)?

161. What is the mean number of automobiles owned?

162. If every family currently not owning a car bought one car, what would be the mean number of automobiles owned?

163. If all the families currently owning three cars bought a fourth car, what would be the mean number of automobiles owned?

Working with the Variance of a Discrete Random Variable

164–165 Use the following formula for the variance of a discrete random variable X as needed to answer the following problems (round each answer to two decimal places):

$$\sigma_x^2 = \sum (x_i - \mu_x)^2 p_i$$

164. If the variance of a discrete random variable X is 3, what is the standard deviation of X?

165. If the standard deviation of X is 0.65, what is the variance of X?

Putting Together the Mean, Variance, and Standard Deviation of a Random Variable

166–169 In the following table, X represents the number of siblings for the 29 students in a first-grade class.

X	$P(X)$
0	0.34
1	0.52
2	0.14

© John Wiley & Sons, Inc.

166. What is the mean number of siblings for these students?

167. What is the variance for the number of siblings for these students?

168. What is the standard deviation for the number of siblings for these students? Round your answer to two decimal places.

169. How would the variance and standard deviation change if the number of siblings, X, doubled in each case but the probabilities, $p(x)$, values stayed the same?

Digging Deeper into the Mean, Variance, and Standard Deviation of a Random Variable

170–173 In the following table, X represents the number of books required for classes at a university.

X	P(X)
0	0.30
1	0.25
2	0.25
3	0.10
4	0.10

© John Wiley & Sons, Inc.

170. What is the mean number of books required?

171. What is the variance of the number of books required? Round your answer to two decimal places.

172. What is the standard deviation of the number of books required? Round your answer to two decimal places.

173. How would the standard deviation and variance change if only 20% of the students required two books, but now 5% of the students require five books (with all other categories unchanged)?

Introducing Binomial Random Variables

174–178 Solve the following problems about the basics of binomial random variables.

174. What condition(s) must a random variable meet to be considered binomial?

(A) fixed number of trials

(B) exactly two possible outcomes on each trial: success and failure

(C) constant probability of success for all trials

(D) independent trials

(E) all of the above

175. You flip a coin 25 times and record the number of heads. What is the binomial random variable (X) in this experiment?

176. You roll a six-faced die ten times and record which face comes up each time (X). Why is X not a binomial random variable?

177. You interview a number of employees selected at random and ask them whether they've graduated from high school. You continue the interviews until you have 30 employees who say they graduated from high school. If X is the number of people you had to ask until you got 30 "yes" responses, why isn't X a binomial random variable?

178. You recruit 30 pairs of siblings and test each individual to see whether they are carrying a particular genetic mutation, and then you add up the total number of people (not pairs) who have the mutation. Why is this not a binomial experiment?

Figuring Out the Mean, Variance, and Standard Deviation of a Binomial Random Variable

179–183 Solve the following problems about the mean, standard deviation, and variance of binomial random variables.

179. What is the mean of a binomial random variable with $n = 18$ and $p = 0.4$?

180. What is the mean of a binomial random variable with $n = 25$ and $p = 0.35$?

181. What is the standard deviation of a binomial distribution with $n = 18$ and $p = 0.4$? Round your answer to two decimal places.

182. What is the variance of a binomial distribution with $n = 25$ and $p = 0.35$? Round your answer to two decimal places.

183. A binomial distribution with $p = 0.14$ has a mean of 18.2. What is n?

Finding Binomial Probabilities with a Formula

184–188 X is a binomial random variable with p = 0.55. Use the following formulas for the binomial distribution for the following problems.

$$P(X = x) = \binom{n}{x} p^x (1-p)^{n-x}$$

$$where \binom{n}{x} = \frac{n!}{x!(n-x)!} \ and$$

$$n! = (n-1)(n-2)(n-3)...(3)(2)(1)$$

184. What is the value of $\binom{n}{x}$ if $n = 8$ and $x = 1$?

185. What is the probability of exactly one success in eight trials? Round your answer to four decimal places.

186. What is the probability of exactly two successes in eight trials? Round your answer to four decimal places.

187. What is the value of $\begin{pmatrix} 8 \\ 0 \end{pmatrix}$?

188. What is the probability of getting at least one success in eight trials? Round your answer to four decimal places.

Digging Deeper into Binomial Probabilities Using a Formula

189–195 Suppose that X has a binomial distribution with $p = 0.50$.

189. What is the probability of exactly eight successes in ten trials? Round your answer to four decimal places.

190. What is the probability of exactly three successes in five trials?

191. What is the probability of exactly four successes in five trials? Round your answer to four decimal places.

192. What is the value of $\begin{pmatrix} 5 \\ 5 \end{pmatrix}$?

193. What is the probability of three or four successes in five trials? Round your answer to four decimal places.

194. What is the probability of at least three successes in five trials? Round your answer to four decimal places.

195. What is the probability of no more than two successes in five trials? Round your answer to four decimal places.

Finding Binomial Probabilities with the Binomial Table

196–200 X is a random variable with a binomial distribution with $n = 15$ and $p = 0.7$. Use the binomial table (Table A-3 in the appendix) to answer the following problems.

196. What is $P(X = 6)$?

197. What is $P(X=11)$?

198. What is $P(X<15)$?

199. What is $P(4<X<7)$?

200. Find $P(4 \leq X \leq 7)$.

Digging Deeper into Binomial Probabilities Using the Binomial Table

201–205 X is a random variable with a binomial distribution with n = 11 and p = 0.4. Use the binomial table (Table A-3 in the appendix) to answer the following problems.

201. What is $P(X=5)$?

202. What is $P(X>0)$?

203. What is $P(X \leq 2)$?

204. What is $P(X>9)$?

205. What is $P(3 \leq X \leq 5)$?

Using the Normal Approximation to the Binomial

206–208 Solve the following problems about the normal approximation to the binomial.

206. Which of the following binomial distributions would allow you to use the normal approximation to the binomial?

(A) $n=10, p=0.3$

(B) $n=10, p=0.4$

(C) $n=20, p=0.9$

(D) $n=25, p=0.3$

(E) $n=30, p=0.4$

207. You're conducting a binomial experiment by flipping a fair coin ($p=0.5$) and recording how many times it comes up heads. What is the minimum size for *n* that will allow you to use the normal approximation to the binomial?

208. You're conducting a binomial experiment by selecting marbles from a large jar and recording how many times you draw a green marble. Overall, 30% of the marbles in the jar are green. What is the minimum size for *n* that will allow you to use the normal approximation to the binomial?

Digging Deeper into the Normal Approximation to the Binomial

209–214 *You conduct a binomial experiment by flipping a fair coin ($p = 0.5$) 80 times and recording how often it comes up heads (X). Use the normal approximation to the binomial to calculate any probabilities in this problem set. You can do so because the two required conditions are met: np and $n(1-p)$ are both at least 10.*

209. How would you formally write the probability that you get at least 50 heads?

210. How would you formally write the probability that you get no more than 30 heads?

211. To use the normal approximation to the binomial, what value will you use for μ?

212. To use the normal approximation to the binomial, what value will you use for σ? Round your answer to two decimal places.

213. What is the z-value corresponding to the x value in $P(X \geq 45)$? Round your answer to two decimal places.

214. What is the probability that you will get at least 45 heads in 80 flips? Round your answer to four decimal places.

Getting More Practice with Binomial Variables

215–223 *X is a binomial variable with $p = 0.45$ and $n = 100$.*

215. What are the mean and standard deviation for the distribution of X? Round your answer to two decimal places.

216. What is the z-value corresponding to the x value in $P(x \leq 40)$? Round your answer to two decimal places.

217. What is $P(x \leq 40)$? Round your answer to four decimal places.

218. What is $P(X \geq 40)$? Round your answer to four decimal places.

219. Suppose you want to find $P(X = 40)$. You can try to solve this by using a normal approximation and converting $x = 40$ to its z-value of $z = -1.00$. What is $P(Z = -1.00)$? Round your answer to four decimal places.

220. What is the z-value corresponding to $x = 56$? Round your answer to two decimal places.

221. What is $P(X \geq 56)$? Round your answer to four decimal places.

222. What is $P(X \leq 56)$? Round your answer to four decimal places.

223. What is $P(56 \leq x \leq 60)$? Round your answer to four decimal places.

Chapter **5**

The Normal Distribution

The normal distribution is the most common distribution of all. Its values take on that familiar bell shape, with more values near the center and fewer as you move away. Because of its nice qualities and practical nature, you see many applications of the normal distribution not only in this chapter but also in many chapters that follow. Understanding the normal distribution from the ground up is critical.

The Problems You'll Work On

In this chapter, you practice in detail everything you need to know about the normal distribution and then some. Here's an overview of what you'll work on:

>> Finding probabilities for a normal distribution (less than, greater than, or in between)

>> Understanding the standard normal (Z-) distribution, its properties, and how its values are interpreted and used

>> Using the Z-table to find probabilities

>> Determining percentiles for a normal distribution

What to Watch Out For

The normal distribution seems easy, but many important nuances exist as well, so stay on top of them. Here are some things to keep in mind:

>> Be very clear about what a z-score is and what it means. You'll use it throughout this chapter.

>> Make sure you understand how to use the Z-table to find the probabilities you want.

>> Know when you're asked for a percent (probability) and when you're asked for a percentile (a value of the variable affiliated with a certain probability). Then know what to do about it.

Defining and Describing the Normal Distribution

224–234 Solve the following problems about the definition of the normal distribution and what it looks like.

224. What are properties of the normal distribution?

(A) It's symmetrical.

(B) Mean and median are the same.

(C) Most common values are near the mean; less common values are farther from it.

(D) Standard deviation marks the distance from the mean to the inflection point.

(E) All of the above.

225. In a normal distribution, about what percent of the values lie within one standard deviation of the mean?

226. In a normal distribution, about what percent of the values lie within two standard deviations of the mean?

227. In a normal distribution, about what percent of the values lie within three standard deviations of the mean?

228. What two parameters (pieces of information about the population) are needed to describe a normal distribution?

229. Which of the following normal distributions will have the greatest spread when graphed?

(A) $\mu = 5, \sigma = 1.5$

(B) $\mu = 10, \sigma = 1.0$

(C) $\mu = 5, \sigma = 1.75$

(D) $\mu = 5, \sigma = 1.2$

(E) $\mu = 10, \sigma = 1.6$

230. For a normal distribution with $\mu = 5$ and $\sigma = 1.2$, 34% of the values lie between 5 and what number? (Assume that the number is above the mean.)

231. For a normal distribution with $\mu = 5$ and $\sigma = 1.2$, about 2.5% of the values lie above what value? (Assume that the number is above the mean.)

232. For a normal distribution with $\mu = 5$ and $\sigma = 1.2$, about 16% of the values lie below what value?

233. A normal distribution with $\mu = 8$ has 99.7% of its values between 3.5 and 12.5. What is the standard deviation for this distribution?

234. What are the mean and standard deviation of the Z-distribution?

Working with z-Scores and Values of X

235–239 *A random variable X has a normal distribution, with a mean of 17 and a standard deviation of 3.5.*

235. What is the z-score for a value of 21.2?

236. What is the z-score for a value of 13.5?

237. What is the z-score for a value of 25.75?

238. What value of X corresponds to a z-score of −0.4?

239. What value of X corresponds to a z-score of 2.2?

Digging Deeper into z-Scores and Values of X

240–245 *All the scores on a national exam have a normal distribution and a range of 0 to 100, but when split up, Form A of the exam has $\mu = 70$ and $\sigma = 10$, while Form B has $\mu = 74$ and $\sigma = 8$.*

240. What are the scores on each exam corresponding to a z-score of 1.5?

241. What are the scores on each exam corresponding to a z-score of −2.0?

242. In terms of standard deviations above or below the mean, what score on Form B corresponds to a score of 80 on Form A?

243. In terms of standard deviations above or below the mean, what score on Form B corresponds to a score of 85 on Form A?

244. In terms of standard deviations above or below the mean, what score on Form A corresponds to a score of 78 on Form B?

245. In terms of standard deviations above or below the mean, what score on Form A corresponds to a score of 68 on Form B?

Writing Probability Notations

246. Write the probability notation for the area shaded in this Z-distribution (where $\mu = 0$ and $\sigma = 1$).

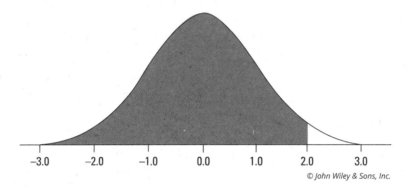

© John Wiley & Sons, Inc.

247. Write the probability notation for the area shaded in this Z-distribution (where $\mu = 0$ and $\sigma = 1$).

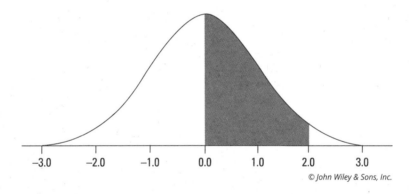

© John Wiley & Sons, Inc.

248. Write the probability notation for the area shaded in this Z-distribution (where $\mu = 0$ and $\sigma = 1$).

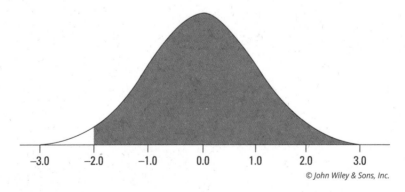

© John Wiley & Sons, Inc.

Introducing the Z-Table

249–253 Use the Z-table (Table A-1 in the appendix) as needed to answer the following problems.

249. What is $P(Z \leq 1.5)$?

250. What is $P(Z \geq 1.5)$?

251. What is $P(Z \geq -0.75)$?

252. What is $P(-0.5 \leq Z \leq 1.0)$?

253. What is $P(-1.0 \leq Z \leq 1.0)$?

Finding Probabilities for a Normal Distribution

254–258 The diameter of a machine part produced by a factory is normally distributed, with a mean of 10 centimeters and a standard deviation of 2 centimeters.

254. What is the z-score for a part with a diameter of 13 centimeters?

255. What is the probability of a part having a diameter of at least 13 centimeters?

256. What is the probability of a part having a diameter no greater than 13 centimeters?

257. What is the probability of a part having a diameter between 10 and 13 centimeters?

258. What is the probability of a part having a diameter between 7 and 10 centimeters?

Digging Deeper into z-Scores and Probabilities

259–270 The weight of adult men in a population is normally distributed, with a mean of 160 pounds and a standard deviation of 20 pounds.

259. What is the z-score for a weight of 135 pounds?

260. What is the z-score for a weight of 170 pounds?

261. What is the z-score for a weight of 115 pounds?

262. What is the z-score for a weight of 220 pounds?

263. What is the z-score for a weight of 205 pounds?

264. What is the probability of a weight greater than 220 pounds?

265. What is the probability of a weight less than 220 pounds?

266. What is the probability of a weight between 135 and 160 pounds?

267. What is the probability of a weight between 205 and 220 pounds?

268. What is the probability of a weight between 115 and 135 pounds?

269. What is the probability of a weight between 135 and 170 pounds?

270. What is the probability of a weight between 170 and 220 pounds?

Figuring Out Percentiles for a Normal Distribution

271–278 In a population of adults ages 18 to 65, BMI (body mass index) is normally distributed with a mean of 27 and a standard deviation of 5.

271. What is the BMI score for which half of the population has a lower value?

272. What BMI marks the bottom 25% of the distribution for this population?

273. What BMI marks the bottom 5% of the distribution for this population?

274. What BMI marks the bottom 10% of the distribution for this population?

275. What BMI value marks the upper 10% of the distribution for this population?

276. What BMI value marks the upper 5% of the distribution for this population?

277. What BMI value marks the upper 30% of the distribution for this population?

278. What two BMI values mark the 1st and 3rd quartiles of the distribution?

Digging Deeper into Percentiles for a Normal Distribution

279–284 Scores on an exam are normally distributed, with a mean of 75 and a standard deviation of 6.

279. What is the value where only 20% of the students scored below it?

280. What is the value where only 5% of the students scored below it?

281. What is the value where only 10% of the students scored above it?

282. What is the value where only 1% of the students scored above it?

283. What is the value where only 2.5% of students scored above it?

284. What is the value where only 5% of the students scored above it?

Getting More Practice with Percentiles

285–290 In a large population of military recruits, the time required to run a mile has a normal distribution with a mean of 360 seconds and a standard deviation of 30 seconds.

285. What time represents the cutoff for the 5% of fastest times?

286. What time represents the cutoff for the fastest 50% of times?

287. What time represents the cutoff for the slowest 10% of times?

288. What time represents the cutoff for the fastest 10% of times?

289. What time represents the cutoff for the fastest 25% of times?

290. What time represents the cutoff for the slowest 25% of times?

Chapter **6**

The *t*-Distribution

The *t*-distribution is a relative of the normal distribution. It has a bell shape with values more spread out around the middle. That is, it's not as sharply curved as the normal distribution, which reflects its ability to work with problems that may not be exactly normal but are close.

The Problems You'll Work On

In this chapter, you'll work on all the ins and outs of the *t*-distribution and its relationship to the standard normal (Z-) distribution. Here's what you'll focus on:

>> Understanding the characteristics that make the *t*-distribution similar to and different from the standard normal (Z-) distribution

>> Finding probabilities for various *t*-distributions, using the *t*-table

>> Finding critical values on the *t*-distribution that are used when calculating confidence intervals

What to Watch Out For

As you work through this chapter, keep the following in mind:

>> Pick out the big four — population, parameter, sample, and statistic — in every situation; they'll follow you wherever you go.

>> Know the circumstances for when you should use the *t*-distribution.

>> The *t*-table (for the *t*-distribution) is different from the Z-table (for the Z-distribution); make sure you understand the values in the first and last rows.

>> Pay special attention to the term *tail probability*, which appears a great deal in this chapter.

Understanding the *t*-Distribution and Comparing It to the Z-Distribution

291–300 *Solve the following problems about the t-distribution, its traits, and how it compares to the Z-distribution.*

291. Which of the following is true of the *t*-distribution, as compared to the Z-distribution? (Assume a low number of degrees of freedom.)

    (A) The *t*-distribution has thicker tails than the Z-distribution.

    (B) The *t*-distribution has a proportionately larger standard deviation than the Z-distribution.

    (C) The *t*-distribution is bell-shaped but has a lower peak than the Z-distribution.

    (D) Choices (A) and (C)

    (E) Choices (A), (B), and (C)

292. Which *t*-distribution do you use for a study involving one population with a sample size of 30?

293. If a particular *t*-distribution from a study involving one population is t_{24}, what is the sample size?

294. If you graphed a standard normal distribution (Z-distribution) on the same number line as a *t*-distribution with 15 degrees of freedom, how would you expect them to differ?

295. If you graphed a standard normal distribution (Z-distribution) and a *t*-distribution with 20 degrees of freedom on the same number line, how would you expect them to differ?

296. Given different *t*-distributions with the following degrees of freedom, which one would you expect to most closely resemble the Z-distribution: 5, 10, 20, 30, or 100?

297. What minimal information do you need to recognize which particular *t*-distribution you have?

298. What statistical procedure would be most appropriate to test a claim about a population mean for a normal distribution when you don't know the population standard deviation?

299. What statistical procedure would be most appropriate for a study testing a claim about the mean of the difference when the subjects (and hence the data) are in pairs?

300. What is the particular *t*-distribution resulting from a paired design involving a total of 50 observations?

Using the *t*-Table

301–335 Use the t-table (Table A-2 in the appendix) as necessary to solve the following problems.

301. For a study involving one population and a sample size of 18 (assuming you have a *t*-distribution), what row of Table A-2 will you use to find the right-tail ("greater than") probability affiliated with the study results?

302. For a study involving a paired design with a total of 44 observations, with the results assuming a *t*-distribution, what row of Table A-2 will you use to find the probability affiliated with the study results?

303. What column of Table A-2 would you use to find an upper-tail ("greater than") probability of 0.025?

304. What column of Table A-2 would you use to find the *t*-value for a two-tailed test with $\alpha = 0.01$?

305. What column of Table A-2 would you use to find the *t*-value for an overall probability of 0.05 for a two-tailed test?

306. For a *t*-distribution with 10 degrees of freedom, what is $P(t \geq 1.81)$?

307. For a *t*-distribution with 25 degrees of freedom, what is $P(t \geq 2.49)$?

308. What is $P(t_{15} \geq 1.34)$?

309. A *t*-value of 1.80, from a *t*-distribution with 22 degrees of freedom, has an upper-tail ("greater than") probability between which two values on Table A-2?

310. A *t*-value of 2.35, from a *t*-distribution with 14 degrees of freedom, has an upper-tail ("greater than") probability between which two values on Table A-2?

311. What is $P(t_{15} \leq -2.60)$?

312. What is $P(t_{27} \leq -2.05)$?

313. What is $P(t_{27} \geq 2.05 \text{ or } t_{27} \leq -2.05)$?

314. What is $P(t_9 \geq 3.25 \text{ or } t_9 \leq -3.25)$?

315. What is the 95th percentile for a t-distribution with 10 degrees of freedom?

316. What is the 60th percentile for a t-distribution with 28 degrees of freedom?

317. What is the 10th percentile for a t-distribution with 20 degrees of freedom?

318. What is the 25th percentile for a t-distribution with 20 degrees of freedom?

319. What value is the 10th percentile for a t-distribution with 16 degrees of freedom?

320. What value is the 5th percentile for a t-distribution with 16 degrees of freedom?

321. Which of the following is a reason to use the t-distribution rather than the Z-distribution to calculate a confidence interval for the mean?

(A) You have a small sample size.

(B) You don't know the population standard deviation.

(C) You don't know the population mean.

(D) Choices (A) and (B)

(E) Choices (A), (B), and (C)

322. Using the last row of Table A-2, what column would you use to find the t-value for a 99% confidence interval?

323. Using the first row (column headings) of Table A-2, what column would you use to find the t-value for a 99% confidence interval?

324. What is the t-value for a 95% confidence interval for a t-distribution with 15 degrees of freedom?

325. What is the t-value for a 99% confidence interval for a t-distribution with 23 degrees of freedom?

326. What is the t-value for a 90% confidence interval for a t-distribution with 30 degrees of freedom?

327. For a t-distribution with 19 degrees of freedom, what is the confidence level of a confidence interval with a t-value of 2.09?

328. For a t-distribution with 26 degrees of freedom, what is the confidence level of a confidence interval with a t-value of 2.78?

329. For a t-distribution with 12 degrees of freedom, what is the confidence level of a confidence interval with a t-value of 1.36?

330. Considering t-distributions with different degrees of freedom, which of the following would you expect to most closely resemble the Z-distribution: 30, 35, 40, 45, or 50?

331. Considering t-distributions with different degrees of freedom, which of the following would you expect to least resemble the Z-distribution: 10, 20, 30, 40, or 50?

332. Considering t-distributions with different degrees of freedom, which of the following would you expect to have the largest t-value for a right-tail probability of 0.05: 10, 20, 30, 40, or 50?

333. Considering t-distributions with different degrees of freedom, which of the following would you expect to have the smallest t-value for a right-tail probability of 0.10: 10, 20, 30, 40, or 50?

334. For a t_{40} distribution, which of the following confidence intervals would you expect to be narrowest: 80%, 90%, 95%, 98%, or 99%?

335. For a t_{50} distribution, which of the following confidence intervals would you expect to be widest: 80%, 90%, 95%, 98%, or 99%?

Using the *t*-Distribution to Calculate Confidence Intervals

336–340 Use Table A-2 in the appendix as needed and the following information to solve the following problems: The mean length for the population of all screws being produced by a certain factory is targeted to be $\mu_0 = 5$ centimeters. Assume that you don't know what the population standard deviation (σ) is. You draw a sample of 30 screws and calculate their mean length. The mean ($\bar{x}$) for your sample is 4.8, and the standard deviation of your sample (s) is 0.4 centimeters.

336. Calculate the appropriate test statistic for the following hypotheses:

$$H_0 : \mu = 5 \text{ cm}$$

$$H_1 : \mu \neq 5 \text{ cm}$$

337. Using Table A-2, what is the probability of getting a t-value of -2.74 or less?

338. What is the 95% confidence interval for the population mean? Round your answer to two decimal places.

339. What is the 90% confidence interval for the population mean? Round your answer to two decimal places.

340. What is the 99% confidence interval for the population mean? Round your answer to two decimal places.

Chapter 7

Sampling Distributions and the Central Limit Theorem

ample results vary — that's a major truth of statistics. You take a random sample of size 100, find the average, and repeat the process over and over with different samples of size 100. Those sample averages will differ, but the question is, by how much? And what affects the amount of difference? Understanding this concept of variability between all possible samples helps determine how typical or atypical your particular result may be. Sampling distributions provide a fundamental piece to answer these problems. One tool that's often needed in the process is the central limit theorem. In this chapter, you work on both.

The Problems You'll Work On

Here are the main areas of focus in this chapter:

» Working with the sample means as a random variable, with their own distribution, mean, and standard deviations (standard errors) and then doing the same with sample proportions

» Understanding the central limit theorem, how to use it, when to use it, and when it's not needed

» Calculating probabilities for $\bar{X}$ and $\hat{p}$

What to Watch Out For

Sampling distributions and the central limit theorem are difficult topics. Here are some main points to keep in mind:

» $\bar{x}$ is the mean of one sample, and $\bar{X}$ represents any possible sample mean from any sample.

» Everything builds on the understanding of what a sampling distribution is — work until you get it.

» The central limit theorem is used only in certain situations — know those situations.

» Finding probabilities for $\bar{X}$ and $\hat{p}$ involve dividing by a quantity involving the square root of n. Don't leave it out.

Introducing the Basics of Sampling Distributions

341–347 Solve the following problems that introduce the basics of sampling distributions.

341. A characteristic of interest that takes on certain values in a random manner is called a _____.

342. Suppose that 10,000 students took the AP statistics exam this year. If you take every possible sample of 100 students who took the AP exam and find the average exam score for each sample and then put all those average scores together, what would it represent?

343. A GPA is the grade point average of a single student. Suppose that you found the GPA for every student in a university and found that the mean of all those GPAs is 3.11. What statistical notation do you use to represent this value of 3.11?

344. If you roll two fair dice, look at the outcomes, and find the average value, you could get any number from 1 (where both your dice came up 1) to 6 (where both dice came up 6). However, in the long run, if you took the average of all possible pairs of dice, you'd get 3.5 (because that's the average value of the numbers 1 through 6). How do you represent 3.5 in this situation using statistical notation?

345. Suppose that you roll several ordinary six-sided dice, choose two of those dice at random, and average the two numbers. Suppose the average of these two dice is 3.5. How would you express the 3.5 in statistical notation?

346. Which of the following features is necessary for something to be considered a sampling distribution?

(A) Each value in the original population should be included in the distribution.

(B) The distribution must consist of proportions.

(C) The distribution can't consist of percentages.

(D) Each of the observations in the distribution must consist of a statistic that describes a collection of data points.

(E) The samples in the distribution must be drawn from a population of normally distributed values.

347. Which of the following would *not* ordinarily be considered a sampling distribution?

(A) a distribution showing the average weight per person in several hundred groups of three people picked at random at a state fair

(B) a distribution showing the average proportion of heads coming up in several thousand experiments in which ten coins were flipped each time

(C) a distribution showing the average percentage daily price change in Dow Jones Industrial Stocks for several hundred days chosen at random from the past 20 years

(D) a distribution showing the proportion of parts found to be deficient in each of several hundred shipments of parts, each of which has the same number of parts in it

(E) a distribution showing the weight of each individual football fan entering a stadium on game day

Checking Out Random Variables and Sample Means

348–352 You have a six-sided die with faces numbered 1 through 6, and each face is equally likely to come up on any given roll.

348. What is the meaning for X in this example?

349. What is the meaning of $\bar{X}$ in this example?

350. Suppose that you rolled the die five times and got the values of 3, 4, 2, 3, and 1. What is the value of $\bar{x}$?

351. Suppose that you rolled the die five times and got the values of 3, 4, 6, 3, and 5. What is the value of $\bar{x}$?

352. Which of the following is the formula for the standard error of a sample mean?

(A) $\sigma = \dfrac{\sigma_{\bar{x}}}{\sqrt{n}}$

(B) $\sigma_{\bar{x}} = \dfrac{\sigma}{\sqrt{n}}$

(C) $\sigma_{\bar{x}} = \dfrac{\sigma_X}{\sqrt{n}}$

(D) $\sigma_{\bar{x}} = \dfrac{\sigma_X}{n}$

(E) $\sigma_X = \dfrac{\sigma_X}{\sqrt{n}}$

Examining Standard Error

353–357 Suppose that you have a population of 100,000 individuals and ask them 200 trivia questions; their score is the number of trivia questions they got right. For this population, the mean score is 158 and the population standard deviation of the scores is 26.

353. Suppose that you draw several samples of 50 people from this particular population, and for each sample, you find $\bar{x}$, their average score. And you do this repeatedly until you have all possible average scores from all possible samples of 50 people. Fill in the blanks with the appropriate statistical terms: The _____ represents the amount of variability in the entire population of trivia test scores, and the _____ represents the amount of variability in all the average trivia scores for all groups of 50.

354. Suppose that you draw 100 samples of size 30 and calculate the mean and standard error. If you then draw 100 samples of size 50 from the same population, you would expect to see the mean _____ and the standard error _____.

(A) be approximately the same; be larger

(B) be approximately the same; be smaller

(C) be smaller; be smaller

(D) be smaller; be larger

(E) be smaller; be approximately the same

355. For this particular population, with its given mean and standard deviation, what is the standard error for a sample of size 50? Round your answer to two decimal places.

356. For this particular population, with its given mean and standard deviation, what is the standard error for a sample of size 60? Round your answer to two decimal places.

357. For this particular population, with its given mean and standard deviation, what is the standard error for a sample of size 30? Round your answer to two decimal places.

Surveying Notation and Symbols

358–364 Solve the following problems about notation and symbols.

358. What is the notation for the average of a set of scores sampled from a larger population of scores?

359. What is the notation for the average value of a random variable X in a population?

360. Which symbol represents the standard error of the sample means?

361. Which symbol represents the standard deviation of individual scores in a population?

362. Suppose that you assign a number value to each card in a standard deck of 52 playing cards (Ace $= 1$, $2 = 2$, $3 = 3$, and so on up to Jack $= 11$, Queen $= 12$, King $= 13$). Then, you draw a hand of five cards, compute the average face value of cards in that hand, return the cards to the deck, shuffle the deck, and do the same thing again. You do this over and over — an infinite number of times — recording the average value of the cards in your hand each time. What symbol would best represent the standard deviation of all the averages you find by this method?

363. Every morning, the crew of a fishing boat pulls up a net full of fish, randomly chooses 50 of those fish, computes the average weight of the fish in the catch, and then records that number in a log. Over several years, the owner of the boat uses the standard deviation of those average weights to make financial projections to be sure to consider random fluctuations in the average size of the fish in a catch when estimating revenues. Which symbol best represents the value that the owner is trying to estimate to make financial projections?

364. A realtor looks at seven houses recently sold that are comparable to a house she'll be trying to sell. Which symbol best represents the sale price of any one of the homes?

Understanding What Affects Standard Error

365–368 Solve the following problems about items that affect standard error.

365. If everything else is held constant, which of the following is most likely to result in a larger standard error?

(A) a larger population mean

(B) a smaller population mean

(C) a larger sample size

(D) a smaller sample size

(E) a smaller population standard deviation

366. What effect would it have on the standard error of the sample mean if you could raise the size of each of the samples without changing the population standard deviation?

367. Population A has a mean of 1,000 and a standard deviation of 70, while Population B has a mean of 1,200 and a standard deviation of 62. For samples of size 35, which will have a smaller standard error for the sample mean $\bar{X}$?

368. If a survey sampling researcher wants to cut the standard error of the mean by half, what should they do regarding the sample size next time?

Digging Deeper into Standard Error

369–374 A quality control expert is studying the length of bolts produced in his factory by drawing a sample of completed parts and studying their dimensions in centimeters (cm).

369. For a sample of size 9, if $\sigma_X = 20$, what is the standard error of the mean, $\sigma_{\bar{X}}$? Round your answer to four decimal places.

370. For a sample of size 16, if $\sigma_X = 20$, what is the standard error of the mean, $\sigma_{\bar{X}}$?

371. Which of the following would be expected to result in a larger standard error of the mean?

(A) a larger sample size

(B) a smaller sample size

(C) a smaller population standard deviation

(D) a larger population standard deviation

(E) Choices (B) and (D)

372. If the standard error of the sample mean is 5, for samples of size 25, what is the population standard deviation?

373. For a sample of size 40, what will be the units for the standard error?

374. To estimate the population mean, which of the following standard errors will give you the most precise estimate?

(A) 0.4856

(B) 0.6818

(C) 0.7241

(D) 1.3982

(E) 8.2158

Connecting Sample Means and Sampling Distributions

375–385 Solve the following problems about sample means and sampling distributions.

375. A researcher draws a series of samples of exam scores from a population of scores that has a normal distribution. What required sample size (if any) is necessary for the distribution of the sample means to also have a normal distribution?

376. Which of the following conditions is enough to ensure that the sampling distribution of the sample means has a normal distribution?

(A) Samples of size $n > 10$ are drawn.

(B) The population of all possible scores is very large.

(C) At least 30 samples are drawn, with replacement, from the distribution of possible scores.

(D) Individual scores x_i are normally distributed.

(E) None of the above.

377. The daily productivity of individual bees has a normal distribution, with each bee producing a slightly different amount of honey. If random samples of ten bees are taken, what is the shape of the sampling distribution of the sample means of honey production?

378. If the population distribution is _____ and the sample size is _____, you need to apply the central limit theorem to assume that the sampling distribution of the sample means is normal.

(A) normal, 10

(B) normal, 50

(C) right-skewed, 60

(D) Choices (B) and (C)

(E) Choices (A), (B), and (C)

379. If individual scores are _____, the sampling distribution of sample means for scores from that population will be normal regardless of sample size n.

380. When individual scores are normally distributed, what can you conclude about the shape of the sampling distribution of the sample means?

381. You have a population of 10,000 test scores with a normal distribution. If you draw all possible samples, each of size 40, and graph their means, what shape do you expect this graph to take?

382. The sizes of adult shoes stocked in a store have a normal distribution. If a shoe store employee randomly pulls four pairs of shoes at a time out of the storeroom, finds the average size of the four pairs, returns them to the shelf, and repeats this process 30 times, how would the distribution of sample means most likely look?

383. Suppose that all possible samples of size 5 are repeatedly drawn from a normally distributed population of 10,000 scores, the average score in each sample is noted, and the distribution of sample averages is examined. What can you confidently say about the distribution of all the possible means in this case?

384. Suppose that samples of size 10 are repeatedly drawn from a population of stones, and the average weight of each sample is recorded. The distribution of weights in the population of stones is normally distributed. What shape do you expect the distribution of sample means to have?

385. A hardware store stocks nails in lengths from 20 to 100 millimeters, with sizes 5 millimeters apart (20, 25, 30, and so on). The store keeps the same number of nails of each length in stock. If a random sample of size 35 is drawn from the stock, the mean length calculated and recorded, the nails returned to the stock, and the experiment repeated 100 times, what shape do you expect the distribution of the sample means to assume?

Digging Deeper into Sampling Distributions of Sample Means

386–392 Use the descriptions of Populations A–D to answer the following problems. The distribution of values in Population A is strongly skewed right. The distribution of values in Population B is flat all the way across (this is called a uniform distribution). The distribution of values in Population C is strongly skewed left. The values in Population D have a normal distribution.

386. Imagine that samples are repeatedly taken from each population. If each sample has $n = 50$ observations in it, for which population will the sampling distribution of the sample means be approximately normal?

    (A) Population A
    (B) Population B
    (C) Population C
    (D) Population D
    (E) all of the above

387. Suppose that samples of size 20 are repeatedly taken from the four populations. Which population is likely to have an approximately normal distribution for the sampling distribution of the sample mean?

388. Suppose that samples of size 40 are repeatedly taken from the four populations. What shape do you expect the sampling distribution of the sample means to take?

389. Suppose that each population has 100,000 units. If samples are repeatedly taken from each, what shape do you expect the sampling distributions of the sample means to have?

390. Suppose that you drew all possible samples of a particular size from Population C. How large would your sample size have to be before you would expect the sampling distribution of sample means to be normal?

391. Imagine that a researcher draws all possible samples of size 3 from a normal distribution. What general shape should the sampling distribution of sample means have?

392. Imagine that a researcher draws all possible samples of size 100 from a population that doesn't have a normal distribution. What general shape should the sampling distribution of sample means have?

Looking at the Central Limit Theorem

393–400 *Solve the following problems that involve the central limit theorem.*

393. Suppose that a researcher draws random samples of size 20 from an unknown distribution. Can the researcher claim that the sampling distribution of sample means is at least approximately normal?

394. As a *general* rule, approximately what is the smallest sample size that can be safely drawn from a non-normal distribution of observations if someone wants to produce a normal sampling distribution of sample means?

395. A researcher draws a sample of 500 values from a large population of scores that has a skewed distribution. The researcher then goes on to do statistical analyses that assume that the sampling distribution of sample means is at least approximately normal. What can you conclude about whether the researcher has violated a condition of his statistical procedure?

396. A researcher draws 150 samples of 10 apiece from a normally distributed population of individual observations. What can the researcher conclude in this case?

397. A researcher is going to draw a sample of 150 values from a population. The researcher wants to find probabilities for the sample mean. Which of the following statements is true?

(A) The central limit theorem can't be used because you don't know the distribution of the population.

(B) The central limit theorem can't be used because the researcher is taking only one sample.

(C) The central limit theorem can be used because the sample size is large enough and the population distribution is unknown.

(D) The central limit theorem isn't needed because you can assume that the population is normal if it isn't stated.

(E) It can't be determined from the information given.

398. The distribution of the length of screws produced by a factory isn't normal. A researcher draws 150 samples of 10 apiece from the day's output of screws. Can the researcher use the central limit theorem in this case to make conclusions about the random variable $\overline{X}$?

399. A researcher draws 40 samples of 150 numbers each apiece from a non-normal skewed population of individual observations. What can the researcher conclude in this case?

400. A researcher draws a sample of size 32 from a population whose individual values are skewed right (not normal). The researcher wants to use the normal distribution to find probabilities regarding his results. Can they use the central limit theorem in this case?

Getting More Practice with Sample Mean Calculations

401–405 Assume that scores on a math exam are normally distributed for students in a particular university, with a mean of 10 and a standard deviation of 3. Samples of differing sizes are drawn from this population.

401. For a sample of 10 students, their scores are as follows: 7, 6, 7, 14, 9, 9, 11, 11, 11, and 11. What is the sample mean?

402. You draw a sample and calculate the sample mean. Which of the following is true?

(A) The sample mean will always be smaller than the population mean because the sample is only part of the population.

(B) The sample mean will always be the same as the population mean.

(C) The sample mean will be larger than the population mean because the sample is only part of the population.

(D) Larger samples tend to yield more precise estimates of the population mean.

(E) Not enough information to tell.

403. Which of the following is true about the relationship between the mean from a particular sample and the population mean?

(A) The sample mean is the same as the population mean.

(B) The sample mean is the best estimate of the population mean.

(C) Larger samples yield more precise estimates of the population mean.

(D) Smaller samples yield more precise estimates of the sample mean.

(E) Choices (B) and (C)

404. Assume that the true population mean is 10, and the population standard deviation is 3. What is the probability of getting a sample average at least as far from the mean as this sample gave, considering that the sample size is $n = 10$?

405. Suppose that in a sample of 500 students, the mean score on the test was found to be 9.7. What is the probability of finding a sample of size 500 whose mean is farther from the population mean than that of this particular sample?

Finding Probabilities for Sample Means

406–411 Find probabilities for sample means in the following problems.

406. If $\mu = 100$, $\sigma = 30$, and $n = 35$, what is the probability of finding a sample average 10 or more units from the population average?

407. If $\mu = 100$, $\sigma = 40$, and $n = 64$, what is the probability of finding a sample average 10 or more units from the population average?

408. If $\mu = 50$, $\sigma = 16$, and $n = 64$, what is the probability of finding a sample average 10 or more units from the population average?

409. If $\mu = 50$, $\sigma = 16$, and $n = 16$, and if the population is normally distributed, what is the probability of finding a sample average 10 or more units from the population average?

410. If $\mu = 0$, $\sigma = 160$, and $n = 100$, what is the probability of finding a sample average 10 or more units from the population average?

411. A medical technician is told that the mean white blood cell count in the population is 7,250 white blood cells per microliter of blood with a standard deviation of 1,375, but the technician doesn't know whether white blood cell counts are normally distributed in the population. The technician takes 40 tubes of blood from different randomly selected people and gets a mean white blood cell count of $\bar{x} = 7,616$ averaging across the 40 tubes. What is the probability of getting a sample mean this high or higher?

Digging Deeper into Probabilities for Sample Means

412–415 The owner of a cookie factory manufactures chocolate chip cookies to very precise specifications. The cookie population weight has a mean of 12 grams and a population standard deviation of 0.1 grams.

412. The statistician who works at the factory pulls a random sample of 36 cookies from the most recent batch one day and discovers that the average weight of a cookie in that sample is 12.011 grams. What is the probability of getting a value this close or closer to the mean?

413. The statistician who works at the factory pulls a random sample of 49 cookies from the most recent batch one day and discovers that the average weight of a cookie in that sample is 12.004 grams. What is the probability of getting a value this close or closer to the mean?

414. The statistician who works at the factory pulls a random sample of 36 cookies from the most recent batch one day and discovers that the average weight of a cookie in that sample is 12.02 grams. What is the probability of getting a value this close or closer to the mean?

415. The statistician who works at the factory pulls a random sample of 49 cookies from the most recent batch one day and discovers that the average weight of a cookie in that sample is 12.02 grams. What is the probability of getting a value this close or closer to the mean?

Adding Proportions to the Mix

416–419 Suppose that a researcher believes that gender bias exists in the presentation of zombies in movies. The researcher believes that exactly 50% of all clearly gendered zombies should be female if no bias is present. The researcher gets a list of all zombie movies ever made, randomly chooses 29 movies from the list, and then chooses a random spot in each movie to begin watching. The researcher codes the apparent gender of the first zombie that appears after the randomly chosen spot to begin watching. The results are that 20 of 29 zombies are apparently female.

416. Assuming that you're interested in the proportion of female zombies, what is $\hat{p}$? Round your answer to two decimal places.

417. What is p? Round your answer to four decimal places.

418. What is $\sigma_{\hat{p}}$? Round your answer to four decimal places.

419. In the research scenario presented, can the central limit theorem be used?

Figuring Out the Standard Error of the Sample Proportion

420–423 Calculate the standard error of the sample proportion in the following problems.

420. If $p = 0.9$ and $n = 100$ in a binomial distribution, what is the standard error of the sample proportion?

421. If $p = 0.1$ and $n = 100$ in a binomial distribution, what is $\sigma_{\hat{p}}$?

422. If $p = 0.5$ and $n = 100$ in a binomial distribution, what is $\sigma_{\hat{p}}$?

423. If $p = 0.67$ and $n = 60$ in a binomial distribution, what is $\sigma_{\hat{p}}$?

Using the Central Limit Theorem for Proportions

424–425 Determine how to use the central limit theorem for proportions in the following problems.

424. If a researcher wants to study a binomial population where $p = 0.1$, what is the minimum size n needed to make use of the central limit theorem?

425. If a researcher wants to study a population proportion where $p = 0.5$, what is the minimum size n needed to make use of the central limit theorem?

Matching *z*-Scores to Sample Proportions

426–428 Calculate z-scores for observed sample proportions in the following problems.

426. If $p = 0.5$, $n = 10$, and $\sigma_{\hat{p}} = 0.1581$, what z-score corresponds to an observed $\hat{p}$ of 0.25? Round your answer to two decimal places.

427. If $p = 0.5$, $n = 25$, and $\sigma_{\hat{p}} = 0.1$, what z-score corresponds to an observed $\hat{p}$ of 0.25?

428. If $p = 0.25$ and $n = 25$, what z-score corresponds to an observed $\hat{p}$ of 0.25?

Finding Approximate Probabilities

429–431 These problems all have a binomial distribution. In each case, find the approximate probability that $\hat{p} \leq 0.25$, using the central limit theorem if the appropriate conditions are met.

429. $p = 0.5, n = 10$

430. $p = 0.5, n = 25$

431. $p = 0.25, n = 40$

Getting More Practice with Probabilities

432–435 Imagine that a fair coin is tossed repeatedly, and the total number of heads across all tosses is noted.

432. If the coin is tossed 36 times, what is the probability of getting at least 21 heads?

433. If the coin is tossed 50 times, what is the probability of getting at least ten heads?

434. If the coin is tossed 100 times, what is the probability of getting more than 60 heads?

435. A large corporation buys computers in bulk. It knows that a certain manufacturer guarantees that no more than 1% of its machines are defective. The corporation has observed a defect rate of 1.5% in 1,000 machines it has bought from that company. If the manufacturer truly achieves the defect rate it claims, what is the probability of observing a defect rate at least as high as 1.5% in a random sample of 1,000 machines?

Digging Deeper into Approximate Probabilities

436–440 A researcher begins with the knowledge that $p = 0.3$. The researcher finds $\hat{p}$ as stated, using n trials given in the question. Find the approximate probability in each case.

436. Given the information provided, what is the approximate probability of observing a proportion $\hat{p}$ outside of the range 0.2 to 0.4, with 36 trials?

437. Given the information provided, what is the probability of observing a proportion $\hat{p} < 2$ in 36 trials?

438. Given the information provided, what is the approximate probability of observing $\hat{p} < 0.4$ with 36 trials?

439. Using the information provided, what is the approximate probability of observing a sample proportion $\hat{p}$ outside the range 0.2 to 0.4 with 81 trials?

440. What is the probability of observing a sample proportion $\hat{p}$ in the range 0.2 to 0.4 with 144 trials?

Chapter **8**

Finding Room for a Margin of Error

A margin of error is the "plus or minus" part you have to add to your statistical results to tell everyone you acknowledge that sample results will vary from sample to sample. The margin of error helps you indicate how much you believe those results could vary, with a certain level of confidence. Anytime you're trying to estimate a number from a population (like the average gas price in the United States), you include a margin of error. And it's never a good idea to assume that a margin of error is small if it's not given.

The Problems You'll Work On

The set of problems in this chapter focus solely on the margin of error. It's related to the confidence interval, and it's being given its own spotlight because it's so important. Here's what you'll be working on:

>> Understanding exactly what margin of error (MOE) means and when it's used

>> Breaking down the components of margin of error and seeing the role of each component

>> Examining the factors that increase or decrease the margin of error and the way they do so

>> Working on margin of error for population means as well as a few population proportions

>> Interpreting the results of a margin of error properly

What to Watch Out For

As you work through the problems in this chapter, pay special attention to the following:

>> Knowing exactly what happens when certain values in a margin of error change

>> Understanding the relationship between values of z and the confidence level needed for a margin of error

>> Accurately calculating the sample size needed to get a particular margin of error you want

>> Keeping your level of understanding high and not getting buried in the calculations

Defining and Calculating Margin of Error

441–446 Solve the following problems about margin of error basics.

441. An opinion survey reports that 60% of all voters who responded to a phone call support a bond measure to provide more money for schools, with a margin of error of ± 4%. What does the term *margin of error* add to your knowledge of the survey results?

442. A sports supplement maker claims that based on a random sample of 1,000 users of its product, 93% of all its users gain muscle and lose fat, with a margin of error of ± 1%. What critical information is missing from this claim?

443. A poll shows that Garcia is leading Smith by 54% to 46% with a margin of error of ± 5% at a 95% confidence level. What conclusion can you draw from this poll?

444. What is the margin of error for estimating a population mean given the following information and a confidence level of 95%?

$$\sigma = 15$$
$$n = 100$$

445. What is the margin of error for estimating a population mean given the following information and a confidence level of 95%?

$$\sigma = 5$$
$$n = 500$$

446. What is the margin of error for estimating a population mean given the following information and a confidence level of 95%?

$$\sigma = \$10,000$$
$$n = 40$$

Using the Formula for Margin of Error When Estimating a Population Mean

447–449 A researcher conducted an Internet survey of 300 students at a particular college to estimate the average amount of money students spend on groceries per week. The researcher knows that the population standard deviation, σ, of weekly spending is $25. The mean of the sample is $85.

447. What is the margin of error if the researcher wants to be 99% confident of the result?

448. What is the margin of error if the researcher wants to be 95% confident in the result?

449. What is the lower limit of an 80% confidence interval for the population mean based on this data?

Finding Appropriate z*-Values for Given Confidence Levels

450–452 Use Table A-4 to find the appropriate z*-value for the confidence levels given except where noted.

450. Table A-1 (the Z-table) and Table A-4 in the appendix are related but not the same. To see the connection, find the z*-value that you need for a 95% confidence interval by using Table A-1.

451. What is the z*-value for a 99% confidence level?

452. What is the z*-value for a confidence level of 80%?

Connecting Margin of Error to Sample Size

453–455 A sociologist is interested in the average age women get married. The sociologist knows that the population standard deviation of age at first marriage is three years for both men and women and hasn't changed for the last 60 years. The sociologist would like to demonstrate with 95% confidence that the mean age of the first marriage for women now is higher than the mean age in 1990, which was 24 years old.

453. If the sociologist can sample only 50 women, what will the margin of error be?

454. If the sociologist can sample 100 women, what will the margin of error be?

455. What is the smallest number of participants the sociologist can sample to estimate the average nuptial age for the population, with a margin of error of only two years?

Getting More Practice with the Formula for Margin of Error

456–460 Two hundred airplane engine screws are to be sampled from a factory run of many thousands of screws to estimate the average length of all the screws (in millimeters).

456. Assuming that $\sigma = 0.01$ millimeters and you're using a 99% level of confidence, find the margin of error for estimating the population mean.

457. Assuming that $\sigma = 0.05$ millimeters and you're using a 99% level of confidence, what is the margin of error for estimating the population mean?

458. Assuming that $\sigma = 0.10$ millimeters and you're using a 99% level of confidence, what is the margin of error for estimating the population mean?

459. Margin of error can help factories determine whether they're manufacturing parts within a satisfactory range of dimensions. Suppose that a factory discovers that the average weights of the parts it's making have a margin of error (with 99% confidence) that is twice as large as it should be when they tried to estimate the average size of the parts. What step could the factory take regarding sample size if it wants to reduce the margin of error by half?

460. An unscrupulous factory owner found that the margin of error for estimating the average size of a factory-produced part is twice as large as it should be. One way of fixing this is to make sure that the manufacturing processes are all very consistent, but doing so is expensive and the factory owner didn't want to go to that much cost and trouble. The factory owner was instantly able to report to the board of directors that they had cut the margin of error by a little more than half. Assuming no changes were made in the manufacturing process, what deceptive action might the factory owner have done to achieve this result?

Linking Margin of Error and Population Proportion

461–464 Solve the following problems related to margin of error and population proportion.

461. A market researcher samples 100 people to find a confidence interval for estimating the average age of their customers. They find that the margin of error is three times larger than they want it to be. How many people should the researcher add to the sample to bring the margin of error down to the desired size?

462. A survey of 10,000 randomly selected adults from across Europe finds that 53% are unhappy with the euro. What is the margin of error for estimating the proportion among all Europeans who are dissatisfied with the euro? (Use a confidence level of 95%.) Give your answer as a percentage.

463. A county election office wants to plan for having enough poll workers for election day, so it conducts a telephone survey to estimate what proportion of those who are eligible to vote intend to do so. The office calls 281 eligible people in the community and finds that 135 of them intend to vote. What is the margin of error for estimating the population proportion using a 99% confidence level? Give your answer as a percentage.

464. A sample of 922 households finds that the proportion with one or more dogs is 0.46. The margin of error is reported as 2.7%. What confidence level is being used in this confidence interval for a proportion?

Chapter **9**

Confidence Intervals: Basics for Single Population Means and Proportions

S uppose that you want to find the value of a certain population parameter (for example, the average gas price in Ohio). If the population is too large, you take a sample (such as 100 gas stations chosen at random) and use those results to *estimate* the population parameter. Knowing sample results vary, you attach a margin of error (plus or minus) to cover your bases. Put it all together, and you get a *confidence interval* — a range of likely values for the population parameter.

The Problems You'll Work On

The problems in this chapter give you practice calculating and interpreting some basic confidence intervals. Here are the highlights:

>> Calculating a confidence interval for a population mean when the population standard deviation is known (involves the *Z*-distribution) or unknown (involves a *t*-distribution)

>> Calculating a confidence interval for a population proportion when the sample is large

>> Finding and interpreting the margin of error of a confidence interval and what factors affect it

>> Determining appropriate sample sizes required to achieve a certain margin of error

What to Watch Out For

When working through these problems, make sure you can do the following:

>> Know when a confidence interval is needed — the word *estimate* is your cue.

>> Know which conditions to check before deciding which formula to use.

>> Understand each part of a confidence interval, what its role is, and how it can be affected.

>> Focus particularly on how the results are *interpreted* (explained in real language) instead of focusing just on the calculations.

Introducing Confidence Intervals

465–468 *A clothing store is interested in the mean amount spent by all of its customers during shopping trips, so it examines a random sample of 100 electronic cash-register records and discovers that, among those who made purchases, the average amount spent was $45 with a 95% confidence interval of $41 to $49.*

465. Which of the following are true statements regarding the 95% confidence interval for this data?

(A) If the same study were repeated many times, about 95% of the time the confidence interval would include the average money spent from the sample, which is $45.

(B) If the same study were repeated many times, about 95% of the time, the confidence interval would contain the average money spent for all the customers.

(C) There is a 95% probability that the average money spent for all the customers is $45.

(D) Choices (A) and (B)

(E) Choices (A), (B), and (C)

466. Which of the following is a reason for reporting a confidence interval as well as a point estimate for this data?

(A) The store studied a sample of sales records rather than the entire population of sales records.

(B) The confidence interval is certain to contain the population parameter.

(C) Because sample results vary, the sample mean is not expected to correspond exactly to the population mean, so a range of likely values is required.

(D) Choices (A) and (B)

(E) Choices (A) and (C)

467. Which of the following statements is a valid argument for drawing a sample of size 500 rather than size 100?

(A) The larger sample will produce a less-biased estimate of the sample mean.

(B) The larger sample will produce a more precise estimate of the population mean.

(C) The 95% confidence interval calculated from the larger sample will be narrower.

(D) Choices (B) and (C)

(E) Choices (A), (B), and (C)

468. Which of the following statements is true regarding the sample mean of $45?

(A) It is the same as the population mean.

(B) It is a good number to use to estimate the population mean.

(C) If you drew another sample of 100 from the same population, you would expect the sample mean to be exactly $45.

(D) Choices (A) and (C)

(E) None of the above

Checking Out Components of Confidence Intervals

469–484 Solve the following problems about confidence interval components.

469. Consider the following samples of $n = 5$ from a population. Without doing any calculations, which would you expect to have the widest 95% confidence interval if you're using the sample to estimate the population mean?

  Sample A: 5, 5, 5, 5, 5

  Sample B: 5, 6, 6, 6, 7

  Sample C: 5, 6, 7, 8, 9, 10

  Sample D: 5, 6, 7, 8, 9, 20

470. When analyzing the same sample of data, which confidence interval would have the widest range of values?

(A) one with a confidence level of 80%

(B) one with a confidence level of 90%

(C) one with a confidence level of 95%

(D) one with a confidence level of 98%

(E) one with a confidence level of 99%

471. How will the confidence interval be affected if the confidence level increases from 95% to 98%?

472. With all factors remaining equal, why does increasing the confidence level increase the width of the confidence interval?

473. A survey of 100 Americans reports that 65% of them own a car. The 95% confidence interval for the percentage of all American households who own one car is 60% to 70%. What is the margin of error for the confidence interval in this example?

474. If all other factors stay the same, which sample size will create the widest confidence interval?

(A) $n = 100$

(B) $n = 200$

(C) $n = 300$

(D) $n = 500$

(E) $n = 1,000$

475. If all other factors are equal, which sample size results in the narrowest confidence interval?

(A) $n = 200$

(B) $n = 500$

(C) $n = 1,000$

(D) $n = 2,500$

(E) $n = 5,000$

476. A hospital is considering what size sample to draw in a study of the accuracy of its clinical records. If it decides to use a random sample of 500 records, rather than a random sample of 200 records, which of the following statements is true?

(A) The sample of 500 will have a wider 95% confidence interval.

(B) The sample of 500 will have a narrower 95% confidence interval.

(C) The width of the 95% confidence interval will not be affected by the sample size.

(D) The sample of 500 will produce a more precise estimate of the population mean.

(E) Choices (B) and (D)

477. What is the relationship between the confidence level and the width of the confidence interval?

478. Assume that Population A has substantially less variability than Population B. Comparing samples of the same size, and confidence intervals of the same confidence level, which of the following statements is true?

(A) The confidence intervals related to Populations A and B are expected to be the same.

(B) The confidence interval related to Population A is expected to be wider.

(C) The confidence interval related to Population A is expected to be narrower.

(D) It depends on how the data was collected.

(E) Not enough information to tell.

479. A university is planning to study student satisfaction with technological services on campus, based on a survey of a random sample of students. Which of the following statements is true?

(A) A confidence level of 80% will produce a wider confidence interval than a confidence level of 90%.

(B) A confidence level of 80% will produce a narrower confidence interval than a confidence level of 90%.

(C) A sample of 300 students will produce a narrower confidence interval than a sample of 150 students.

(D) Choices (A) and (C)

(E) Choices (B) and (C)

480. The following samples are drawn from different populations. Assuming that the samples are accurate reflections of the variability of the populations and the same confidence level is used for each, which sample will have the widest confidence interval?

Sample A: 10, 20, 30, 40, 50, 60, 70, 80, 90, 100

Sample B: 1, 2, 3, 40, 50, 60, 70, 800, 900, 1000

Sample C: 41, 42, 43, 44, 45, 46, 47, 48, 49, 50

Sample D: 10, 15, 20, 21, 22, 23, 24, 25, 30, 35

Sample E: 510, 520, 530, 540, 550, 560, 570, 580, 590, 600

481. A random sample of 100 people taken from which of the following populations will yield the widest confidence interval for the mean income?

(A) workers ages 22 to 30 who live in Denver

(B) all U.S. workers ages 22 to 30

(C) all U.S. workers ages 16 to 22

(D) all workers ages 22 to 30 who live in North America (Canada, the U.S., and Mexico), adjusted in U.S. dollars

(E) all workers ages 22 to 30 who live in U.S. cities with populations of fewer than 10,000 people

482. A random sample of 100 taken from which of the following populations will yield the narrowest confidence interval for average height?

(A) children ages 1 to 5

(B) children ages 5 to 10

(C) children ages 10 to 16

(D) teenagers ages 13 to 19

(E) adults ages 55 to 65

483. Which of the following samples will yield the widest confidence interval for the same population mean?

(A) one with confidence level 95%, $n = 200$, and $\sigma = 8.5$

(B) one with confidence level 95%, $n = 200$, and $\sigma = 12.5$

(C) one with confidence level 95%, $n = 400$, and $\sigma = 8.5$

(D) one with confidence level 80%, $n = 200$, and $\sigma = 8.5$

(E) one with confidence level 80%, $n = 400$, and $\sigma = 8.5$

484. Which of the following will decrease the margin of error of a confidence interval?

(A) Increase the sample size from 200 to 1,000 subjects.

(B) Decrease the confidence level from 95% to 90%.

(C) Increase the confidence level from 95% to 98%.

(D) Choices (A) and (B)

(E) Choices (A) and (C)

Interpreting Confidence Intervals

485–489 Solve the following problems about interpreting confidence intervals.

485. A sample of the heights of boys in a school class shows the mean height is 5 feet 9 inches. The margin of error is ± 4 inches for a 95% confidence interval. Which of the following statements is true?

(A) The 95% confidence interval for the mean height of all the boys is between 5 feet 5 inches and 6 feet 1 inch.

(B) The mean of any one sample has a 95% chance of being between 5 feet 5 inches and 6 feet 1 inch.

(C) Based on the data, the sample mean height is 5 feet 9 inches, and 95% of all the boys will be between 5 feet 5 inches and 6 feet 1 inch.

(D) It means that based on the sample data, 95% of the heights are calculated to have a mean of 5 feet 9 inches, 5% of the boys are more than 4 inches shorter and 5% are at least 4 inches taller than 5 feet 9 inches.

(E) It means that a boy randomly selected from the class has a 95% chance of having a height between 5 feet 5 inches and 6 feet 1 inch.

486. A sample of college students showed that they earned a mean summer income of $4,500. The margin of error is ± $400 for a 95% confidence interval. What does this mean?

487. Which of the following correctly describes the margin of error?

(A) The margin of error is the percentage of errors that were made in taking the sample.

(B) The margin of error is an estimate that adjusts for the false reporting by the people surveyed.

(C) The margin of error is used to calculate a range of likely values for a population parameter, based on a sample.

(D) The margin of error identifies the quality of the sampling methods. A margin of error ± 5% indicates a well-designed study.

(E) The margin of error shows the distance of the sample results from the population mean.

488. A sample of college students found that the mean amount that students spend on books, supplies, and lab fees was $450 per semester, with a margin of error of ± $50. The confidence level for these results is 99%. Based on these results, which of the following statements is true about the margin of error?

(A) The margin of error means that 99% of the fees on books are within $50 of each other.

(B) The margin of error measures the amount by which your sample results could change, with 99% confidence.

(C) The margin of error means you have to adjust your results by $50 to account for inaccurate reporting by the people surveyed.

(D) The margin of error identifies the quality of the sampling methods. A margin of error ± $50 indicates a poorly designed study.

(E) The margin of error shows that the total range of all the purchases on average was $400 to $500, and the mean was $450.

489. A poll of 1,000 likely voters showed that Candidate Smith had 48% of the vote, and Candidate Jones had 52% of the vote. The margin of error was ± 3%, and the confidence level was 98%. Who is most likely to win the election?

Spotting Misleading Confidence Intervals

490–494 Solve the following problems about misleading confidence intervals.

490. In a survey, 6,500 of the first 10,000 fans at a football game chose chocolate ice cream as their favorite flavor. The ice-cream company running the survey then claims that 65% of all Americans prefer chocolate ice cream over other flavors, based on this survey. Which of the following choices best describes the conclusions of the ice-cream company?

(A) The survey has a built-in bias.

(B) The survey will have valid results because the sample size is high.

(C) The results are biased because the confidence interval is too wide when only 10,000 people respond.

(D) The survey isn't valuable because it doesn't list the choices of flavors made by the other fans.

(E) The results of the survey are biased because it didn't account for people who don't like ice cream.

491. A company took a random sample of 30 first-year employees and asked them their level of satisfaction with their jobs. It found that 80% of those sampled were "very happy" with their employment, ± 3% at a confidence level of 95%. The company took this information and reported that 80% of all its employees were very happy with their jobs, ± 3%. There is at least one problem with the company's reported results. Choose the answer(s) that best describes the problem(s).

(A) The survey is accurate because it's based on a random sample.

(B) The survey is biased because it was based only on first-year employees, who may feel differently about their jobs than other employees.

(C) The survey is misleading because it doesn't report the results of the other first-year employees.

(D) The sample size is only 30. The margin of error must be higher than 3% based on the size of the sample and the confidence level.

(E) Choices (B) and (D)

492. A company conducted a random online survey of 1,000 visitors to its website during the past three months. The sample showed that the visitors had a mean of five online purchases at all Internet sites during the past 12 months. The margin of error was ± 0.6 purchases with a confidence level of 95%. So the company concludes that the average number of online purchases for all visitors to its website during the past three months is 5, plus or minus 0.6. Which of the following choices best describes the survey?

(A) The survey can be used by the company to help predict the spending habits of all its website visitors during the year.

(B) The survey can be used only by the company as part of its analysis of the Internet spending habits of all visitors to its website during the last three months.

(C) The survey can be used by the company as part of its analysis of the Internet spending habits of all of its customers (in-store and website) during the last three months.

(D) The sample is faulty and unusable because it surveys visitors to its website only, not necessarily those who make purchases.

(E) The survey is faulty because it's based on too small of a sample.

493. Over a three-month period, a random sample of teenagers visiting a local movie theater was asked how many times they went to a movie in the last year. The purpose of the survey is to estimate the average number of movies a teenager attended in the last year. The results of the survey were as follows:

$n = 1,000$ teenagers

$\bar{x} = 4.5$ visits

MOE $= 0.7$ visits

Confidence level $= 95\%$

Which of the following choices best describes the survey?

(A) The results are invalid because the survey was done at a movie theater.

(B) The 95% confidence interval for the average number of movies attended by any teenager is between 3.1 and 5.9 visits per year and is valid because that's what the formula tells you to calculate.

(C) This survey isn't a valid method for estimating how many times teenagers visited movie theaters in the past 12 months because the survey was taken only over a three-month period.

(D) This survey isn't valid because it was based only on a sample and not the population of all teenagers.

(E) Choices (A) and (C)

494. A Colorado fashion magazine that has 2 million readers found in a mail-in survey with 5,800 responses that 56% of respondents chose Colorado as their favorite place to live, ± 2% with a confidence level of 98%. Which of the following choices best describes the survey?

(A) The sample is probably based on a representative sample of the magazine's readers because it's so large.

(B) The sample isn't based on a representative sample of the magazine's readers.

(C) Because the magazine is based on Colorado, more readers are likely to buy the magazine who are from Colorado; therefore, they'd be more likely to vote it as the best place to live.

(D) The sample results are likely biased because the respondents had to make the effort to mail back the survey.

(E) Choices (B), (C), and (D)

Calculating a Confidence Interval for a Population Mean

495–522 *Calculate confidence intervals for population means in the following problems.*

495. In a random sample of 50 intramural basketball players at a large university, the average points per game was 8, with a standard deviation of 2.5 points and a 95% confidence level. Which of the following statements is correct?

   (A) With 95% confidence, the average points scored by all intramural basketball players is between 7.3 and 8.7 points.

   (B) With 95% confidence, the average points scored by all intramural basketball players is between 7.7 and 8.4 points.

   (C) With 95% confidence, the average points scored by all intramural basketball players is between 5.5 and 10.5 points.

   (D) With 95% confidence, the average points scored by all intramural basketball players is between 7.2 and 8.8 points.

   (E) With 95% confidence, the average points scored by all intramural basketball players is between 7.6 and 8.4 points.

496. On the SAT Math test, a random sample of the scores of 100 students in a high school had a mean of 650. The standard deviation for the population is 100. What is the confidence interval if 99% is the confidence level?

497. An apple orchard harvested ten trees of apples. From a random sample of 50 apples, the mean weight of an apple was 7 ounces. The population standard deviation is 1.5 ounces. What is the confidence interval if 99% is the confidence level?

498. A random sample of 200 students at a university found that they spent an average of three hours a day on homework. The standard deviation for all the university's students is one hour. What is the confidence interval if 90% is the confidence level?

499. Analysis of a random sample of 200 people ages 18 to 22 showed that a person spent an average of $32.50 on a typical outing with a friend. The population standard deviation for this age group is $15.00. What is the confidence interval if 95% is the confidence level?

500. A random sample of 150 people over age 17 showed that a person spent an average of 30 minutes a day on vigorous exercise. The population standard deviation for exercise for this age group is 15 minutes. What is the confidence interval if 90% is the confidence level?

501. A random sample of 200 college graduates showed that a person made an average of $36,000 income in the first year after graduation. The standard deviation for income for all first-year college graduates is $8,000. What is the confidence interval if 95% is the confidence level?

502. A random sample of 300 trips by a city bus along a specific route showed that the average time to complete the bus route was 45 minutes. The standard deviation for all trips on this bus route is 3 minutes. What is the confidence interval if 95% is the confidence level?

503. A random sample of 1,100 travel itinerary requests on an airline ticket website showed that the average itinerary request took 4.5 seconds to be calculated and displayed to the traveler. The standard deviation for all itinerary requests is 2 seconds. What is the confidence interval if 98% is the confidence level?

504. A random sample of 200 MP3 players on an assembly line showed that the average amount of time to assemble an MP3 player was 12.25 minutes. The population standard deviation for assembly is 2.15 minutes. What is the confidence interval if 95% is the confidence level?

505. A random sample of 300 university students found that the average distance to the hometown of a student was 125 miles. The standard deviation for the distance for all students at the university is 40 miles. What is the confidence interval if 90% is the confidence level?

506. A random sample of 75 data entry specialists at a data center found that specialists made an average of 2.7 errors out of 10,000 data items. The standard deviation for the errors made by all specialists at the bank is 0.75 per 10,000 items. What is the confidence interval if 95% is the confidence level?

507. A random sample of 500 bats (of the type that has a standard length of 38 inches) made for major league baseball players found that the average bat had a length of 38.01 inches. The standard deviation for all 38-inch bats is 0.01 inches. What is the confidence interval if 99% is the confidence level?

508. A random sample of 2,000 special valve parts for an engine resulted in an average length of 3.2550 centimeters. The standard deviation for the population of special valve parts is 0.025 centimeters. What is the confidence interval if 99% is the confidence level?

509. A random sample of 40 purchases of medium-quality hardwood purchased over a 12-month period by a furniture maker from different suppliers showed a mean cost of $0.78 per board foot. The standard deviation for the year for all wood purchased was $0.12 per board foot. What is the confidence interval if 95% is the confidence level?

510. A standardized math test was given to a random sample of 25 first-year college students. Their mean score was 84% with a sample standard deviation of 5%. What is the confidence interval if 95% is the confidence level?

511. A random sample of 25 households found that the mean size household had 3.4 people, with a sample standard deviation of 0.8 people. What is the confidence interval if 90% is the confidence level?

512. A random sample of 30 teenagers found that the average number of social network friends each person had was 85, with a sample standard deviation of 50 friends. What is the confidence interval if 95% is the confidence level?

513. A random sample of 24 first-year college students found that the students traveled an average of 400 miles on the longest trip they took last year, with a sample standard deviation of 300 miles. What is the confidence interval if 95% is the confidence level?

514. A random sample of 20 shoppers leaving a mall found that they spent an average of $78.50 that day, with a sample standard deviation of $50.75. What is the confidence interval if 95% is the confidence level?

515. A random sample of 20 visitors leaving a museum found that they spent an average of three hours in the museum, with a sample standard deviation of one hour. What is the confidence interval if 90% is the confidence level?

516. A random sample of 50 1-pound loaves of bread at a bakery found that the mean weight was 18 ounces, with a sample standard deviation of 1.5 ounces. What is the confidence interval if 90% is the confidence level?

517. A random sample of ten people shopping at a grocery store found that they visited the store an average of 2.8 times per month, with a sample standard deviation of 2 visits. You know from previous research that the number of shopping visits each month is approximately normally distributed. What is the confidence interval if 80% is the confidence level?

518. A random sample of 18 first-year students at a university found that they watched an average of five movies a month (whether in theaters, online, or on DVD), with a population standard deviation of three movies. What is the confidence interval if 90% is the confidence level?

519. A random sample of 25 visitors to an amusement park found that they spent an average of $32.00 that day while at the park. The population standard deviation is $6.00. What is the confidence interval if 99% is the confidence level?

520. In calculating the sample sizes needed for a particular margin of error, which of the following results will be rounded to 118 (the number of subjects required)?

(A) 117.2

(B) 117.6

(C) 118.1

(D) Choices (A) and (B)

(E) Choices (A), (B), and (C)

521. In calculating the sample sizes needed for a particular margin of error, how would the following results be rounded: 121.1, 121.5, 131.2, and 131.6?

522. Although in general terms researchers like to have larger rather than smaller sample sizes, to get more precise results, what are some limiting factors on the sample size used in a study?

(A) A larger sample often means greater costs.

(B) It may be difficult to recruit a larger sample (for example, if you're studying people with a rare disease).

(C) At some point, increasing the sample size may not significantly improve precision (for instance, increasing the sample size from 3,000 to 3,500).

(D) Choices (A) and (B)

(E) Choices (A), (B), and (C)

Determining the Needed Sample Size

523–529 Figure out the sample size needed in the following problems.

523. A physician wants to estimate the average BMI (*body mass index*, a measure that combines information about height and weight) for their adult patients. They decide to draw a sample of clinical records and retrieve this information from them. They want an estimate with a margin of error of 1.5 units of BMI, with 95% confidence, and believe that the national population standard deviation of adult BMI of 4.5 also applies to their patients. They knows that BMI is approximately normally distributed for adults. How large a sample do they need to draw?

524. A physician wants to estimate the height of 6-year-old boys in their community using a random sample drawn from administrative records. They want an estimate with a margin of error of 0.5 inches with 95% confidence, and they believe that the population standard deviation of 1.8 inches applies to their population. They also know that height is approximately normally distributed for this population. How large a sample do they need to draw?

525. You want to estimate the average height of 10-year-old boys in your community. The population standard deviation is 3 inches. What size sample do you need for a margin of error of no more than ±1 inch and a confidence level of 95% when constructing a confidence interval for the mean height of all 10-year-old boys?

526. You want to take a sample that measures the weekly job earnings of high-school students during the school year. The population standard deviation is $20. What size sample do you need for a margin of error of no more than ± $5 and a confidence level of 99% when constructing a confidence interval for the mean weekly earnings of all high-school students?

527. You want to take a sample that measures the weekly job earnings of university students during the school year. The population standard deviation is $55. What size sample do you need for a margin of error of no more than ± $10 and a confidence level of 90% when constructing a confidence interval for the mean weekly earnings of all university students?

528. You want to take a sample that measures the amount of sleep university students get each night. The population standard deviation is 1.2 hours. What size sample (number of students) do you need for a margin of error of no more than ± 0.25 hours and a confidence level of 95% when constructing a confidence interval for the mean amount of sleep of all university students?

529. You want to take a sample that measures the attendance at women's Division I university basketball games. The population standard deviation is 2,300. What size sample (number of games) do you need for a margin of error of no more than ± 800 and a confidence level of 95% when constructing a confidence interval for the mean attendance at all such games?

Introducing a Population Proportion

530–536 *A random sample of 100 students at a university found that 38 students were thinking of changing their major.*

530. You were asked to report both a point estimate (a single number) and a confidence interval (range of values) for your survey. Why would the confidence interval be requested?

(A) You may make a mistake with your calculations.

(B) You are using sample data to estimate a parameter.

(C) If you drew a different sample of the same size, you would expect the results to be slightly different.

(D) Choices (B) and (C)

(E) Choices (A), (B), and (C)

531. What does 0.38 represent in this example?

(A) the proportion of students at the university who are thinking of changing their major

(B) the number of students at the university who are thinking of changing their major

(C) an estimate of the proportion of all students at the university who are thinking of changing their major

(D) the proportion of students in the sample of 100 who are thinking of changing their major

(E) Choices (C) and (D)

532. Can you use the normal approximation to the binomial to calculate a confidence interval for this data? Why or why not?

533. What is the standard error of $\hat{p}$?

534. With all other relevant values being fixed, which of the following confidence levels will result in the widest confidence interval?

(A) 80%

(B) 90%

(C) 95%

(D) 98%

(E) 99%

535. At a confidence level of 90%, what is the confidence interval for the proportion of all students thinking of changing their major?

536. At a confidence level of 95%, what is the confidence interval for the proportion of all students thinking of changing their major?

Connecting a Population Proportion to a Survey

537–539 A website ran a random survey of 200 customers who purchased products online in the past 12 months. The survey found that 150 customers were "very satisfied."

537. What are the sample proportion and the standard error for the sample proportion based on this data?

538. With a 95% confidence level, what is the margin of error for the estimate of the proportion of all customers who purchased products online in the past 12 months?

539. With a 99% confidence level, what is the confidence interval for the proportion of all customers who purchased products online in the past 12 months? Round to two decimal places.

Calculating a Confidence Interval for a Population Proportion

540–545 Assume that you have a random sample of 80 with a sample proportion of 0.15.

540. Can you use the normal approximation to the binomial for this data?

541. With an 80% level of confidence, what is the confidence interval for the population proportion? Round your answer to four decimal places.

542. With a 90% level of confidence, what is the confidence interval for the population proportion? Round your answer to four decimal places.

543. With a 95% level of confidence, what is the confidence interval for the population proportion? Round your answer to four decimal places.

544. Assuming a 98% confidence level, what is the confidence interval for the population proportion based on this data? Round your answer to four decimal places.

545. Assuming a 99% confidence level, what is the confidence interval for the population proportion based on this data? Round your answer to four decimal places.

Digging Deeper into Population Proportions

546–547 *Solve the following problems about confidence intervals and population proportions.*

546. Assume that the 95% confidence interval for a population proportion based on a certain data set is 0.20 to 0.30. Which of the following could be the 98% confidence interval for the population proportion using the same data?

(A) 0.15 to 0.35

(B) 0.21 to 0.29

(C) 0.22 to 0.38

(D) 0.23 to 0.27

(E) 0.24 to 0.26

547. Assume that the 95% confidence interval for a population proportion based on a certain data set is 0.20 to 0.30. Based on the same data and pertaining to the same population proportion, what level of confidence could the interval 0.22 to 0.28 represent?

(A) 80%

(B) 90%

(C) 99%

(D) Choice (A) or (B)

(E) Choice (A), (B), or (C)

Getting More Practice with Population Proportions

548–552 In a random sample of 160 adults in a large city, 88 were in favor of a new 0.5% sales tax. Assume that you can use the normal approximation to the binomial for this data.

548. If you had only one number to use to estimate the proportion of all the adults in the city who favor the new tax, what number would you use?

549. What is the standard error for $\hat{p}$?

550. With a 95% confidence level, what is the margin of error for estimating the proportion of all adults in the city who favor the new tax?

551. With a 99% confidence level, what is the margin of error for estimating the proportion of all adults in the city who favor the new tax?

552. With an 80% level of confidence, what is the margin of error for estimating the proportion of all adults in the city who favor the new tax?

Chapter **10**

Confidence Intervals for Two Population Means and Proportions

M any real-world scenarios are looking to compare two populations. For example, what is the difference in survival rates for cancer patients taking a new drug compared to cancer patients on the existing drug? What is the difference in the average salary for males versus females? What's the difference in the average gas price this year compared to last year? All these questions are really asking you to compare two populations in terms of either their averages or their proportions to see how much difference exists (if any). The technique you use here is confidence intervals for two populations.

The Problems You'll Work On

The problems in this chapter give you practice with the following:

>> Calculating and interpreting confidence intervals for the difference in two population means when the population standard deviations are known (involves the *Z*-distribution) or unknown (involves a *t*-distribution)

>> Calculating and interpreting confidence intervals for the difference in two population proportions when the samples are large

What to Watch Out For

Keep the following in mind as you work through this chapter:

>> This chapter is about the *difference in the means,* not *the mean of the differences.*

>> When you look at the difference between two means (or two proportions), keep track of what populations you're calling Population 1 and Population 2. Subtracting two numbers in the opposite order changes the sign of the results!

>> If a confidence interval for a difference in means (or proportions) contains negative values, positive values, and zero, you can't conclude that a difference exists.

Working with Confidence Intervals and Population Proportions

553–557 A random poll of 100 males and 100 females who were likely to vote in an upcoming election found that 55% of the males and 25% of the females supported candidate Johnson. Call the population of males "Population 1" and the population of females "Population 2" while working these problems.

553. If you could use only one number to estimate the difference in the proportions of all males and all females supporting candidate Johnson among all likely voters, what number would you use?

554. What is the standard error for the estimate of the difference in proportions in the male and female populations?

555. With a 95% confidence level, what is the confidence interval for the difference in the percentage of males and females favoring Johnson among all likely voters?

556. With a 90% confidence level, what is the confidence interval for the difference in the percentage of males and females favoring Johnson among all likely voters?

557. With a 99% confidence level, what is the confidence interval for the difference between the percentage of males and females favoring Johnson among all likely voters?

Digging Deeper into Confidence Intervals and Population Proportions

558–560 A random survey found that 220 out of 300 adults living in large cities (with populations of more than 1 million) wanted more state funding for public transportation. In small cities (with populations fewer than 100,000), 120 of 300 adults surveyed wanted more state funding. While working these problems, call the population of adults living in large cities "Population 1" and the population of adults living in small cities "Population 2."

558. If you had only one number to estimate the difference between the proportion of adults in large cities and the proportion of adults in small cities favoring increased state funding for public transportation, what number would you use?

559. With a 90% confidence level, what is the confidence interval for the difference in the population proportions? Give your answer to four decimal places.

560. With an 80% confidence level, what is the confidence interval for the difference in the population proportions?

Working with Confidence Intervals and Population Means

561–565 *A random sample of 70 12th-grade boys in a high school showed a mean height of 71 inches. (Assume that among all boys in the high school, their heights have a population standard deviation of 2 inches.) A random sample of 60 12th-grade girls in the same school showed a mean height of 67 inches. (Assume that among all girls in the high school, their heights have a population standard deviation of 1.8 inches.) While working these problems, call the population of boys "Population 1" and the population of girls "Population 2."*

561. Using a 95% confidence level, what is the confidence interval for the population difference in the mean heights of all 12th-grade boys compared to 12th-grade girls at this school? Round to the nearest hundredth.

562. Using an 80% confidence level, what is the confidence interval for the population difference in the mean heights of all 12th-grade boys compared to 12th-grade girls at this school? Round to the nearest hundredth.

563. Using a 99% confidence level, what is the confidence interval for the population difference in the mean heights of all 12th-grade boys compared to 12th-grade girls at this school? Round to the nearest hundredth.

564. Using a 98% confidence level, what is the confidence interval for the population difference in the mean heights of all 12th-grade boys compared to 12th-grade girls at this school? Round to the nearest hundredth.

565. Suppose that you want a confidence interval for the difference in the mean heights of all 12th-grade boys compared to all 12th-grade girls at this school. In one case, you treat the population of boys as Population 1 and the population of girls as Population 2. In another case, you switch the order and treat the population of girls as Population 1 and the population of boys as Population 2. How would the resulting confidence intervals differ?

Making Calculations When Population Standard Deviations Are Known

566–571 *A random sample of 120 college students who were physics majors found that they spent an average of 25 hours a week on homework; the standard deviation for the population was 7 hours. A random sample of 130 college students who were English majors found that they spent an average of 18 hours a week on homework; the standard deviation for the population was 4 hours. Call the population of physics majors "Population 1" and the population of English majors "Population 2" while working these problems.*

566. Using a 90% confidence level, what is the margin of error for the estimated difference in average time spent on homework for college physics majors versus college English majors? Round to one decimal place.

567. Using an 80% confidence level, what is the margin of error for the estimated difference in average time spent on homework for college physics majors versus college English majors? Round to one decimal place.

568. Using a 95% confidence level, what is the confidence interval for the true difference in average time spent on homework for college physics majors versus college English majors? Round your answer to the nearest tenth.

569. Using a 99% confidence level, what is the confidence interval for the true difference in average time spent on homework for college physics majors versus college English majors? Round your answer to the nearest tenth.

570. If you did not know the population standard deviations, how would your calculation of confidence intervals differ?

(A) You would use $t*$ from a t-distribution rather than $z*$ from the standard normal distribution.

(B) You would use the sample standard deviations rather than the population standard deviations.

(C) You would combine the sample sizes and divide the sum of the standard deviations by $(n_1 + n_2)$.

(D) Choices (A) and (B)

(E) Choices (A), (B), and (C)

571. If your sample sizes were 35 English majors and 20 physics majors, how, if at all, would your calculation of confidence intervals differ?

Digging Deeper into Calculations When Population Standard Deviations Are Known

572–574 A random sample of 200 men in North America found that their average age at first marriage was 29 years old; the standard deviation for the population was 6 years. A random sample of 220 women in North America found that their average age at first marriage was 26 years old; the standard deviation for the population was 4 years. While working these problems, call the group of men "Population 1" and the group of women "Population 2."

572. For an 80% confidence level, what is the margin of error for the estimate of the difference in average age at first marriage for men and women? Round your answer to the nearest tenth.

573. For a 90% confidence level, what is the margin of error for the estimate of the difference in average age at first marriage for men and women? Round your answer to the nearest tenth.

574. For a 95% confidence level, what is the confidence interval for the true difference in mean age at first marriage for men and women in North America? Round your answer to the nearest tenth.

Working with Unknown Population Standard Deviations and Small Sample Sizes

575–580 A random sample of 20 12th-grade boys in a high school showed a mean weight of 170 pounds with a sample standard deviation of 18 pounds. A random sample of 20 9th-grade boys in the same school showed a mean weight of 140 pounds with a sample standard deviation of 12 pounds. While working these problems, call the population of 12th graders "Population 1" and the population of 9th graders "Population 2."

575. What degrees of freedom will you use to calculate the confidence interval for the difference in weights?

576. Using a 99% level of confidence, what is the margin of error for the estimated difference in mean weights between 12th-grade boys and 9th-grade boys at this school? Round to the nearest whole number of pounds.

577. Using an 80% level of confidence, what is the margin of error for the estimated difference in mean weights between 12th-grade boys and 9th-grade boys at this school? Round to the nearest whole number of pounds.

578. Using a 90% level of confidence, what is the margin of error for the estimated difference in mean weights between 12th-grade boys and 9th-grade boys at this school? Round to the nearest whole number of pounds.

579. Using a 95% confidence level, what is the confidence interval for the true difference in the mean weights of all the 12th-grade and 9th-grade boys at this school? Round the endpoints to the nearest whole number.

580. Using a 98% confidence level, what is the confidence interval for the true difference in the mean weights of the 12th-grade and 9th-grade boys at this school? Round the endpoints to the nearest whole number.

Digging Deeper into Unknown Population Standard Deviations and Small Sample Sizes

581–584 A random sample of 20 men in North America found that their average annual income after five years of employment, in thousands of dollars, was 37, with a standard deviation for the sample of 3.5. A random sample of 25 women in North America found that their average income after five years of employment, in thousands of dollars, was 30, with a standard deviation for the sample of 3. While working these problems, call the population of men "Population 1" and the population of women "Population 2."

581. You want to calculate a confidence interval for the true difference in average salaries for all men and women after five years of employment (in thousands of dollars). What degrees of freedom will you use?

582. With a 90% level of confidence, what is the margin of error for the estimated difference between average male and female salaries after five years of employment (in thousands of dollars)? Round your answer to one decimal place.

583. With a 95% level of confidence, what is the margin of error for the estimated difference between average male and female salaries after five years of employment? Round your answer to one decimal place.

584. With a 99% level of confidence, what is the confidence interval for the true difference in average income between all North American men and women after five years of employment? Round the confidence limits in your answer to one decimal place.

Chapter **11**

Claims, Tests, and Conclusions

Hypothesis testing is a scientific procedure for asking and answering questions. Hypothesis tests help people decide whether existing claims about a population are true, and they're also commonly used by researchers to see whether their ideas have enough evidence to be declared statistically significant. This chapter is about understanding the basics of hypothesis testing.

The Problems You'll Work On

In this chapter, you break down the basic elements of a hypothesis test. Here's an overview of the problems you'll work on:

>> Setting up a pair of hypotheses: the current claim (null hypothesis) and the one challenging it (alternative hypothesis)

>> Using data to form a test statistic, which determines how far apart the two hypotheses are

>> Measuring the strength of the new evidence through a probability (*p*-value)

>> Making your decision as to whether the current claim can be overturned

>> Understanding that your decision can be wrong and what errors you can commit

What to Watch Out For

Hypothesis testing has two levels at each step: how to do it and figuring out what it means. Here are some things to pay special attention to as you work through this chapter:

>> Realizing it's all about the null hypothesis and whether you have enough evidence to overturn it

>> Knowing the role of the test statistic, beyond calculating it and looking it up on a table

>> Understanding how to interpret a *p*-value properly and why a small *p*-value means you reject H_0

>> Being clear on Type I and Type II errors in the real sense — as false alarms and missed opportunities

Knowing When to Use a Hypothesis Test

585–586 *Solve the following problems about using a hypothesis test.*

585. Which of the following statements, as currently written, could be tested using a hypothesis test?

(A) An automobile factory claims 99% of its parts meet stated specifications.

(B) An automobile factory claims that it produces the best-quality cars in the country.

(C) An automobile factory claims that it can assemble 500 automobiles an hour when the assembly line is fully staffed.

(D) Choices (A) and (B)

(E) Choices (A) and (C)

586. Which of the following scenarios, as currently stated, could *not* involve a hypothesis test without further clarification?

(A) A political party conducts a survey in an attempt to contradict published claims of the proportion of voter support for a proposed law.

(B) A commercial laboratory does sample tests on a hand sanitizer to see whether it kills the percentage of bacteria claimed by the manufacturer.

(C) A school gives its students standardized tests to measure levels of achievement compared to prior years.

(D) A laboratory takes samples of a yogurt to see whether the manufacturer has met its published standard of being 99% fat-free.

(E) A university evaluation group gives random surveys to students to see whether university claims regarding the proportion of students who are satisfied with student life are valid.

Setting Up Null and Alternative Hypotheses

587–604 *Determine null and alternative hypotheses in the following problems.*

587. You decide to test the published claim that 75% of voters in your town favor a particular school bond issue. What will your null hypothesis be?

588. You decide to test the published claim that 75% of voters in your town favor a particular school bond issue. What will your alternative hypothesis be?

589. Given the null hypothesis H_0: $\mu = 132$, what is the correct alternative hypothesis?

590. A university claims that work-study students earn an average of $10.50 per hour. What is the null hypothesis for a hypothesis test of this statement?

591. The manufacturer of the new GVX Hybrid car claims that it gets an average of 52 miles per gallon of gas. What is the null hypothesis for this statement?

592. Suppose that μ is the average number of songs on an MP3 player owned by a college student. Write down the description of the null hypothesis H_0: $\mu = 228$.

593. A think tank announces that 78% of teen-agers own cell phones. What is the null hypothesis for a hypothesis test of this statement?

594. A travel agency claims that people from States 1 and 2 are equally likely to have taken a vacation in Hawaii. What is the null hypothesis for this statement?

595. According to a newspaper report, seven out of ten Americans think Congress is doing a good job. What alternative hypothesis would you use if you believe this stated proportion is too high?

596. Amtrak claims that a train trip from New York City to Washington, D.C., takes an average of 2.5 hours. What alternative hypothesis would you use if you think the average trip length is actually longer?

597. An airline company claims that its flights arrive early 92% of the time. What alterna-tive hypothesis would you use if you think this statistic is too high?

598. A car manufacturer advertises that a new car averages 39 miles per gallon of gasoline. What alternative hypothesis would you use if you think this statistic is too high?

599. A company claims that only 1 out of every 200 computers it sells has a mechanical malfunction. What alternative hypothesis would you use if you think this statistic is too low?

600. A hospital claims that only 5% of its patients are unhappy with the care pro-vided. What is the alternative hypothesis if you think this statistic is too low?

601. A health study states that American adults consume an average of 3,300 calories per day. What alternative hypothesis would you use if you think this statistic is incorrect?

602. A study claims that adults watch televi-sion an average of 1.8 hours per day. What alternative hypothesis would you use if you think this figure is too low?

603. An investment company claims that its clients make an average of 8% return on investments every year. What alternative hypothesis would you use if you think this figure is too high?

604. Someone claims that high-school students living in cities with populations more than 1 million (Population 1) are 25% more likely to attend college than high-school students living in cities with populations less than 1 million (Population 2). Write the alternative hypothesis if you think this statistic is incorrect.

Finding the Test Statistic and the *p*-Value

605–612 You're conducting an experiment with the following hypotheses:

$$H_0: \mu = 4$$
$$H_a: \mu \neq 4$$

The standard error is 0.5, and the alpha level is 0.05. The population of values is normally distributed.

605. If $\bar{x} = 3$ in your sample, what is the test statistic?

606. If $\bar{x} = 4.5$ in your sample, what is the test statistic?

607. If $\bar{x} = 5.2$ in your sample, what is the test statistic?

608. If $\bar{x} = 3.6$ in your sample, what is the test statistic?

609. Suppose that your test statistic is 1.42. What is the *p*-value for this result?

610. Suppose that your test statistic is –1.56. What is the *p*-value for this result?

611. Suppose that your test statistic is 0.75. What is the *p*-value for this result?

612. Suppose that your test statistic is –0.81. What is the *p*-value for this result?

Making Decisions Based on Alpha Levels and Test Statistics

613–618 Suppose that you're conducting a study with the following hypotheses:

$$H_0: p = 0.45$$
$$H_a: p > 0.45$$

613. If your alpha level (significance level) is 0.05 and your test statistic is 1.51, what will your decision be? Assume that n is large enough to use the central limit theorem.

614. If your alpha level is 0.10 and your test statistic is 1.51, what will your decision be? Assume that n is large enough to use the central limit theorem.

615. If your alpha level is 0.01 and your test statistic is 1.98, what will your decision be? Assume that n is large enough to use the central limit theorem.

616. If your alpha level is 0.05 and your test statistic is 1.98, what will your decision be? Assume that n is large enough to use the central limit theorem.

617. If your alpha level is 0.05 and your test statistic is −1.98, what will your decision be? Assume that n is large enough to use the central limit theorem.

618. If your alpha level is 0.01 and your test statistic is −3.0, what will your decision be? Assume that n is large enough to use the central limit theorem.

Making Conclusions

619–633 Make conclusions after reading the information in the following problems.

619. What does it mean if a test statistic has a p-value of 0.01?

620. You are conducting a statistical test with an alpha level of 0.10. Which of the following is true?

(A) There is a 10% chance that you will reject the null hypothesis when it is true.

(B) There is a 10% chance that you will fail to reject the null hypothesis when it is false.

(C) You should reject the null hypothesis if your test statistic has a p-value of 0.10 or less.

(D) Choices (A) and (C)

(E) Choices (B) and (C)

621. The alpha level of a test is 0.05. The p-value for your test statistic is 0.0515. What is your decision?

622. A test was done with a significance level (α level) of 0.05, and the p-value was 0.001. Write down the best description of this result.

623. A test is done to challenge the statistic that 70% of people spend their summer vacations at home. The significance level is $\alpha = 0.05$.

$$H_0: p = 0.70$$
$$H_a: p < 0.70$$
$$p\text{-value} = 0.03$$

What can you conclude about the results?

624. If the significance level α is 0.02, which p-value for a test statistic will result in a test conclusion to reject H_0?

(A) 0.03

(B) 0.01

(C) 0.05

(D) 0.97

(E) 0.98

625. If the significance level α is 0.05, which p-value for a test statistic will result in a test conclusion to reject H_0?

(A) 0.95

(B) 0.10

(C) 0.06

(D) 0.055

(E) 0.04

626. Based on the following information, what do you conclude?

$$H_0: p = 0.03$$
$$H_a: p \neq 0.03$$
$$\alpha = 0.01$$
$$p\text{-value} = 0.007$$

627. Based on the following information, what do you conclude?

$$H_0: p = 0.65$$
$$H_a: p > 0.65$$
$$\alpha = 0.03$$
$$p\text{-value} = 0.02$$

628. Based on the following information, what do you conclude?

$$H_0: \mu = 220$$
$$H_a: \mu < 220$$
$$\alpha = 0.05$$
$$p\text{-value} = 0.06$$

629. Based on the following information, what do you conclude?

$$H_0: p = 0.42$$
$$H_a: p > 0.42$$
$$\alpha = 0.05$$
$$p\text{-value} = 0.42$$

630. Based on the following information, what do you conclude?

$$H_0: \mu = 0.2$$
$$H_a: \mu > 0.2$$
$$\alpha = 0.02$$
$$p\text{-value} = 0.2$$

631. Based on the following information, what do you conclude?

$$H_0: \mu = 10$$
$$H_a: \mu \neq 10$$
$$\alpha = 0.01$$
$$p\text{-value} = 0.018$$

632. Based on the following information, what do you conclude?

$$H_0: \mu = 9.65$$
$$H_a: \mu > 9.65$$
$$\alpha = 0.05$$
$$\text{Test statistic: } -1.88$$
$$p\text{-value} = 0.03$$

633. Based on the following information, what is your conclusion?

$$H_0: \mu = 348$$
$$H_a: \mu > 348$$
$$\alpha = 0.05$$
$$p\text{-value} = 0.07$$

Understanding Type I and Type II Errors

634–640 Solve the following problems about Type I and Type II errors.

634. Which of the following describes a Type I error?

(A) accepting the null hypothesis when it is true

(B) failing to accept the alternative hypothesis when it is true

(C) rejecting the null hypothesis when it is true

(D) failing to reject the alternative hypothesis when it is false

(E) none of the above

635. Which of the following describes a Type II error?

(A) accepting the alternative hypothesis when it is true

(B) failing to accept the alternative hypothesis when it is true

(C) rejecting the null hypothesis when it is true

(D) failing to reject the null hypothesis when it is false

(E) none of the above

636. If the alpha level is 0.01, what is the probability of a Type I error?

637. If the alpha level is 0.05, what is the probability of a Type II error?

638. Which of the following is a description of the power of the test?

(A) the probability of accepting the alternative hypothesis when it is true

(B) the probability of failing to accept the alternative hypothesis when it is true

(C) the probability of rejecting the null hypothesis when it is true

(D) the probability of rejecting the null hypothesis when it is false

(E) none of the above

639. What is the key to avoiding a Type II (missed detection) error?

(A) having a low significance level

(B) having a random sample of data

(C) having a large sample size

(D) having a low p-value

(E) Choices (B) and (C)

640. What is the key to avoiding a Type I error?

(A) having a low significance level

(B) having a random sample of data

(C) having a large sample size

(D) having a low p-value

(E) Choices (A), (B), and (C)

Chapter **12**

Hypothesis Testing Basics for a Single Population Mean: *z*- and *t*-Tests

Conducting a hypothesis test is somewhat like doing detective work. Every population has a mean, and it's usually unknown. Many people claim that they know what it is; others assume that it hasn't changed from a past value; and in many cases, the population mean is supposed to follow certain specifications. Your modus operandi is to challenge or test that value of the population mean that's already assumed, given, or specified and use data as your evidence. That's what hypothesis testing is all about.

The Problems You'll Work On

In this chapter, you'll work out the basic ideas of hypothesis testing in the context of the population mean, including the following areas:

>> Setting up the original, or null, hypothesis (the assumed or specified value) and the alternative hypothesis (what you believe it to be)

>> Working through the details of doing a hypothesis test

>> Making conclusions and assessing the chance of being wrong

What to Watch Out For

Hypothesis testing can seem like a plug-and-chug operation, but that can take you only so far. To truly master the materials in this chapter, keep the following in mind:

>> Pay careful attention to the problem to determine how to set up the alternative hypothesis.

>> Make sure you can calculate a test statistic and, more importantly, know what it's telling you.

>> Remember that a small *p*-value comes from a large test statistic, and both mean rejecting H$_0$.

>> Get an intuitive feeling about what Type 1 and Type 2 errors are — don't simply memorize!

Knowing What You Need to Run a *z*-Test

641–642 Figure out what you need to know to run a z-test in the following problems.

641. Dr. Thompson, a health researcher, claims that a teenager in the United States drinks an average of 30 ounces of sugary carbonated soda per day. A high-school statistics class decides to test this claim and is open to soda consumption being, on average, either higher or lower than the claimed 30 ounces per day. They conduct a random survey of 15 of their classmates and find self-reported soda consumption is, on average, 25 ounces per day. What additional information would you need to run a *z*-test to determine whether the students in this school drink significantly more or less soda than Dr. Thompson claims is consumed by U.S. teens in general?

642. You want to run a *z*-test to determine whether the sample from which a sample mean is drawn differs significantly from a population mean. You have the sample mean and sample size ($n = 20$); what other information do you need to know?

(A) whether the characteristic of interest is normally distributed in the population

(B) the population size

(C) the population mean and standard deviation

(D) the sample standard deviation

(E) Choices (A) and (C)

Determining Null and Alternative Hypotheses

643–646 Figure out null and alternative hypotheses in the following problems.

643. A researcher believes that people who smoke have lower shyness scores than the population average on a shyness scale, which is 25. What is the null hypothesis in this case, given that μ is the mean shyness score for all smokers?

644. Suppose a researcher has heard that children watch an average of ten hours of TV per day. The researcher believes this is wrong but has no theory about whether it's an overestimate or an underestimate of the truth. If the researcher wants to do a *z*-test of one population mean, what will the researcher's alternative hypothesis be?

645. A computer store owner reads that its consumers buy five flash drives each year, on average. The owner feels that, on average, their customers actually buy more than that, so the owner does a random survey of the customers on their mailing list. What is the store owner's alternative hypothesis?

646. A person reads that the average cost of dry-cleaning a shirt is 3 dollars, but in their city, it seems cheaper. So, they randomly choose ten dry cleaners in town and asks them the price to dry-clean a shirt. What will their alternative hypothesis be?

Introducing *p*-Values

647–650 Calculate p-values in the following problems.

647. A researcher has a *less than* alternative hypothesis and wants to run a single sample mean z-test. The researcher calculates a test-statistic of $z = -1.5$ and then uses a Z-table (such as Table A-1 in the appendix) to find a corresponding area of 0.0668, which is the area under the curve to the left of that value of z. What is the p-value in this case?

648. Suppose that a researcher has a *not equal to* alternative hypothesis and calculates a test statistic that corresponds to $z = -1.5$ and then finds, using a Z-table (such as Table A-1 in the appendix), a corresponding area of 0.0668 (the area under the curve to the left of that value of z). What is the p-value in this case?

649. A researcher has a *not equal to* alternative hypothesis and calculates a test statistic that corresponds to $z = -2.0$. Using a Z-table (such as Table A-1 in the appendix), the researcher finds a corresponding area of 0.0228 to the left of -2.0. What is the p-value in this case?

650. A scientist with a *not equal to* alternative hypothesis calculates a test statistic that corresponds to $z = 1.1$. Using a Z-table (such as Table A-1 in the appendix), the scientist finds that this corresponds to a curve area of 0.8643 (to the left of the test statistic value). What is the p-value in this case?

Calculating the *z*-Test Statistic

651–652 Determine the z-test statistic in the following problems.

651. A coffee shop manager reads that the preferred temperature for coffee among the U.S. populous is 110 degrees Fahrenheit with a standard deviation of 10 degrees. However, the manager doesn't believe this is true of their customers. Through complex and extensive testing, the manager finds that a random sample of 50 of their customers prefer their coffee, on average, to be 115 degrees Fahrenheit. Calculate the z-test statistic for this case and give your answer to two decimal places.

652. A very mathematically oriented musician has read studies showing that the average piece of popular music has 186.39 chord changes with a standard deviation of 26.52. The musician examines a random sample of 40 of their favorite songs and finds a mean of 172.12 chord changes. Calculate the z-statistic for this case and give your answer to four decimal places.

Finding *p*-Values by Doing a Test of One Population Mean

653–654 Calculate p-values in the following problems.

653. A pen company surveys the market and finds that people, on average, hope to use a pen for 40 days before having to replace it, with a standard deviation of 9 days. The company director of research believes that customers are actually less ambitious and will hope to use a pen for fewer than 40 days. The research director conducts a customer survey of 25 customers and finds that its customers, on average, expect to replace a pen after 36 days. What is the *p*-value if the director uses a *Z*-distribution to do a test of one population mean?

654. A berry farmer decided to grow blackberries after reading that bushes give an average of 3 pounds of fruit per year, with a standard deviation of 1 pound. The farmer suspects they won't get quite so much fruit because growing conditions aren't entirely right. The farmer identifies a random sample of 100 bushes and keeps careful track of how much fruit each bush gives. At the end of the year, the farmer finds that the average yield for each of the bushes in the sample was 2.9 pounds of fruit. Assume that the farmer conducts a *z*-test of a single population mean. What *p*-value will the farmer get?

Drawing Conclusions about Hypotheses

655–657 Figure out the conclusions that can be drawn in the following problems.

655. A psychiatrist reads that the average age of someone who's first diagnosed with schizophrenia is 24 years, with a standard deviation of 2 years. The psychiatrist suspects that the schizophrenic patients in their clinic were diagnosed at a younger age than 24. The psychiatrist examines records for a random sample of clinic patients and finds that the age at first diagnosis in the sample is 23.5 years on average. If age at first diagnosis is normally distributed and the *p*-value found in this case is 0.02, and the psychiatrist wants to run a test with a 0.05 significance level, what conclusion can the psychiatrist draw?

656. Airplane passengers carry an average of 45 pounds of luggage on a flight, with a standard deviation of 10 pounds. A researcher suspects that business travelers carry less than that on average. The researcher randomly samples 250 business travelers and finds that they carry an average of 44.5 pounds of luggage when traveling. The researcher wants a significance level of 0.05. What conclusion can the researcher draw based on this data pattern if the researcher performs a *z*-test for one population mean?

657. In a particular city, a square foot of office space averages $2.00 per month in rent, with a standard deviation of $0.50. A shopkeeper hopes to open a store in a neighborhood with significantly cheaper rent than that, using a 0.05 level of significance. They sample 49 offices at random from the neighborhood they're most interested in. The shopkeeper finds that the average rent in their sample is $3.00 per month. The shopkeeper conducts a z-test and finds their p-value to be 0.10. What can the shopkeeper conclude about their average rent?

Digging Deeper into *p*-Values

658–659 *Calculate p-values in the following problems.*

658. Assume that the average temperature of a healthy human being is 98.6 degrees Fahrenheit with a standard deviation of 0.5 degrees. A doctor believes that their patients average a higher temperature than that, so they randomly select 36 of their patients and find that their temperature is 98.8 degrees on average. If the doctor carries out a z-test for one population mean, what is the p-value for these results? Give your answer to four decimal places.

659. The average satisfaction rating for a company's customers is a 5 (on a scale of 0 to 7), with a standard deviation of 0.5 points. A researcher suspects that the Northeastern division has customers who are more satisfied. After randomly sampling 60 customers from the Northeastern division and asking them about their satisfaction, the researcher finds an average satisfaction level of 5.1. What is the p-value in this case if the researcher does a z-test for a single sample mean?

Digging Deeper into Conclusions about Hypotheses

660–665 *Figure out the conclusions that can be drawn in the following problems.*

660. In a company, employees type an average of 20 words per minute. Typing rates are normally distributed with a standard deviation of 3. The manager of a large branch of the company believes that their employees do better than that. They randomly sample 30 employees from their branch and find an average typing rate of 20.5 words per minute. If the manager wants a significance level of 0.05, what can they conclude?

661. A potter believes that their workshop assistants can cover a certain size of vase with only 2 ounces of glaze, which is the industry standard. The potter knows that the amount of glaze required to cover a vase of the specified size follows a normal distribution with a standard deviation of 0.8 ounces, and they believe their workshop is still putting too much glaze on each vase. The potter samples 30 vases from a large production run and finds a sample mean of 2.3 ounces of glaze. Using a significance level of 0.01, what can they conclude?

662. A researcher believes that the tissue cultures in their lab are significantly denser than average. The researcher takes a random sample of 40 specimens of tissue from their lab and finds the average weight is 0.005 grams per cubic millimeter. Textbooks claim that such tissues should weigh 0.0047 grams per cubic millimeter with a standard deviation of 0.00047 grams. What can this researcher conclude if they use a one-sample z-test and a significance level of 0.001?

663. On average, hens lay 15 eggs per month with a standard deviation of 5. A farmer tests this claim on their hens. They sample 30 hens and find that the hens lay an average of 16.5 eggs per month. Can the farmer reject the null hypothesis that their hens, on average, lay the same number of eggs as the larger hen population? Use a one-sample z-test and a significance level of 0.05.

664. A nationally known moving company knows that the typical family uses 110 boxes in a long-distance move, with a standard deviation of 30 boxes. A company from Chicago wants to see how it compares. The company randomly samples 80 families from the Chicago area and finds that they used, on average, 103 boxes in their most recent move. Using a one-sample z-test and a significance level of 0.05, what is your decision?

665. A travel magazine claims that the American family who uses a car to go on vacation travels an average of 382 miles from home. A researcher believes that families with dogs who vacation by car on average travel a different distance. The researcher draws a sample of 30 families with dogs who drive on their vacations and finds that they drive an average of 398 miles. (Assume that the population standard deviation is 150 miles.) At a 0.05 significance level and a one-sample z-test, what can you conclude from this data?

Digging Deeper into Null and Alternative Hypotheses

666–667 Figure out null and alternative hypotheses in the following problems.

666. A magazine reports that the average number of minutes that U.S. teenagers spend texting each day is 120. You believe it's less than that. What are your null and alternative hypotheses?

667. The average number of calories that a pizza slice contains is 250, with a standard deviation of 35 calories. A nutrition researcher suspects that slices of pizza on the college campus where they work contain a higher number of calories. That researcher randomly samples 35 pizza slices in the area of their campus and finds a mean of 265 calories. What are the null and alternative hypotheses?

Knowing When to Use a *t*-Test

668–670 Solve the following problems about knowing when to use a t-test.

668. Which of the following conditions indicate that you should use a *t*-test instead of using the *Z*-distribution to test a hypothesis about a single population mean? (Assume that the population has a normal distribution.)

(A) The population standard deviation is unknown.

(B) The population standard deviation is known.

(C) The sample standard deviation is unknown.

(D) The sample standard deviation is known.

(E) None of the above conditions are related to a *t*-test.

669. A researcher is trying to decide whether to use a *Z*-distribution or a *t*-test to evaluate a hypothesis about a single population mean. Which of the following conditions would indicate that the researcher should use the *t*-test?

(A) The population standard deviation isn't known.

(B) The sample size is only $n = 50$.

(C) The sample mean is less than the population mean.

(D) The alternative hypothesis is a *not equal to* hypothesis.

(E) The sampling distribution of sample means is normal.

670. A student finds that a sample of 50 of their friends reports, on average, that they spend 43 hours per week using social networking sites, with a sample standard deviation of 8 hours. The student believes that this is significantly less than journalists' claims that students spend 50 hours per week using social networking sites. The student has a *less than* alternative hypothesis about a single population mean. How should the student test this hypothesis?

Connecting Hypotheses to *t*-Tests

671–674 Solve the following problems about hypotheses and t-tests.

671. A student believes that their friends spend less time on social networking sites, on average, than is claimed in the media. If μ_1 is the average amount of time spent by the student's friends and μ_0 is the amount of time claimed in the media, what is the null hypothesis the student would use to do a *t*-test on a single population mean?

672. A student believes that their friends spend less time on social networking sites, on average, than is claimed in the media. If μ_1 is the average amount of time spent by the student's friends and μ_0 is the amount of time claimed in the media, what is the alternative hypothesis the student would use to do a *t*-test on a single population mean?

673. A national retail chain says in its ad that the average price of a certain hair product it sells across the United States is $10 per bottle (the null hypothesis). A manager of one of the stores in the chain believes that it's more than that. The manager samples 30 bottles of the product at random from their own store and conducts a t-test. The manager's p-value is smaller than 0.05 (pre-specified significance level), so they reject the null hypothesis. (Assume that the population standard deviation is unknown.) Based on the manager's data, can they conclude that the average price of this product across the United States is actually more than $10 per bottle?

674. Suppose that a dentist believes their patients experience less pain than does the average dental patient. The dentist samples 40 of their patients and receives pain ratings from them after they receive a filling. The dentist then compares their sample mean and standard deviation with a population value they found in a medical journal. The dentist's sample average pain rating was 3.2 (on a 10-point scale), and the medical journal reported an average pain rating of 3.5. What are the null and alternative hypotheses in this case?

Calculating Test Statistics

675–676 *Figure out test statistics in the following problems.*

675. Use the following information to calculate a t-value (test statistic).

> Sample mean: 30
>
> Claimed population mean: 35
>
> Sample standard deviation: 10
>
> Sample size: 16

What is the value for the test statistic, t?

676. It's believed that the average amount of sleep a person in the United States gets per night is 6.3 hours. A mom believes that mothers get far less sleep than that. They contact a random sample of 20 other moms on a mom social networking site and find that they get an average of 5.2 hours of sleep per night, with a standard deviation of 1.8 hours. Using this data, what is the value of the test statistic t from a t-test for a single population mean?

Working with Critical Values of t

677–680 *Solve the following problems about the critical value of t.*

677. A researcher hypothesizes that a population of interest has a mean greater than 6.1. The researcher uses a sample of 15. What is the critical value of t needed to reject the null hypothesis, using a significance level of 0.05?

678. Suppose that a researcher believes that a sampled population of interest has a mean that differs from a claimed value of 100 but is unsure of the direction of the difference. The researcher wants a 0.05 significance level with a sample size of $n = 10$. What critical value of t should the researcher use for a t-test involving a single population mean?

679. It's believed that the average amount that a person in the United States sleeps is 6.3 hours per night. A psychologist believes that mothers get far less sleep than that. The psychologist contacts a random sample of other moms on a social networking site for moms, and they find among the sample of 20 moms an average of 5.2 hours of sleep per night with a standard deviation of 1.8 hours. If this psychologist wants a significance level of 0.05 and sets up a *less than* alternative hypothesis, what would they conclude based on the t-value of −2.733 in this case?

680. A researcher wants to use a 90 percent confidence interval to determine whether their sample of 17 ball bearings differs in either direction from an average target value of 0.0112 grams. In the sample, the standard deviation is 0.0019 grams, and the mean is 0.0123 grams. How does the critical value of t compare with the observed value of t in this case?

Linking *p*-Values and *t*-Tests

681–685 Solve the following problems about *p*-values and *t*-tests.

681. A study finds a test statistic t-value of 1.03 for a t-test on a single population mean. The sample size is 11, and the alternative hypothesis is $H_a: \mu \neq 5$. Using Table A-2 in the appendix, what range of values is sure to include the p-value for this value of t?

682. A researcher believes that their sample mean is smaller than 90 and finds a sample mean of 89.8 with a sample standard deviation of 1. Using the t-table (Table A-2 in the appendix), and given that the sample had 29 observations, what is the approximate p-value?

683. Imagine that it costs an average of $50,000 per year to incarcerate someone in the United States. A prison warden wants to know how the costs in their prison compare to the population average, so they randomly sample 12 of their inmates, reviews their records carefully, and finds an average cost of $58,660 per inmate, with a standard deviation of $10,000. If they perform a t-test on a single population mean, what is the approximate p-value for this data?

684. A researcher has a *greater than* alternative hypothesis and observes a sample mean greater than the claimed value. The test statistic t for a t-test for a single population mean is 2.5, with 14 degrees of freedom. What is the p-value associated with this test statistic?

685. A researcher with a *not equal to* alternative hypothesis observes a sample mean less than the claimed value. The test statistic is found to be −2.5 with 20 degrees of freedom. What is the p-value associated with this test statistic value of t if the researcher runs a t-test for a single population mean?

Drawing Conclusions from *t*-Tests

686–692 *Make conclusions from t-tests in the following problems.*

686. An instructor claims their students take an average of 45 minutes to complete their exams. You think that the average time is higher than that. You decide to investigate this using a significance level of 0.01. You take a sample, find the test statistic, and find the *p*-value is 0.0001 (small by anyone's standards). What is your conclusion?

687. Suppose it's claimed that people who play musical instruments have average skills in verbal ability. A scientist, however, has a hypothesis that people who play musical instruments are below average in verbal ability. On a verbal ability test with a population mean score of 100, a random sample of 8 musicians yields an average score of 97.5, with a standard deviation of 5. Verbal ability, as measured by this test, is normally distributed in the population. What should the scientist conclude if the significance level for their test is 0.05?

688. The president of a large corporation believes their employees donate less money to charity than the corporate target of $50 per year, per employee. They conduct a survey of ten randomly selected employees and find an average annual donation of $43.40 with a standard deviation of $5.20. Donations are normally distributed in this population. If the president does a *t*-test for a single population mean, with a 0.05 level of significance, what should they conclude?

689. A coat maker promises that their heavy winter coats feel warm down to a temperature of −5 degrees Centigrade, but they believe that their coats actually protect people well at even colder temperatures. They survey 15 of their customers, and they report, on average, that the coats feel warm down to a temperature of −6.5 degrees Centigrade, with a standard deviation of 1.0 degrees Centigrade. If the coat maker wants a 0.10 significance level ($\alpha = 0.10$), what should the coat maker conclude if they run a one-sample *t*-test, assuming that temperatures are normally distributed?

690. A teacher believes that other teachers in their school get lower evaluations compared to teachers in other schools in the district. They know that the average teaching evaluation in the district is a rating of 7.2 on a 10-point scale, with scores normally distributed. They survey six randomly selected teachers in their school and finds an average teaching rating among them of 6.667 with a standard deviation of 2. What should they conclude from a single population *t*-test, using their data and a significance level of 0.05?

691. A popsicle company tries to keep its products at a temperature of −1.92 degrees Centigrade. The company president believes that the freezers are set too low, thus wasting money to keep products colder than necessary. They randomly sample five freezers and find that they're running at an average temperature of −2.25 degrees Centigrade, with a sample standard deviation of 1.62 degrees. Overall, temperature is normally distributed for this brand of freezer. Using a significance level of 0.01, what should they conclude?

692. You're given the following information and asked to perform a t-test on a single population mean. The null hypothesis is that the mean weight of all parts of an object of a certain type is 50 grams per object: H_0: $\mu = 50$. The alternative hypothesis is that the objects are heavier on average: H_a: $\mu > 50$. The sample mean is 54 grams, and the sample standard deviation is 8 grams. The sample size is 16. Weight is normally distributed in this population. Using a significance level of $\alpha = 0.05$, what is your conclusion?

Performing a t-Test for a Single Population Mean

693–694 Suppose you expect that the average number of beads in a 1-pound bag coming from the factory is 1,200. However, the retailer believes that the average number is larger. You draw a random sample of 30 bags and find that the mean of that sample is 1,350 beads, with a standard deviation of 500. Assuming that the population of values is normally distributed, you use this information to perform a t-test for a single population mean.

693. Using a significance level of 0.01, what is your decision?

694. Using a significance level of 0.05, what is your decision?

Drawing More Conclusions from t-Tests

695–700 Determine the conclusions that should be drawn in the following problems.

695. A popular financial advisor states that, on average, a person should spend no more than $100 per month on entertainment. A researcher believes that the average person spends more than that and conducts a survey of 25 randomly chosen people. In that survey, the average monthly spending on entertainment was found to be $118.44, with a standard deviation of $35.00. Assuming that spending follows a normal distribution in the population and using a significance level of $\alpha = 0.01$, what should the researcher conclude, based on a t-test of a single mean?

696. A Wisconsin cheese company read that, on average, a person in Europe consumes 25.83 kilograms of cheese per year. The company's sales director believes that Americans consume even more cheese. The company does a survey of a random sample of Americans and finds that they eat, on average, 27.86 kilograms of cheese per year with a standard deviation of 6.46 kilograms. Assume that cheese consumption among Americans is normally distributed. If there were 30 people in the sample and the significance level is 0.05, what can the cheese company managers conclude about their sales director's hypothesis?

697. A meditation teacher reads that the ideal amount of time to meditate each day is 20 minutes. They want to know whether the practice of their students differs significantly from the ideal amount of meditation by running a t-test on a single mean, using a significance level of $\alpha = 0.10$. The teacher surveys nine people randomly chosen from their meditation classes and finds that they meditate for an average of 24 minutes per day with a standard deviation of 5 minutes. Assuming that meditation time among all such students is normally distributed, what should the teacher conclude?

698. Suppose that a heavy-duty laser printer is claimed to produce an average of 20,000 pages of print before needing to be serviced. Also, assume that the page output of such printers until service is needed is normally distributed. A company that uses a lot of heavy-duty laser printers randomly samples 16 of its printers to see how many pages it gets, per printer, before needing service. The study finds an average of 18,356 pages between servicing and a standard deviation of 2,741 pages. Using a significance level of 0.05 and a *not equal to* alternative hypothesis, what is your decision?

699. A doctoral student has heard that dissertations are, on average, 90 pages in length, but they believe that dissertations written by students in their program may have a different average length. They randomly select ten dissertations completed by people who went through the doctoral student's program and find a mean length of 85.2 pages with a standard deviation of 7.59 pages. If they assume that dissertation page length is normally distributed and runs a t-test on a single mean with a significance level of 0.05, what should they conclude?

700. A logger walks through a forest with their spouse. The logger's spouse estimates that the average tree in the forest is 30 years old. Because all the trees in a certain area of the forest are to be cleared anyway, the logger randomly chooses five trees, cuts them down, and counts the rings. The logger is interested in knowing only whether these trees differ from theirspouse's estimated mean of 30 years on average, so they choose a significance level of $\alpha = 0.50$. Their sample averaged 33 years of age with a standard deviation of 5.6 years. In this forest, tree age is approximately normally distributed. What is the logger's conclusion, based on a significance test for the true average age of all trees planned to be cleared?

Chapter **13**

Hypothesis Tests for One Proportion, Two Proportions, or Two Population Means

This chapter gives you practice doing hypothesis tests for three specific scenarios: testing one population proportion; testing for a difference between two population proportions; and testing for a difference between two population means. Most hypothesis tests use a similar framework, so patterns will develop, but each hypothesis test has its own special elements, and you work on those here.

The Problems You'll Work On

In this chapter, you practice setting up and carrying out three different hypothesis tests, focusing specifically on the following:

>> Knowing what hypothesis test to use and when

>> Setting up and understanding the null and alternative hypotheses correctly in all cases

>> Working through the appropriate formulas and knowing where to get the needed numbers

>> Calculating and interpreting the test statistics and *p*-values

What to Watch Out For

As you go through more hypothesis tests, knowing their similarities and differences is important. Here are a few notes about this chapter:

>> Notation plays a big role — make a list for yourself to keep it all straight.

>> The formulas go up a notch in this chapter. Be sure you label and understand the parts of the formula and what clues you need to set them up.

>> If you do find a statistically significant difference, make it clear which population is the one with the larger proportion or mean.

Testing One Population Proportion

701–715 Solve the following problems about testing one population proportion.

701. A bank will open a new branch in a particular neighborhood if it can be reasonably sure that at least 10% of the residents will consider banking at the new branch. The bank will use a significance level of 0.05 to make its decision. The bank does a survey of residents of a particular neighborhood and finds that 19 out of 100 random people surveyed said they'd consider banking at the new branch. Run a z-test for a single proportion and determine whether the bank should open the new branch, considering its standard policy.

702. A corporate call center hopes to resolve 75% or more of customer calls through an automated computer voice recognition system. It randomly surveys 50 recent customers; 45 customers report that their issue was resolved. Can the management of the corporation conclude that the computer system is hitting its minimum target, using a significance level of 0.05? Use a z-test for a single proportion to provide an answer.

703. A used bookstore will buy a collection of books if it's reasonably sure that it can sell at least 50% of the books within six months. A customer comes in with 30 books; the clerk evaluates them and determines that 17 are likely to sell within six months. Using the threshold of a 0.05 significance level, should the clerk make an offer for the collection of books? Use the z-test for a single proportion to decide.

704. A factory owner hopes to maintain a standard of less than 1% defects. The owner randomly samples 1,000 ball bearings and finds that 6 are defective. Can the factory owner conclude that the process is producing a defect rate of 1% or less? Use a z-test for a single proportion and a level of significance of 0.05 to decide.

705. The two symbols $\hat{p}$ and p_0 appear throughout a hypothesis test for one proportion (for example, the formula for the test statistic is $\frac{\hat{p} - p_0}{SE}$). What is the difference between these two symbols?

706. A computer manufacturer is willing to buy components from a supplier only if it can be reasonably sure that the defect rate is less than 1%. If it inspects a shipment of 10,000 randomly selected components and finds that 90 are defective, should the computer manufacturer work with that supplier? Perform a hypothesis test using a z-test on a single proportion. The significance level is 0.01.

707. What is the difference between the symbol p and the term "p-value" in a hypothesis test for a proportion?

(A) p is the actual population proportion, and a p-value is a sample proportion.

(B) A p-value is the actual population proportion, and p is a sample proportion.

(C) A p-value is the claimed value for the population, and p is the actual value of the population proportion.

(D) p is the claimed value for the population proportion, and p-value is the actual value of the population proportion.

(E) None of the above.

708. An antique dealer is willing to buy collections of antiques as long as no more than 5% of the items in the entire collection are found to be fakes. At a recent estate sale, the dealer randomly samples 10 items from 200 items for sale and finds that 2 are fakes. Should the dealer buy that entire collection, based on a z-test for a single proportion, using a significance level of 0.01?

709. A blood bank wants to ensure that none of its blood carries diseases to the recipients. It tests a sample of 1,000 specimens and finds that 2 of them have potentially deadly diseases. The director of research has asked that a z-test on a single proportion be done. With what confidence can the blood bank say that the true population proportion of illnesses in the blood specimens is greater than 0?

710. A European city council likes to ensure that no more than 3% of the pigeons that inhabit the city carry disease that humans can catch from droppings. As long as there is significant evidence that fewer than 3% of pigeons carry potential human diseases, the council leaves the pigeons alone. If there isn't enough evidence, it aggressively works to control the population of pigeons by trapping and moving them or shooting them.

The city council randomly samples 200 pigeons and finds that 6 have droppings with potential for human disease transmission. Should the council try to control the pigeon population or leave them alone, considering its guidelines and a threshold of $\alpha = 0.05$?

711. A fashion designer likes to use small defects to make their pieces more interesting. They send designs off to manufacturing plants abroad and reject shipments if the true defect rate is anything other than 25%. The designer doesn't want rates much lower or much greater than 25%. In a recent large shipment, a random sample of 50 pieces found a defect rate of 12%. Should the shipment be accepted or rejected, if a z-test for a single population proportion and a significance level of $\alpha = 0.05$ is used to make the decision?

712. Suppose that you flip a coin 100 times, and you get 55 heads and 45 tails. You want to know whether the coin is fair, so you conduct a hypothesis test to help you decide. What are the null and alternative hypotheses in this situation? (*Note:* p represents the overall probability of getting a heads, also known as the proportion of heads over an infinite number of coin flips.)

713. A person claims to have has ESP and aims to prove it. The person asks you to shuffle five regular decks of cards. One by one, you pick a card, note its suit (diamonds, spades, hearts, or clubs), and return it to the deck. As you select each card, the person tells you what they think its suit is. You repeat this process 100 times and then look at the proportion of correct answers they have given (designated as $\hat{p}$). Suppose you want the person to be at least 20 percentage points above the accuracy you would expect by chance. In this scenario, what are the null and alternative hypotheses?

714. A biologist wants to study cell lines where nearly 25% of the cells in a sample have a particular phenotype. As a result, the biologist wants to reject any sample of cells that differ from the target 25% phenotype present based on a significance level of 0.10. In a sample of 1,000,000 cells, 250,060 are found to have the phenotype. Should that sample be studied or rejected? Use a z-test for a single proportion to answer.

715. Suppose a grocery store owner has been watching customers as they stand in the checkout aisle waiting to pay for their groceries. Based on the owner's observations, they believe that more than 30% of the customers purchase at least one item in the checkout aisle. You believe the percentage is even higher than that. You conduct a hypothesis test to find out what the story is. Your null and alternative hypotheses are $H_0: p = 0.30$ and $H_a: p > 0.30$. You take a random sample of 100 customers and observe their behavior in the checkout aisle. Your sample finds that only 20% of the customers purchased items. What conclusion can you make regarding H_0 at this point (if any)?

Comparing Two Independent Population Means

716–720 A manager of a large grocery store chain believes that happy employees are more productive than unhappy ones. The manager randomly samples 60 of their grocery checkout clerks and classifies them into one of two groups: those who smile a lot (Group 1) and those who don't (Group 2). After sorting the random sample, there just happens to be 30 in each group. The manager then examines their productivity scores (based on how quickly and accurately they're able to assist customers) and gets the following data for each group:

> Group 1 score mean: 33.3
>
> Group 2 score mean: 14.4

The population of productivity scores is normally distributed with a standard deviation of 17.32, which is assumed to be the population standard deviation that applies both to Group 1 and Group 2.

Conduct an appropriate test to see whether a difference in productivity levels exists between the two groups among all employees, using $\alpha = 0.05$.

716. What are the appropriate null and alternative hypotheses for this test?

717. What is the critical value for a z-test for this hypothesis?

718. What is the standard error for this test?

719. What is the test statistic for this data?

720. What is your decision, given this data?

Digging Deeper into Two Independent Population Means

721–725 A psychologist read a claim that average intelligence differs between smokers and nonsmokers and decided to investigate. They sampled 30 smokers and 30 nonsmokers and gave them an IQ test. The psychologist used an alpha level of 0.10. The mean for the smokers is 51.9, and the mean for the nonsmokers is 52.6. The population variance for each group is 5.

721. What are the null and alternative hypotheses for this data?

722. What is the critical value for a z-test for this hypothesis?

723. What is the test statistic for this data?

724. What is your decision, given this data?

725. Suppose that you were using an alpha level of 0.05. What would your decision be regarding this data?

Getting More Practice with Two Independent Population Means

726–730 A sleep researcher investigates performance following a night with two different levels of sleep deprivation. In Group 1, 40 people were permitted to sleep for five hours. In Group 2, 35 people were permitted to sleep for three hours. The outcome measure was performance on a memory test with a population standard deviation of 6 in each group. The mean memory test score in the group that slept three hours was 58; in the group that slept five hours, it was 62.

726. If the researcher is interested in whether performance differs according to the amount of sleep, what are the null and alternative hypotheses?

727. If the researcher is interested in whether more sleep is associated with better performance, what are the null and alternative hypotheses?

728. Using a z-test for two population means, what is the test statistic for this data?

729. Using a z-test for two population means, a *not equal to* alternative hypothesis, and an alpha level of 0.01, what is the researcher's decision regarding this data?

730. Using a z-test for two population means, a *greater than* alternative hypothesis, and an alpha level of 0.05, what is the researcher's decision regarding this data?

Using the Paired *t*-Test

731–736 A city power company wants to encourage households to conserve energy. It decides to test two advertising campaigns. The first emphasizes the potential to save money by saving energy (the "wallet" appeal). The second emphasizes the humanitarian value of reducing the carbon footprint (the "moral" appeal). The company randomly chooses ten households and presents one randomly chosen appeal of the two to them, observes their energy consumption for a month, and then presents the other appeal and observes their energy consumption for a month. Energy consumption is expressed in kilowatts.

With the "wallet" appeal as Group 1 and the "moral" appeal as Group 2, the results are as follows:

$$\bar{d} = -83.5$$
$$s_d = 46.39$$
$$n = 10$$
$$df = 9$$

Assume that power consumption is normally distributed at both time points. The research question is whether energy consumption differs following one appeal rather than the other.

731. What is the appropriate test to determine if one appeal is more successful than the other?

732. Given the information provided, which of the following is a true statement regarding this sample data?

   (A) The wallet appeal was more successful.

   (B) The moral appeal was more successful.

   (C) There was greater variability in the wallet appeal data.

   (D) There was greater variability in the moral appeal data.

   (E) On average, there was no difference between the wallet and moral appeals.

733. What is the degrees of freedom for the appropriate test for this data?

734. What is the standard error for a paired t-test on this data? Round your answer to two decimal places.

735. What is the test statistic for this data? Round to two decimal places.

736. Using a *not equal to* alternative hypothesis and an alpha level of 0.10, what is your conclusion regarding this data?

Digging Deeper into the Paired *t*-Test

737–745 *A precocious child wants to know which of two brands of batteries tends to last longer. The child finds seven toys, each of which requires one battery. For each toy, the child then randomly chooses one of the brands, puts a fresh battery of that brand in the toy, turns the toy on, and records the time before the battery dies. The child repeats the experiment with a battery of the brand not used in the first trial for each toy.*

The data is listed in the following table (in terms of hours of battery life before failure):

	Brand 1	Brand 2	Difference
Toy 1	11.1	12.8	−1.7
Toy 2	10.2	12.4	−2.2
Toy 3	10.3	10.8	−0.5
Toy 4	7.9	8.7	−0.8
Toy 5	10.9	12.0	−1.1
Toy 6	14.2	13.5	0.7
Toy 7	7.1	8.0	−0.9

Illustration by Ryan Sneed

Assume that the difference scores are normally distributed. The sample standard deviation of difference scores (s_d) is 0.9214.

737. What is the appropriate test to determine whether the two brands have a different battery life?

738. If the child claims that one brand lasts longer than the other, what are the null and alternative hypotheses for this research question?

739. What is $\bar{d}$ for this data? Round your answer to four decimal places.

740. What is the standard error for this data? Round your answer to four decimal places.

741. At an alpha level of 0.05, given a *not equal to* alternative hypothesis, what is the critical value of *t*?

742. At an alpha level of 0.01, given a *not equal to* alternative hypothesis, what is the critical value of *t*?

743. What is the test statistic for this data? Round your answer to four decimal places.

744. At the $\alpha = 0.01$ level, what is your decision regarding this data?

745. At the $\alpha = 0.05$ level, what is your decision regarding this data?

Comparing Two Population Proportions

746–752 A computer security consultant is interested in determining which of two password generation rules is more secure. One rule requires users to include at least one special character (*, @, !, %, or \$); the other doesn't.

The consultant creates 100,000 phantom accounts and observes them for six months, monitoring carefully for inappropriate logins. Among the 50,000 accounts with passwords that used the first rule (requiring special characters), 1,055 were breached by security threats. Among the 50,000 who followed the second rule, 2,572 security breaches occurred.

746. What is the appropriate statistical test to address this research question?

747. Assume that the researcher claims that the password systems differ in strength. What are the appropriate null and alternative hypotheses for this research question?

748. What are the values of $\hat{p}_1$ and $\hat{p}_2$ for this data?

749. What is the value of $\hat{p}$ for this data?

750. What is the standard error for this data? Round your answer to four decimal places.

751. What is the z-statistic for this data? Round your answer to two decimal places.

752. Using a significance level of 0.05 and a *not equal to* alternative hypothesis, what is your decision based on this data?

Digging Deeper into Two Population Proportions

753–760 A prison warden wants to see whether it's beneficial or harmful to give prisoners access to the Internet. The warden randomly assigns prisoners in 100 cells to receive such access and 100 cells to be observed as controls. The outcome variable is whether each cell had a report of a behavioral problem during the week following the introduction of Internet access.

Of those in the Internet group (Group 1), 50 of the cells had a reported episode of behavioral disturbance, and of those in the non–Internet group (Group 2), 70 cells had such incidents. If appropriate, conduct a z-test for independent proportions and report your results. Use a 95% confidence level.

753. What are the values of $\hat{p}_1$ and $\hat{p}_2$ for this data?

754. What is the value of $\hat{p}$ for this data?

755. What is the standard error for this data? Round your answer to four decimal places.

756. For the following hypotheses and an alpha level of 0.01, what is the critical value for the test statistic?

$$H_0: p_1 = p_2$$
$$H_a: p_1 \neq p_2$$

757. For the following hypotheses and an alpha level of 0.01, what is the critical value for the test statistic?

$$H_0: p_1 = p_2$$
$$H_a: p_1 > p_2$$

758. What is the test statistic for this data, given the following hypotheses? Round your answer to four decimal places.

$$H_0: p_1 = p_2$$
$$H_a: p_1 \neq p_2$$

759. What is your decision regarding this data, given an alpha level of 0.05 and the following hypotheses?

$$H_0: p_1 = p_2$$
$$H_a: p_1 \neq p_2$$

760. What is your decision regarding this data, given an alpha level of 0.01 and the following hypotheses?

$$H_0: p_1 = p_2$$
$$H_a: p_1 \neq p_2$$

Chapter 14

Surveys

Surveys are everywhere, and their quality can range from good to bad to ugly. The ability to critically evaluate surveys is the focus of this chapter.

The Problems You'll Work On

The problems in this chapter focus on the following big ideas:

>> Understanding the basics and nuances of good and bad surveys (issues involved in designing the survey, selecting a sample of participants, and collecting the data)

>> Identifying problems that commonly occur in surveys and samples that create bias

>> Using the proper terminology at the proper time to describe a problem that's been found with a survey or a sample

What to Watch Out For

Pay particular attention to the following:

>> The way a sample is selected depends on the situation — know the differences.

>> Specific terms are used to describe problems with surveys and samples. Make sure you pick up on subtle differences. Here are the most common terms:

- **Sampling frame:** A list of all the members of the target population.

- **Census:** Getting desired information from everyone in the target population.

- **Random sample:** Each member of the population has an equal chance of being selected for the sample.

- **Bias:** Systematic unfairness in sample selection or data collection.

- **Convenience sample:** Chosen solely for convenience, not based on randomness.

- **Volunteer/self-selected sample:** Sample where people determine on their own to be involved.

- **Non-response bias:** Occurs when someone in the sample doesn't return or doesn't finish the survey.
- **Response bias:** When the respondent takes the survey but doesn't give correct information.
- **Undercoverage:** Sampling frame doesn't include adequate representation from certain groups within the target population.

Planning and Designing Surveys

761–766 You're interested in the willingness of adult drivers (age 18 and over) in a metropolitan area to pay a toll to travel on less-congested roads. You draw a sample of 100 adult drivers and administer a survey on this topic to them.

761. What is the target population for this study?

762. Suppose that you collect your data in a way that makes it likely that the survey respondents aren't representative of the target population. What is this called?

763. If you were to select your sample by drawing numbers at random from the published phone directory and calling during daytime hours on weekdays, how could these actions bias the results?

   (A) Not everyone has a phone or a listed phone number.

   (B) Not everyone is at home during the day on weekdays.

   (C) Not everyone is willing to participate in telephone polls.

   (D) Choices (A) and (B)

   (E) Choices (A), (B), and (C)

764. What is the main problem with this survey question: "Don't you agree that drivers should be willing to pay more for less-congested roads?"

765. Of the 100 people in your sample, 20 choose not to participate. You later discover that they're in a lower income bracket than those who did participate. What problem does this introduce into your study?

766. One of your questions asks whether the respondent voted in the last election. You find a much higher proportion of individuals claiming to have voted than is indicated by public records. What is this an example of?

Selecting Samples and Conducting Surveys

767–775 Solve the following problems about selecting samples and conducting surveys.

767. Why is bias particularly problematic in surveys?

768. Which of the following is an example of a good census of 2,000 students in a high school?

(A) calculating the mean age of all the students by using their official records

(B) asking the first 25 students who arrive at school on a given day their age and calculating the mean from this information

(C) sending an e-mail to all students asking them to respond with their age and calculating the mean from those who respond

(D) Choices (A) and (C)

(E) Choices (A), (B), and (C)

769. You want to survey students at a high school and calculate the mean age. Which of the following procedures will result in a simple random sample?

(A) classifying the students as male or female and drawing a random sample from each

(B) using an alphabetized student roster and selecting every 15th name, starting with the first one

(C) selecting three tables at random from the cafeteria during lunch hour and asking the students at those tables for their age

(D) selecting one student at random, asking them to suggest three friends to participate and continuing in this fashion until you have your sample size

(E) numbering the students by using the school's official roster and selecting the sample by using a random number generator

770. Suppose that you intend to conduct a survey among employees at a firm. You use a file supplied by the personnel department for your sampling file, not realizing that it excludes people hired in the past six months. What kind of bias is likely to result?

771. Suppose that you conduct a survey by showing an 800 number on the television screen during a popular program and inviting people to call in with their response to a posted question. What type of bias is likely to result?

772. Suppose that you conduct a survey by interviewing the first 50 people you see in a shopping mall. What type of bias is likely to result?

773. Sometimes questions are rephrased, and scales are reversed during the course of a survey. For example, with Likert scale items, 1 may sometimes mean *strongly agree* and sometimes *strongly disagree*. What is one reason for this reversal?

774. What is the biggest problem that makes the question "Should everyone go to college and seek gainful employment?" less than ideal?

775. Why do some surveys include "don't know" as a response?

(A) because the respondent may be uninformed about the topic

(B) because the respondent may not remember enough information to answer the question

(C) because the respondent may find the question offensive

(D) Choices (A) and (B)

(E) Choices (A), (B), and (C)

Chapter 15

Correlation

I n this chapter, you explore possible linear relationships between a pair of quantitative (numerical) variables, X and Y. Your basic question is this: As the X variable increases in value, does the Y variable increase with it, does it decrease in value, or does it just basically not react at all? Answering this question requires the use of graphs as well as the calculation and interpretation of a certain numerical measure of togetherness — correlation.

The Problems You'll Work On

Your job in this chapter is to look for, describe, and quantify possible linear relationships between two quantitative variables using the following methods:

>> Graphing pairs of data on a scatter plot and describing what you see

>> Measuring the strength and direction of a linear relationship, using correlation

>> Interpreting correlation properly and knowing its properties

>> Understanding what elements can affect correlation

What to Watch Out For

Correlation is more than just a number; it's a way of describing relationships in a universal way.

>> Understand that correlation applies to quantitative variables (like age and height), even though the "street definition" of correlation relates to any variables (like gender and voting pattern).

>> Know the many properties of correlation — some are counterintuitive (for example, you may think that switching the values of X and Y will change the correlation, but it doesn't).

>> Always remember: "Humans do not live by a correlation alone." You always need to look at a scatter plot of the data as well. (Neither is foolproof by itself.)

Interpreting Scatter Plots

776–781 *This scatter plot represents the high-school and freshman college GPAs of 24 students.*

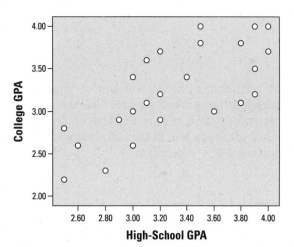

Illustration by Ryan Sneed

776. How would you describe the linear relationship between high-school GPA and college GPA?

   (A) strong

   (B) weak

   (C) positive

   (D) negative

   (E) Choices (A) and (C)

777. Looking at the following high-school GPAs for five students, which one would you predict to have the highest college GPA?

   (A) 2.5

   (B) 2.8

   (C) 3.1

   (D) 3.4

   (E) 4.0

778. How do you know this scatter plot displays a positive linear relationship between the two variables?

779. How do you know this scatter plot displays a relatively strong linear relationship between the two variables?

780. If these two quantitative variables had a correlation of 1, how would the scatter plot be different?

781. If these two quantitative variables had a correlation of –1, what would the scatter plot look like?

   (A) The points would all lie on a straight line.

   (B) All the points would have to be between –1 and 0.

   (C) All the points would slope downward from left to right.

   (D) Choices (A) and (C)

   (E) Choices (A), (B), and (C)

Creating Scatter Plots

782–783 *Solve the following problems about making scatter plots.*

782. For a group of adults, suppose you want to create a scatter plot between height and one other variable. Which variable(s) would be the appropriate candidate?

   (A) gender

   (B) race/ethnicity

   (C) weight

   (D) zip code of residence

   (E) Choices (C) and (D)

783. You have a data set containing four variables collected from a group of children ages 5 to 21 years. The variables are height, age, weight, and gender. Which pair(s) of variables is appropriate for making a scatter plot?

(A) height and age

(B) height and gender

(C) gender and age

(D) height and weight

(E) Choices (A) and (D)

Understanding What Correlations Indicate

784–787 Figure out what the correlations in the following problems indicate.

784. Which of the following correlations indicates a strong, negative linear relationship between two quantitative variables?

(A) −0.2

(B) −0.8

(C) 0

(D) 0.4

(E) 0.8

785. Which of the following correlations indicates a weak, positive linear relationship between two quantitative variables?

(A) −0.2

(B) −0.6

(C) 0.2

(D) 0.75

(E) 0.9

786. Which of the following correlations indicates a very strong, positive linear relationship between two quantitative variables?

(A) −0.7

(B) −0.1

(C) 0.2

(D) 0.4

(E) 0.9

787. Which of the following correlations indicates a weak, negative linear relationship between two quantitative variables?

(A) −0.2

(B) −0.8

(C) −1

(D) 0.4

(E) 0.8

Digging Deeper into Scatter Plots

788–790 In this scatter plot, X and Y both represent measurement variables.

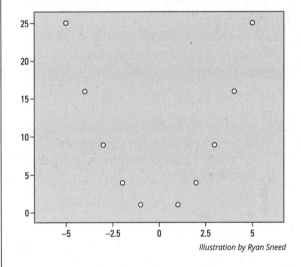

Illustration by Ryan Sneed

788. How do you describe the relationship between the two quantitative variables in this scatter plot?

789. What is your best guess for the correlation between these two quantitative variables?

790. Why is correlation a poor choice to describe the relationship between these two particular variables?

Digging Deeper into What Correlations Indicate

791–793 Figure out what the correlations in the following problems indicate.

791. Which of the following correlations represents the data with the strongest linear relationship between two quantitative variables?

(A) −0.85

(B) −0.56

(C) 0.23

(D) 0.45

(E) 0.6

792. Which of the following correlations represents the data with the weakest linear relationship between two quantitative variables?

(A) −0.63

(B) −0.23

(C) 0.1

(D) 0.45

(E) 0.73

793. Which of the following represents the strongest possible linear relationship between two quantitative variables?

(A) −1

(B) 0

(C) 1

(D) Choices (A) and (C)

(E) Choices (A), (B), and (C)

Calculating Correlations

794–799 Use the following data set to help with calculating the correlation in the following problems.

X	Y
1	2
2	2
3	4
4	3

Illustration by Ryan Sneed

794. What is the value of $\bar{x}$ for this data?

795. What is the value of $\bar{y}$ for this data?

796. What is the value of $n - 1$ in this case?

797. What is the value of s_x for this data?

798. What is the value of s_y for this data?

799. Suppose you have the following information about two variables, X and Y:

$$\sum_x \sum_y (x - \bar{x})(y - \bar{y}) = 2.5$$

Using the formula for correlation, calculate the correlation between these two variables.

Noting How Correlations Can Change

800–803 Use this information to answer the following problems: The following statistics describe two variables, X and Y:

$\bar{x} = 8.00; \bar{y} = 8.53$

$s_x = 4.47; s_y = 5.36$

$n = 15$

$\sum_x \sum_y (x - \bar{x})(y - \bar{y}) = 274$

800. What is the correlation of X and Y in this case?

801. If the sample size is changed to 20, with all the other given values the same, how will the correlation change?

802. If s_y is changed to 4.82, with all the other given values the same, how will the correlation of X and Y change?

803. If $\sum_x \sum_y (x - \bar{x})(y - \bar{y})$ is changed to 349, how does the correlation of X and Y change?

Looking at the Properties of Correlations

804–806 Solve the following problems about correlation properties.

804. Suppose that you calculate the correlation between the heights of fathers and sons, measured in inches, and then you convert the data to centimeters; how does the correlation change?

805. Which of the following values isn't possible for a correlation?

(A) −2.64

(B) 0.99

(C) 1.5

(D) Choices (A) and (C)

(E) Choices (A), (B), and (C)

806. If you compute the correlation between the heights and weights of a group of students and then compute the correlation between their weights and heights, how will the two correlations compare?

Digging Deeper into How Correlations Can Change

807–810 *You conduct a study to see whether changing the font size of the text displayed on a computer screen influences reading comprehension.*

807. Which variable corresponds to "reading comprehension" in this study?

(A) the X variable

(B) the Y variable

(C) the response variable

(D) Choices (A) and (C)

(E) Choices (B) and (C)

808. Which variable corresponds to "font size" in this study?

(A) the X variable

(B) the Y variable

(C) the response variable

(D) Choices (A) and (C)

(E) Choices (B) and (C)

809. How will the correlation change if you switch the designation of the two variables — that is, if you make the X variable the Y variable and make the Y variable the X variable?

810. How does the correlation change if text font size is measured in centimeters rather than inches?

Making Conclusions about Correlations

811–813 *Draw conclusions about the correlations in the following problems.*

811. If a correlation between two quantitative variables is calculated to be 1.2, what do you conclude?

812. If you compute a correlation of −0.86 between two quantitative variables, what do you conclude?

813. If you compute a correlation of 0.27 between two quantitative variables, what do you conclude?

Getting More Practice with Scatter Plots and Correlation Changes

814–817 *You conduct a study to see whether the amount of time spent studying per week is related to GPA for a group of college computer science majors.*

814. How do you designate the "time spent studying" variable on a scatter plot of your data?

815. How do you designate the variable "GPA" on a scatter plot of your data?

   (A) the X variable

   (B) the Y variable

   (C) the response variable

   (D) Choices (A) and (C)

   (E) Choices (B) and (C)

816. How does the correlation change if you switch the measurement of study time from minutes to hours?

817. How does the correlation change if you switch the designation of the two variables — that is, make the X variable the Y variable and make the Y variable the X variable?

Making More Conclusions about Correlations

818–820 *Draw conclusions about the correlations in the following problems.*

818. If you compute a correlation of −0.23 between two quantitative variables, what do you conclude?

819. If you compute the correlation between two quantitative variables to be 1.05, what do you conclude?

820. If you compute a correlation of −0.87 between two quantitative variables, what do you conclude?

Chapter **16**

Simple Linear Regression

With simple linear regression, you look for a certain type of relationship between two quantitative (numerical) variables (like high-school GPA and college GPA.) This special relationship is a *linear relationship* — one whose pairs of data resemble a straight line. After you find that right relationship, you fit a line to it and use the line to make predictions for future values. Sounds romantic, doesn't it?

The Problems You'll Work On

Your job in this chapter is to find and interpret the results of a regression line and its elements and to carefully check exactly how well your line fits. *Note:* Regression assumes you've found that a strong relationship exists (see Chapter 15 for the details of correlation and scatter plots).

>> Find the best fitting (regression) line to describe a linear relationship.

>> Understand and/or interpret the slope and *y*-intercept of the regression line.

>> Use the regression line to make predictions for one variable given another, where appropriate.

>> Assess line fit and look for anomalies, such as offbeat patterns or points that stand out.

What to Watch Out For

It's easy to get caught up in all the calculations of regression. Always remember that understanding and interpreting your results is just as important as calculating them!

>> Slope measures change and are used all over the place in regression — be sure to know them cold.

>> Equations aren't smart — you have to know when they can be used and applied and when they can't.

>> Many techniques exist for determining how well a line fits and for pointing out problems. Know all the tools available and the specifics of what their results tell you.

Introducing the Regression Line

821 Solve the following problem about regression line basics.

821. What conditions must be met before it's appropriate to find the least-squares regression line between two quantitative variables?

(A) Both variables are numeric.

(B) The scatter plot indicates a linear relationship.

(C) The correlation is at least moderate.

(D) Choices (A) and (C)

(E) Choices (A), (B), and (C)

Knowing the Conditions for Regression

822–823 In this scatter plot, the two variables plotted are quantitative (numerical). The correlation is $r = 0.75$.

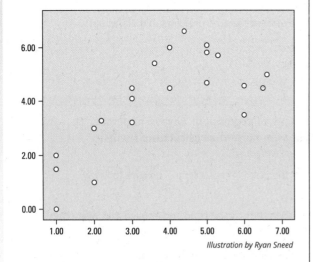

Illustration by Ryan Sneed

822. In looking at this scatter plot, which of the following violations of a necessary condition for fitting a regression line is observed?

(A) The variables aren't numeric.

(B) Their correlation isn't strong enough.

(C) Their relationship isn't linear.

(D) Choices (B) and (C)

(E) None of the above.

823. The equation for calculating the least-squares regression line is $y = mx + b$. If two variables have a negative relationship, which letter will be preceded by the negative sign?

Examining the Equation for Calculating the Least-Squares Regression Line

824–825 Solve the following problems about the equation for the least-squares regression line.

824. The equation for calculating the least-squares regression line is $y = mx + b$. Which letter in this equation represents the slope of the regression line?

825. The equation for calculating the least-squares regression line is $y = mx + b$. Which letter in this equation represents the y-intercept for the regression line?

Finding the Slope and *y*-Intercept of a Regression Line

826–829 *The linear relationship between two variables is described by the regression line $y = 3x + 1$. (Assume that the correlation is strong and that the scatter plot shows a strong linear relationship.)*

826. What is the slope of the regression line?

827. What is the *y*-intercept of the regression line?

828. If $x = 3.5$, what is the expected value of *y*?

829. If $x = 0.4$, what is the expected value of *y*?

Seeing How Variables Can Change in a Regression Line

830–832 *Suppose that the linear relationship between two quantitative variables is described by the regression line $y = -1.2x + 0.74$. (Assume that the correlation is strong and that the scatter plot shows a strong linear relationship.)*

830. If *x* increases by 1.5, how does the value of *y* change?

831. If *x* decreases by 2.3, how does the value of *y* change?

832. Where does this regression line intersect the *y*-axis?

Finding a Regression Line

833–842 *This scatter plot shows the relationship between the GRA verbal (GRA_V) and GRA math (GRA_M) scores of a group of high-school seniors.*

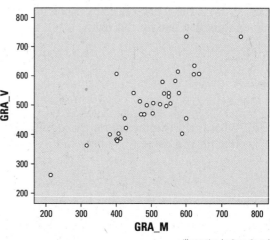

Illustration by Ryan Sneed

For the purpose of this analysis, consider GRA_M to be the X variable and GRA_V to be the Y variable. These variables and their relationship are characterized by the following statistics:

$$\bar{x} = 502.9; \bar{y} = 506.1$$
$$s_x = 103.2; s_y = 103.2$$
$$r = 0.792$$

833. How would you characterize the linear relationship between X and Y in this case?

834. What is the approximate range of values for each of these variables?

835. Without doing any calculations, what is the best estimate for the slope of the regression line for these variables?

836. What is the actual calculated slope for the regression line for this data?

837. What is the y-intercept for the regression line for this data?

838. What is the actual calculated equation of the regression line for this data?

839. The linear relationship between two quantitative variables is described by the equation $y = 0.792x + 107.8$. Using this equation, if $x = 230$, what is the expected value of y?

840. Student A has a math score 210 points higher than Student B. How much higher do you expect Student A's verbal score to be, compared to Student B?

841. Student C has a math score 50 points lower than Student D. Using the regression equation, how do you expect their verbal scores to compare?

842. There is a strong, positive correlation between GRA_M and GRA_V in this data. Why can't you assume that this is a cause-and-effect relationship?

Digging Deeper into Finding a Regression Line

843–852 The scatter plot represents data on home size (in square feet) and selling price (in thousands of dollars) for 35 recent sales in an American community. Home size is the X variable, and selling price is the Y variable.

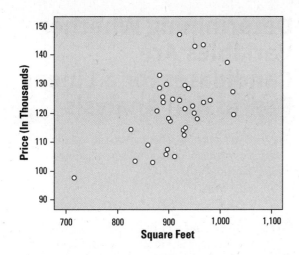

The following statistics describe these two variables and their relationship in the data set:

$\bar{x} = 915.1; \bar{y} = 121.1$

$s_x = 58.5; s_y = 11.8$

$r = 0.527$

843. How do you describe the linear relationship between these two variables?

844. Given the following home sizes in square feet, which size home do you expect to sell for the highest price?

   (A) 800 square feet

   (B) 850 square feet

   (C) 870 square feet

   (D) 890 square feet

   (E) 910 square feet

845. What is the slope of the regression line for this data?

846. What is the y-intercept of the regression line for this data?

847. What is the equation of the regression line for this data?

848. What do you expect a house of 1,000 square feet to sell for (in dollars)?

849. What do you expect a house of 1,500 square feet to sell for (in dollars)?

850. What do you expect a house of 890 square feet to sell for (in dollars)?

851. House A is 90 square feet larger than House B. What do you expect their price difference to be (in dollars)?

852. House C is 54 square feet smaller than House D. What do you expect their price difference to be (in dollars)?

Connecting to Correlation and Linear Relationships

853–856 Use the following scatter plot to answer the following problems.

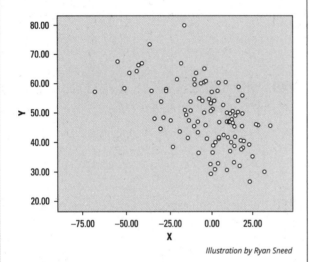

Illustration by Ryan Sneed

853. In words, how would you describe the general linear relationship between the X and Y variables in this scatter plot?

854. In terms of numbers, what is the most plausible value for the correlation between X and Y?

855. If the variables X and Y were switched in this scatter plot, how would the correlation be affected?

856. Does the scatter plot suggest that X and Y are good candidates for a linear regression analysis?

Determining Whether Variables Are Candidates for a Linear Regression Analysis

857–859 Figure out whether you can use the variables in the following problems in a linear regression analysis.

857. Suppose that you're considering running a linear regression with two variables, X and Y. As a preliminary check, you compute the correlation and find that it's 0.54. At this point, can you say that these variables are good candidates for a linear regression analysis? Why or why not?

858. Suppose that you're considering running a linear regression with two variables, X and Y. As a preliminary check, you compute the correlation and find that it's 0.05. Are these variables good candidates for a linear regression analysis? Why or why not?

859. Which of the following correlations indicates the strongest linear relationship between two variables? (Assume that the scatter plots match the correlations, respectively.)

(A) −0.9

(B) −0.5

(C) 0.0

(D) 0.9

(E) Choices (A) and (D)

Digging Deeper into Correlations and Linear Relationships

860–865 As part of a health behaviors study, you collect data on heights and weights for a group of adults. The scatter plot of the data indicates a possible linear relationship.

860. If the correlation between height and weight is 0.65, what is the correlation between weight and height?

861. Originally, you collected height in inches and weight in pounds. You decide to convert your measurements to centimeters and kilograms, respectively, and recalculate the correlation. How will it change?

862. For females in this data set, the correlation coefficient between weight and height is 0.50. For males, it's 0.80. What do these two correlations tell you about the linear relationship between weight and height for females and males in this data set? (Assume that the scatter plots agree with the correlations in both cases.)

863. Which of the following doesn't represent a possible value for a correlation in general?

(A) −1.5

(B) −0.6

(C) 0.0

(D) 0.2

(E) 0.8

864. Do a strong correlation and a strong linear pattern on a scatter plot allow you to conclude a cause-and-effect relationship between two variables X and Y? Why or why not?

865. If you want to predict weight using height, which variable is designated as the X variable: height or weight?

Describing Linear Relationships

866–872 As part of a study of academic performance, you collect data about GPA, major, minutes spent studying per week, and minutes spent watching TV per week from a random sample of college juniors. A scatter plot of minutes studying and GPA suggests a linear relationship.

866. For the entire sample, the correlation between minutes studying and GPA is 0.74. How would you describe the linear relationship between these two variables?

867. Suppose that the responses for minutes spent studying per week range from 0 to 480 minutes. Further, the responses average 250 minutes with a standard deviation of 60 minutes. A scatter plot between minutes studying and GPA demonstrates a linear trend, and the correlation coefficient of the two variables is 0.84. Among the five choices given, how many minutes spent studying is the most likely prediction for students with the highest GPAs, on average? Choose the best answer.

(A) 20 minutes

(B) 60 minutes

(C) 250 minutes

(D) 260 minutes

(E) 450 minutes

868. For the entire sample, the correlation between minutes watching TV and GPA is −0.38. How would you describe the linear relationship between these two variables?

869. Given a correlation of −0.68 between minutes watching TV and GPA for five students, which of the following minutes would you expect the student with the highest GPA to have spent watching TV, on average? (Assume that the scatter plot suggests a linear relationship.)

(A) 30

(B) 80

(C) 120

(D) 250

(E) 500

870. For the purpose of a regression analysis, which is the more logical choice for the Y variable: time spent studying or GPA?

871. Among English majors, the correlation between minutes spent studying and GPA is 0.48. Among engineering majors, it's 0.78. Given this information, which of the following statements about the linear relationship between time spent studying and GPA is correct for this data set? Assume that scatter plots suggest some type of linear relationship between the variables in each case.

(A) The linear relationship is stronger for English majors.

(B) The linear relationship is stronger for engineering majors.

(C) The linear relationships are equal between English and engineering majors.

(D) Correlation can't be used to compare two groups.

(E) Not enough information to tell.

872. You calculate a correlation of −2.56 between minutes spent studying per week and minutes spent watching TV per week. What do you conclude?

Getting More Practice with Finding a Regression Line

873–877 *You conduct a survey including variables for current income (measured in thousands of dollars) and life satisfaction (measured on a scale of 0 to 100, with 100 being most satisfied and 0 being least satisfied). A scatter plot of the data suggests a linear relationship.*

873. If you want to predict life satisfaction using income, which variable would you designate as X?

874. For this data, the mean of satisfaction is 60.4, and the standard deviation of satisfaction is 12.5. The mean of income is 80.5, and the standard deviation of income is 16.7. The correlation coefficient between satisfaction and income is 0.77. What is the slope for the regression line predicting satisfaction from income?

875. For this data, the mean of satisfaction is 60.4, and the standard deviation of satisfaction is 12.5. The mean of income is 80.5, and the standard deviation of income is 16.7. The correlation coefficient between satisfaction and income is 0.77. What is the y-intercept for the regression line when predicting satisfaction from income?

876. The slope and y-intercept for the linear relationship between income (X) and life satisfaction (Y) are 0.58 and 13.7, respectively. What is the equation describing the linear relationship between income and life satisfaction based on this data?

877. In this data set, incomes (x values) range from \$50,000 to \$150,000. Which of the following would qualify as extrapolation when attempting to predict life satisfaction (Y) using income (X)?

(A) predicting satisfaction for someone with an income of \$75,000

(B) predicting satisfaction for someone with an income of \$45,000

(C) predicting satisfaction for someone with an income of \$200,000

(D) Choices (A) and (B)

(E) Choices (B) and (C)

Making Predictions

878–884 *A building contractor examines the cost of having carpentry work done in some of his buildings in the current year. He finds that the cost for a given job can be predicted by this equation:*

$$y = 50x + 65$$

Here, y is the cost of a job (in dollars), and x is the number of hours a job takes to complete. So, the cost of a given job can be predicted by a base fee of \$65 per job plus a cost of \$50 per hour. Assume that the scatter plot and correlation both indicate strong linear relationships.

878. What is the predicted cost of a job that takes 2.5 hours to complete?

879. What is the predicted cost for a job that takes 4.75 hours to complete?

880. How much more money do you predict a job taking 3.75 hours to complete will cost, as compared to a job taking 3.5 hours to complete?

881. Suppose that in a different city, a similar equation predicts carpentry costs, but the intercept is $75 (the slope remains the same). What is the predicted cost for a job taking 2 hours in this city?

882. Suppose you're comparing the predicted cost of a job lasting 3.5 hours in the two cities (the first with an intercept of $65 and the second with an intercept of $75). How much more do you predict a job lasting 3.5 hours will cost in the first city versus the second city?

883. In a third city, carpentry costs are predicted by an equation with a slope of $60 and an intercept of $65. How much do you predict a job lasting 2.3 hours will cost?

884. Looking at data from an earlier year, the equation predicting costs in the first city has an intercept of $65 and a slope of $60. Compared to the current year, how much more did a job lasting 3.6 hours cost in the previous year?

Figuring Out Expected Values and Differences

885–893 You conduct a survey of job satisfaction rating (measured on a scale of 0 to 100 points, with 100 being the most satisfied) and years of experience among employees in a large company. The linear relationship between these two variables is described by the regression line $y = 1.4x + 62$. Here, y is job satisfaction rating, and x is years of employment. Assume that the scatter plot and correlation project a linear relationship.

885. What are the y-intercept and slope for this equation?

886. What does the slope mean in a simple regression equation?

887. What does the y-intercept mean in a simple regression equation?

888. For this data, what is the expected job satisfaction rating for someone with 20 years of experience, on average?

889. For this data, what is the expected job satisfaction rating for someone with two years of experience, on average?

890. What is the expected difference in job satisfaction ratings when comparing someone with 15 years of experience to someone with 8 years of experience?

891. What is the expected difference in job satisfaction ratings, on average, comparing someone with 15 years of experience to a new hire (with 0 years of experience)?

892. What is the expected difference, on average, in job satisfaction ratings, comparing someone with 11.5 years of experience to a new hire (0 years of experience)?

893. For a second company, the equation predicting job satisfaction from years of employment, on average, is $y = 1.4x + 67$. Comparing employees with 10 years of experience, how much does the predicted job satisfaction rating for an employee from the second company differ from that of the first, on average?

Digging Deeper into Expected Values and Differences

894–902 You collect data on square footage and market value for homes in two communities, and you want to use square footage to predict market value. The variables x_1 and x_2 represent the square footage of homes in Community 1 and Community 2, respectively, and y_1 and y_2 represent the market value of homes in Community 1 and Community 2, respectively. The scatter plots and correlations both indicate strong linear relationships for each community. The regression equations describing these linear relationships are

$$y_1 = 77x_1 - 15,400$$
$$y_2 = 74x_2 - 11,300$$

894. What is the expected market value for a home of 1,500 square feet in Community 1, on average?

895. What is the expected market value for a home of 1,840 square feet in Community 1, on average?

896. What is the expected market value for a home of 1,500 square feet in Community 2, on average?

897. What is the expected market value for a home of 980 square feet in Community 2, on average?

898. Comparing two homes with 1,000 square feet each, one in Community 1 and one in Community 2, how do their expected values differ, on average?

899. Comparing two homes with 1,620 square feet each, one in Community 1 and one in Community 2, how do their expected values differ, on average?

900. Comparing two homes with 1,930 square feet each, one in Community 1 and one in Community 2, how do their expected values differ, on average?

901. Which has a larger expected market value: a home in Community 1 with 1,100 square feet or a home in Community 2 with 1,200 square feet, on average?

902. Suppose that the home sizes for Community 1 in your sample range from 800 square feet to 2,780 square feet. Which of the following would constitute extrapolation?

(A) estimating the market value of a home in Community 1 with 2,700 square feet

(B) estimating the market value of a home in Community 1 with 2,900 square feet

(C) estimating the market value of a home in Community 1 with 750 square feet

(D) Choices (A) and (B)

(E) Choices (B) and (C)

Digging Deeper into Predictions

903–905 *For a group of 100 high-school students, the following equation relates their SAT math score (ranging from 200 to 800) with minutes per week watching TV (denoted by x, ranging from 0 to 720): $SAT = 725 - 0.5x$*

903. What is the predicted SAT math score for a student who watches 360 minutes of TV per week?

904. What is the predicted SAT math score for a student who watches 600 minutes of TV per week?

905. Student A watches 60 minutes of TV per week, Student B watches 800 minutes of TV per week, and Student C watches 220 minutes of TV per week. Rank the students from high to low in terms of their expected SAT math scores.

Getting More Practice with Expected Values and Differences

906–915 *You collect data on income (in thousands of dollars) and years of experience of part-time employees working for two companies, and you calculate separate regression equations to explore these linear relationships. The regression equations describing these linear relationships are*

$$y_1 = 6.7x_1 + 2.5$$
$$y_2 = 7.2x_2 + 1.2$$

Here, y_1 and x_1 are the expected salary and years of experience for an employee in Company 1, and y_2 and x_2 are the expected salary and years of experience for an employee in Company 2.

906. What is the expected salary for a part-time employee in Company 1 with 6 years of experience?

907. What is the expected salary for a part-time employee in Company 1 with 17 years of experience?

908. What is the expected salary for a part-time employee in Company 2 with 2.5 years of experience?

909. How much more do you expect a part-time employee in Company 2 with 13 years of experience to make when compared to a part-time employee in Company 2 with 7 years of experience?

910. How much more do you expect a part-time employee in Company 2 with 6.5 years of experience to make when compared to a part-time employee in Company 2 with 1.2 years of experience?

911. Which company has the higher expected starting salary for its part-time employees?

912. Which company has the higher rate of expected increase of salary for its part-time employees?

913. Compare expected salaries between an employee at Company 1 with 3 years of experience and an employee at Company 2 with 5.5 years of experience. Which employee has the higher expected salary?

914. Compare expected salaries between an employee at Company 1 with 3.8 years of experience and an employee at Company 2 with 4.3 years of experience. Which employee has the higher expected salary?

915. Under which circumstances could you declare that a strong relationship between x and y leads you to conclude that a change in x *causes* a change in y?

(A) replication of this study at other companies

(B) a longitudinal study tracking growth in individual employee salaries as years of employment increase

(C) adding additional variables to the model to control for other influences on salary

(D) Choices (A) and (B)

(E) Choices (A), (B), and (C)

Chapter 17

Two-Way Tables and Independence

M any applications of statistics involve categorical variables, such as gender (male/female), opinion (yes/no/undecided), home ownership (yes/no), or blood type. One common statistical application is to look for relationships between two categorical variables. In this chapter, you focus on pairs of categorical variables: how to organize and interpret their data (the two–way tables part), and how to describe findings and look for relationships (the checking for independence part).

The Problems You'll Work On

In this chapter, you work on all the ins and outs of two–way tables and how to interpret and use them, including

>> Being able to read and interpret all parts of a two-way table by using either counts or percentages

>> Finding marginal, joint, and conditional probabilities from a two-way table or related graph

>> Using probabilities from a two-way table to look for and describe relationships between two categorical variables

What to Watch Out For

Numbers sitting in a little table seem easy enough, but you'd be surprised at all the information you can get out of a table and how many equations, formulas, and notations that you can squeeze out of them. Be ready and keep the following in mind:

>> The numbers within a two-way table represent intersections of two characteristics.

>> What goes into the denominator is the key for all two-way table probabilities. Pay attention to what group you're looking at; the total number in that group becomes your denominator.

> » The wording of the problems for two-way tables can be extremely tricky; one small change in wording can lead to a totally different answer. Practice as many problems as you can.

Introducing Variables and Two-Way Tables

916–920 Solve the following problems about variables in two-way tables.

916. Which of the following variables are categorical (that is, their possible values fall into nonnumerical categories)?

(A) blood type

(B) country of origin

(C) annual income

(D) Choices (A) and (B)

(E) Choices (A), (B), and (C)

917. Which of the following variables is categorical?

(A) gender

(B) hair color

(C) zip code

(D) Choices (A) and (B)

(E) Choices (A), (B), and (C)

918. Which of the following variables would be suitable for a two-way table?

(A) years of education

(B) height in centimeters

(C) homeownership (yes/no)

(D) gender

(E) Choices (C) and (D)

919. For a two-way table including education as one of the variables, which of the following would be an appropriate way to categorize the data?

(A) whether someone is a high-school graduate

(B) whether someone is a college graduate

(C) highest level of school completed

(D) Choices (A) and (B)

(E) Choices (A), (B), and (C)

920. How many cells does a 2-x-2 table contain? (A cell is any of the possible combinations of the two variables being studied.)

Reading a Two-Way Table

921–933 This 2-x-2 table displays results from a poll of randomly selected male and female college students at a certain college, asking whether they were in favor of increasing student fees to expand the college's athletics program. The results of their opinions are broken down by gender in the following table.

	Favor Fee Increase	Do Not Favor Fee Increase
Male	72	108
Female	48	132

Illustration by Ryan Sneed

921. What does the value 72 represent in this table?

922. What does the value 132 represent in this table?

923. How many of the students are female and favor the fee increase?

924. How many male students were included in this poll?

925. How many female students were included in this poll?

926. How many students favor the fee increase?

927. How many students do not favor the fee increase?

928. What is the total number of students who took part in the poll?

929. What proportion of male students favors the fee increase?

930. What proportion of female students does not favor the fee increase?

931. What proportion of all students does not favor the fee increase?

932. The following pie chart was calculated based only on the 180 females who were polled. In the poll, the females were asked for their opinions on whether they favored a fee increase. What does the 27% represent?

Females

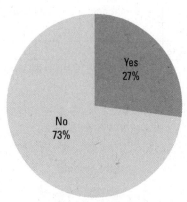

Illustration by Ryan Sneed

933. In this poll, 360 students were asked for their opinions on whether they favor a fee increase. A breakdown of the results is shown in the following pie chart. What does the 33% represent?

Opinion

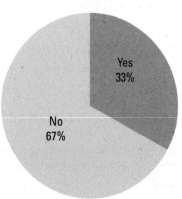

Illustration by Ryan Sneed

Interpreting a Two-Way Table by Using Percentages

934–937 *This 2-x-2 table displays results from a poll of 360 randomly selected male and female college students at a certain college, asking whether they were in favor of increasing student fees to expand the college's athletics program. The results of their opinions are broken down by gender in the following table.*

	Favor Fee Increase	Do Not Favor Fee Increase
Male	72	108
Female	48	132

Illustration by Ryan Sneed

The following pie chart breaks down the gender and opinion of all 360 students polled.

All Students

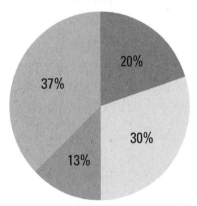

Illustration by Ryan Sneed

934. What is the most appropriate way to describe the 37% represented in this pie chart?

935. What is the most appropriate label for the section in the pie chart representing 20% of the data?

936. What is the most appropriate label for the section in the pie chart representing 30% of the data?

937. What is the most appropriate label for the section in the pie chart representing 13% of the data?

Interpreting a Two-Way Table by Using Counts

938–950 *The following table displays information about cigarette smoking and diagnosis with hypertension for a group of patients at a medical clinic.*

	Hypertension Diagnosis	No Hypertension Diagnosis
Smoker	48	24
Nonsmoker	26	50

Illustration by Ryan Sneed

938. How many of the patients are smokers?

939. How many of the patients have a hypertension diagnosis?

940. How many of the patients are both non-smokers and have a hypertension diagnosis?

941. How many of the patients are smokers and have a hypertension diagnosis?

942. What is the total number of patients in this study?

943. How many of the patients do not have a hypertension diagnosis and are smokers?

944. How many of the patients do not have a hypertension diagnosis and are nonsmokers?

945. What proportion of patients with a hypertension diagnosis are smokers?

946. What proportion of patients with a hypertension diagnosis are nonsmokers?

947. What proportion of nonsmokers has a hypertension diagnosis?

948. What proportion of nonsmokers does not have a hypertension diagnosis?

949. What proportion of all the patients in this study are smokers and have no hypertension diagnosis?

950. What proportion of all the patients in this study are nonsmokers and have no hypertension diagnosis?

Connecting Conditional Probabilities to Two-Way Tables

951–955 *The following table displays information about cigarette smoking and diagnosis with hypertension for a group of patients at a medical clinic.*

	Hypertension Diagnosis	No Hypertension Diagnosis
Smoker	48	24
Nonsmoker	26	50

Illustration by Ryan Sneed

The following stacked bar graph displays the smoking and hypertension data for the group of patients at the medical clinic.

Note: This graph shows two bars, one for the group with a hypertension diagnosis and one for the group with no hypertension diagnosis. The total percentages within each bar sum to 100%. Any percentage within a group represents a conditional probability for that group — that is, the percentage within that group with a certain characteristic. Use conditional probability terms and notation to answer these problems.

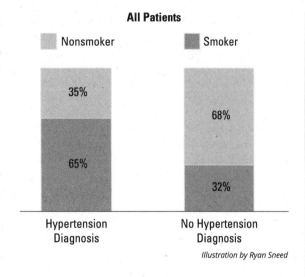

All Patients

Nonsmoker   Smoker

35%

65%

68%

32%

Hypertension Diagnosis   No Hypertension Diagnosis

Illustration by Ryan Sneed

951. What does the area labeled 35% represent?

952. What does the area labeled 65% represent?

953. What does the area labeled 68% represent?

954. What does the area labeled 32% represent?

955. Based on this data and understanding that you are working with only a single sample of data, which of the following statements appears to be true?

(A) Patients with a hypertension diagnosis are more likely to be smokers than nonsmokers.

(B) Patients with a hypertension diagnosis are less likely to be smokers than nonsmokers.

(C) Patients without a hypertension diagnosis are more likely to be smokers than nonsmokers.

(D) Patients without a hypertension diagnosis are more likely to be nonsmokers than smokers.

(E) Choices (A) and (D)

Investigating Independent Variables

956–960 Solve the following problems about independent variables.

956. If variables *A* and *B* are independent, which of the following must be true?

(A) $P(A) = P(B)$

(B) $P(A) \neq P(B)$

(C) $P(A)$ does not depend on whether or not *B* occurs.

(D) $P(A)$ depends on $P(B)$.

(E) Choices (A) and (C)

957. You collect data on the choice of major among a random sample of students from a large university. You find that among engineering majors, 70% are male, and among English majors, 80% are female. Which of the following statements is true?

(A) Gender and choice of major are independent.

(B) Gender and choice of major are not independent.

(C) There are more males than females enrolled in the university.

(D) There are fewer males than females enrolled in the university.

(E) Impossible to say anything without further information.

958. Suppose that in a population of high-school seniors, the choice to enroll in higher education after graduation is independent of gender. Which of the following statements would be true?

(A) The same number of males and females choose to enroll in higher education.

(B) The same proportion of males and females choose to enroll in higher education.

(C) More males enlist in the military, and more females go directly to full-time work.

(D) Choices (B) and (C).

(E) None of the above.

959. A small town has 300 male registered voters and 350 female registered voters. Overall, 60% of voters voted for a bond initiative. If voting is independent of gender in this sample, how many women voted for the bond initiative?

960. A primary school has 200 male students and 190 female students. Overall, 40% of students participate in after-school activities. If participation is independent of gender in this sample, how many boys did not participate in after-school activities?

Calculating Marginal Probability and More

961–970 *The following table represents data from a survey on the type of diet and cholesterol level among a group of adults ages 50 to 75 years.*

Note: A vegetarian diet excludes meat, and a vegan diet excludes any animal-derived products in addition to meat. Regular diet in this table refers to those who are neither vegan nor vegetarian.

	High Cholesterol	No High Cholesterol	
Vegetarian			100
Vegan			100
Regular Dieter			100
	100	200	300

Illustration by Ryan Sneed

961. What is the marginal probability of being a vegetarian?

962. What is the marginal probability of not being a vegan?

963. What is the marginal probability of having high cholesterol?

964. What is the marginal probability of not having high cholesterol?

965. If diet and cholesterol level are independent, which of the following would you expect to be true in this sample?

(A) The same percentage of vegetarians, vegans, and regular dieters will have high cholesterol.

(B) The percentage of vegans, vegetarians, and regular dieters with high cholesterol will differ.

(C) Among those with high cholesterol, equal numbers will be vegetarians, vegans, and regular dieters.

(D) Different numbers of people with high cholesterol will be vegetarians, vegans, and regular dieters.

(E) Choices (A) and (C).

966. If diet and cholesterol level are independent, how many adults in this data set would you expect to be vegetarians without high cholesterol?

967. If being a vegetarian and having high cholesterol are independent, how many adults in this data set would you expect to be vegetarians with high cholesterol?

968. Suppose that in this data, 10 vegetarians and 20 vegans have high cholesterol. How many regular dieters have high cholesterol?

969. Suppose that in this data, ten vegetarians have high cholesterol. How many vegetarians do not have high cholesterol?

970. Suppose that in this data, 35 regular dieters do not have high cholesterol. How many regular dieters do have high cholesterol?

Adding Joint Probability to the Mix

971–978 This table contains data from a survey that asked a sample of adults what type of phone they used most commonly. The adults were classified into three age categories: 18 to 40 years old, 41 to 65 years old, and 66 years old or older.

	Smartphone	Other Mobile Phone	Landline
Age 18–40	60	30	10
Age 41–65	40	40	20
Age 66 or older	20	30	50

Illustration by Ryan Sneed

971. How many people took part in this survey?

972. What is the marginal probability that an individual in the sample is 41 to 65 years old?

973. What is the marginal probability of a respondent's most commonly used type of phone not being a landline?

974. What is the joint probability of being between the ages of 18 and 40 and most commonly using a smartphone?

975. What is the joint probability of being age 66 or older and most commonly using a landline phone?

976. Assume that the marginal frequencies are correct but that the cell entries of the table are unknown. If age and phone preference were independent, how many people ages 18 to 40 would prefer a smartphone?

977. Assume that the marginal frequencies are correct but that the cell entries of the table are unknown. If age and phone preference were independent, how many people age 66 or older would prefer a mobile phone?

978. Comparing the observed values in this table and the expected values under independence, what can you conclude about age and phone preference?

(A) Age and phone preference are independent.

(B) People ages 18 to 40 are less likely to prefer smartphones than would be expected if age and phone preference were independent.

(C) People age 66 or older are less likely to prefer smartphones than would be expected if age and phone preference were independent.

(D) People age 66 or older are more likely to prefer smartphones than would be expected if age and phone preference were independent.

(E) Choices (B) and (D).

Digging Deeper into Conditional and Marginal Probabilities

979 *Answer the following problem.*

979. Suppose that you have a 2-x-2 table displaying values on gender (male or female) and laptop computer ownership (yes or no) for 100 male and 100 female college students. Overall, 75% of the students own laptops; 85% of the male students own laptops, as do 65% of the female students. Are gender and laptop ownership independent in this data set?

(A) Yes, because the overall ownership rate is 75%.

(B) Yes, because the conditional and marginal probabilities are equal.

(C) No, because the ownership rates differ by gender.

(D) No, because the marginal ownership rate differs from the conditional ownership rates.

(E) Choices (C) and (D).

Figuring Out the Number of Cells in a Two-Way Table

980 *Determine how many cells a two-way table will have in the following problem.*

980. You are constructing a two-way table showing the responses to two questions in a survey: type of residence (private home, rented apartment, or condominium) and annual income category ($20,000 or less; $21,000 to $45,000; $46,000 to $75,000; and $76,000 or more). How many cells will this table have?

Including Conditional Probability

981–992 *This bar chart displays frequency results from a survey conducted with a random sample of adults, asking their gender and whether they own a car. All four combinations are shown in the graph, similar to a two-way table.*

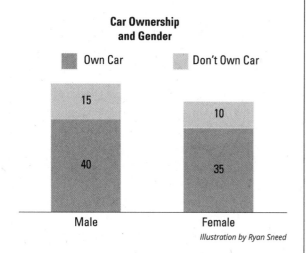

Car Ownership and Gender

Own Car   Don't Own Car

Male: 15, 40
Female: 10, 35

Illustration by Ryan Sneed

981. What was the total sample size for this survey?

982. If a person is male, what is the probability of him owning a car?

983. Given a person is female, what is the probability of her owning a car?

984. What is the marginal probability of car ownership?

985. What is the conditional probability of being male, given car ownership?

986. What is the conditional probability of being female, given car ownership?

987. Assuming that the marginal probability of car ownership among all adults also applies to the group of males, how many of the males in this sample would be car owners?

988. Assuming that the probability of car ownership among all adults also applies to the group of females, how many of the females in this sample would be car owners?

989. What is the marginal probability of being male?

990. What is the marginal probability of being female?

991. Which of the following statements is true when you compare the prevalence of car ownership between females and males?

(A) In this sample, more males than females own cars.

(B) In this sample, more females than males own cars.

(C) In this sample, the conditional probability of car ownership is higher for males.

(D) In this sample, the conditional probability of car ownership is higher for females.

(E) Choices (A) and (D).

992. Which of the following would bolster your ability to draw conclusions about the relationship between car ownership and gender?

(A) replication of the survey in other locations

(B) replication of the survey with a larger sample

(C) replicating the survey with a nationally representative sample

(D) Choices (A) and (B)

(E) Choices (A), (B), and (C)

Digging into Research Designs

993 Solve the following problem about research designs.

993. Which of the following research designs offers the best justification for making cause-and-effect statements from statistical results?

(A) a survey

(B) a randomized clinical trial

(C) an observational study

(D) Choices (A) and (B)

(E) Choices (A) and (C)

Digging Deeper into Two-Way Tables

994–1,001 This table displays results from a study evaluating the effectiveness of a new screening instrument for depression. All participants in the study were given the screening test and were also clinically evaluated for depression.

	Evaluated Positive	Evaluated Negative
Screened Positive	25	20
Screened Negative	10	60

Illustration by Ryan Sneed

994. Suppose that a third category, labeled "Inconclusive," was added to each of the two variables. How many cells would the table then have?

995. What is the marginal probability of screening positive for depression?

996. What is the marginal probability of evaluating positive for depression?

997. What is the conditional probability of evaluating negative for depression, given a positive screening result?

998. What is the conditional probability of evaluating positive for depression, given a negative screening result?

999. Of the four possible outcomes (screened positive and evaluated positive; screened positive and evaluated negative; screened negative and evaluated positive; screened negative and evaluated negative), which has the highest joint probability?

1000. Do the results from this sample suggest that screening positive for depression and being diagnosed for depression are independent? Why or why not?

1001. Which of the following could strengthen your confidence in drawing causal conclusions from this study?

(A) The study sample was randomly selected from the population.

(B) The people doing the evaluation had no knowledge of the screening results.

(C) This study replicated an earlier study that produced similar results.

(D) Choices (A) and (B).

(E) Choices (A), (B), and (C).

2

The Answers

This part provides fully worked-out solutions and discussions for each of the problems in Part I. Be sure to work through a problem as best as you can before peeking at the solutions. That way, you'll be able to gauge which ideas you've already got under your belt and which ones you need to focus more time on.

As you're working through the practice problems and reading the answer explanations, you may decide that you could use a little brushing up in certain areas or a little more supporting material to give you an edge. No worries! The following books are easily available courtesy of your *For Dummies* friends (written by Deborah J. Rumsey, PhD, and published by Wiley):

Statistics For Dummies, 2nd Edition

Statistics Workbook For Dummies

Statistics Essentials For Dummies

Visit www.dummies.com for more information.

Chapter 18

Answers

1. **all households in the city**

   A *population* is the entire group you're interested in studying. The goal here is to estimate what percent of *all households in a large city* have a single woman as the head of the household. The population is all households, and the variable is whether a single woman runs the household.

2. **the 200 households selected**

   The *sample* is a subset drawn from the entire population you're interested in studying. So in this example, the subset is the 200 households selected out of all the households in the city.

3. **the percent of households headed by single women in the city**

   A *parameter* is some characteristic of the population. Because directly studying a population isn't usually possible, parameters are usually estimated by using statistics (numbers calculated from sample data).

4. **the percent of households headed by single women among the 200 selected households**

   The *statistic* is a number describing some characteristic that you calculate from your sample data; the statistic is used to estimate the parameter (the same characteristic in the population).

5. **D. a person's height, recorded in inches**

   Quantitative variables are measured and expressed numerically, have numeric meaning, and can be used in calculations. (That's why another name for them is numerical variables.) Although zip codes are written in numbers, the numbers are simply convenient labels and don't have numeric meaning (for example, you wouldn't add together two zip codes).

6. **E. Choices (B) and (C) (college major; high-school graduate or not)**

   A categorical variable doesn't have numerical or quantitative meaning but simply describes a quality or characteristic of something. College major (such as English or mathematics) and high-school graduate (yes or no) both describe non-numerical qualities.

The numbers used in categorical or qualitative data designate a quality rather than a measurement or quantity. For example, you can assign the number 1 to a person who's married and the number 2 to a person who isn't married. The numbers themselves don't have meaning — that is, you wouldn't add the numbers together.

7. E. All of these choices are true.

Bias is systematic favoritism in the data. You want to get data that represents all customers at the store, no matter what day or what time they shop, whether they shop in couples or alone, and so on. You can't assume that the people who shopped during those three hours on that Saturday morning are representative of the store's total clientele. This sample wasn't drawn randomly — everyone who walked in was counted.

8. E. Choices (C) and (D) (the gender of each shopper who comes in during the time period; the number of men entering the store during the time period)

A variable is a characteristic or measurement on which data is collected and whose result can change from one individual to the next. That means gender is a variable, and the number of men entering the store is also a variable. The day you collect data and the store you observe are just part of the design of your study and were determined beforehand.

9. gender; number of shoppers

Gender is a categorical variable (the categories are male or female), and *number of shoppers* is a quantitative variable (because it represents a count). The day you collect data and the store you observe are just part of the design of your study and were determined beforehand.

10. D. Choices (A) and (C) (a bar graph; a pie chart)

Gender is a categorical variable, so both bar graphs and pie charts are appropriate to display the proportion of males versus females among the shoppers. You could use a time plot only if you knew how many males and how many females were in the store at each individual time period.

11. Add together the total shoppers from each observer and divide this total by 3.

The mean number of shoppers per hour is calculated by dividing the total number of shoppers (found by adding together the total from each observer) and dividing by the number of hours (3).

12. E. Choices (A) and (C) (3, 3, 3, 3, 3; 1, 2, 3, 4, 5)

To find the median, put the data in order from lowest to highest, and find the value in the middle. It doesn't matter how many times a number is repeated. In this case, the data sets 3, 3, 3, 3, 3 and 1, 2, 3, 4, 5 each have a median of 3.

13. **It means that 90% of students who took the exam had scores less than or equal to Susan's.**

A *percentile* shows the relative standing of a score in a population by identifying the percent of values below that score. Susan scored in the 90th percentile, so 90% of the students' scores are less than or equal to Susan's.

14. **E. all of the above**

All of the choices are correct. A distribution is basically a list of all possible values of the variable and how often they occur. In a sentence, you can say, "60% of the 100 people surveyed said they like chocolate, and 40% said they don't" — this sentence gives the distribution. You can also make a table with rows labeled "Like chocolate" and "Don't like chocolate" and show the percents, or you can use a pie chart or a bar graph to visually describe the distribution.

15. **Your score is 0.70 standard deviations above the mean.**

A *z*-score tells you how many standard deviations a data value is below or above the mean. If your *z*-score is 0.70, your exam score is 0.70 standard deviations above the mean. It doesn't tell you your actual score or how many students scored above or below you, but it does tell you where a data value stands, compared to the *average* exam score.

16. **It means that it's likely that between 59% and 71% of all Americans approve of the president.**

The *margin of error* tells you how much your sample results are likely to change from sample to sample. It's measured as "plus or minus a certain amount." In this case, the margin of error of 6% tells you that the result from this sample (65% approving of the president) could change by as much as 6% on either side. Therefore, in using the sample results to draw conclusions about the whole population, the best you can say is, "Based on the data, the percentage of all Americans who approve of the president is likely between 59% (65% − 6%) and 71% (65% + 6%)."

17. **a confidence interval**

You use a confidence interval when you want to estimate a population parameter (a number describing the population) when you have no prior information about it. In a confidence interval, you take a sample, calculate a statistic, and add/subtract a margin of error to come up with your estimate.

18. **a hypothesis test**

You use a hypothesis test when someone reports or claims that a population parameter (such as the population mean) is equal to a certain value and you want to challenge that claim. Here, the claim is that the percentage of all high-school graduates who participated in sports is equal to 60%. You think it's higher than that, so you're challenging that claim.

19. **E.** $p = 0.001$

A p-value measures how strong your evidence is against the other person's reported value. A small p-value means your evidence is strong against them; a large p-value says your evidence is weak against them. In this case, the smallest p-value is 0.001, which in any statistician's book is deemed highly significant, meaning your data and test results show strong evidence against the report.

20. **C. 1, 1, 4, 4**

The standard deviation measures how much variability (diversity) is in the data set, compared to the mean. If all the data values are the same, the standard deviation is 0. To increase the standard deviation, move the values farther and farther away from the mean. The choice that moves them the farthest from the mean here is 1, 1, 4, 4.

21. **25.7**

Use the formula for calculating the mean

$$\bar{x} = \frac{\sum x}{n}$$

where $\bar{x}$ is the mean, $\sum$ represents the sum of the data values, and n is the number of values in the data set.

In this case, $x = 14 + 14 + 15 + 16 + 28 + 28 + 32 + 35 + 37 + 38 = 257$, and $n = 10$. So, the mean is

$$\frac{257}{10} = 25.7$$

22. **52.4**

Use the formula for calculating the mean

$$\bar{x} = \frac{\sum x}{n}$$

where $\bar{x}$ is the mean, $\sum$ represents the sum of the data values, and n is the number of values in the data set.

In this case, $x = 15 + 25 + 35 + 45 + 50 + 60 + 70 + 72 + 100 = 472$, and $n = 9$. So, the mean is

$$\frac{472}{9} = 52.4444$$

The question asks for the nearest tenth, so you round to 52.4.

23. **7.5**

Use the formula for calculating the mean

$$\bar{x} = \frac{\sum x}{n}$$

where $\bar{x}$ is the mean, $\sum$ represents the sum of the data values, and n is the number of values in the data set.

In this case, $x = 0.8 + 1.8 + 2.3 + 4.5 + 4.8 + 16.1 + 22.3 = 52.6$, and $n = 7$. So, the mean is

$$\frac{52.6}{7} = 7.5143$$

The question asks for the nearest tenth, so you round to 7.5.

24. **4.525**

Use the formula for calculating the mean

$$\bar{x} = \frac{\sum x}{n}$$

where $\bar{x}$ is the mean, $\sum$ represents the sum of the data values, and n is the number of values in the data set.

In this case, $x = 0.003 + 0.045 + 0.58 + 0.687 + 1.25 + 10.38 + 11.252 + 12.001 = 36.198$, and $n = 8$. So, the mean is

$$\frac{36.198}{8} = 4.52475$$

The questions asks for the nearest thousandth, so you round to 4.525.

25. **15.0**

To find the median, put the numbers in order from smallest to largest:

4, 5, 6, 12, 15, 16, 18, 20, 22

Because this data set has an odd number of values (nine), the median is simply the middle number in the data set: 15.

26. **17.0**

To find the median, put the numbers in order from smallest to largest:

10, 12, 15.5, 16, 17, 17, 17, 18, 18, 21, 21

Because this data set has an odd number of values (11), the median is simply the middle number in the data set: 17.

27. **9.0**

To find the median, put the numbers in order from smallest to largest:

1, 2, 6, 7, 8, 10, 14, 15, 21, 30

Because this data set has an even number of values (ten), the median is the average of the two middle numbers:

$$\frac{8 + 10}{2} = 9.0$$

28. **4.04**

To find the median, put the numbers in order from smallest to largest.

0.001, 0.1, 0.25, 1.22, 6.85, 8.2, 13.2, 25.2

Because this data set has an even number of values (eight), the median is the average of the two middle numbers:

$$\frac{1.22 + 6.85}{2} = 4.035$$

The question asks for the nearest hundredth, so round to 4.04.

29. **The mean will have a higher value than the median.**

A data set distribution that is skewed right is asymmetrical and has a large number of values at the lower end and few numbers at the high end. In this case, the median, which is the middle number when you sort the data from smallest to largest, lies in the lower range of values (where most of the numbers are). However, because the mean finds the average of all the values, both high and low, the few outlying data points on the high end cause the mean to increase, making it higher than the median.

30. **The mean will have a lower value than the median.**

A data set distribution that is skewed left is asymmetrical and has a large number of values at the high end and few numbers at the low end. In this case, the median, which is the middle number when you sort the data from smallest to largest, lies in the upper range of values (where most of the numbers are). However, because the mean finds the average of all the values, both high and low, the few outlying data points on the low end cause the mean to decrease, making it lower than the median.

31. **The mean and median will be fairly close together.**

When a data set has a symmetrical distribution, the mean and the median are close together because the middle value in the data set, when ordered smallest to largest, resembles the balancing point in the data, which occurs at the average.

32. **median**

The median is the middle value of the data points when ordered from smallest to largest. When the data is ordered, it no longer takes into account the values of any of the other data points. This makes it resistant to being influenced by outliers. (In other words, outliers don't really affect the median.) In contrast, the mean takes every specific data value into account. If the data points contain some outliers that are extreme values to one side, the mean will be pulled toward those outliers.

33. **how concentrated the data is around the mean**

A standard deviation measures the amount of variability among the numbers in a data set. It calculates the typical distance of a data point from the mean of the data. If the standard deviation is relatively large, it means the data is quite spread out away from the mean. If the standard deviation is relatively small, it means the data is concentrated near the mean.

34. **approximately 68%**

According to the empirical rule, the bell-shaped curve of a normal distribution will have 68% of the data points within one standard deviation of the mean.

35. **E. Choice (A) or (C) (standard deviation or variance)**

The standard deviation is a way of measuring the typical distance that data is from the mean and is in the same units as the original data. The variance is a way of measuring the typical *squared* distance from the mean and isn't in the same units as the original data. Both the standard deviation and variance measure variation in the data, but the standard deviation is easier to interpret.

36. **E. Choices (A) and (C) (margin of error; standard deviation)**

The standard deviation measures the typical distance from the data to the mean (using all the data to calculate). Outliers are far from the mean, so the more outliers there are, the higher the standard deviation will be. You calculate the margin of error by using the sample standard deviation so it's also sensitive to outliers. The interquartile range is the range of the middle 50% of the data, so outliers won't be included, making it less sensitive to outliers than the standard deviation or margin of error.

37. **5 years**

The formula for the sample standard deviation of a data set is

$$s = \sqrt{\frac{\sum (x - \bar{x})^2}{n-1}}$$

where x is a single value, $\bar{x}$ is the mean of all the values, $\sum (x - \bar{x})^2$ represents the sum of the squared differences from the mean, and n is the sample size.

First, find the mean of the data set by adding together the data points and then dividing by the sample size (in this case, $n = 10$):

$$\bar{x} = \frac{0+1+2+4+8+3+10+17+2+7}{10}$$
$$= \frac{54}{10} = 5.4$$

Then, subtract the mean from each number in the data set and square the differences, $(x - \bar{x})^2$:

$$(0 - 5.4)^2 = (-5.4)^2 = 29.16$$
$$(1 - 5.4)^2 = (-4.4)^2 = 19.36$$
$$(2 - 5.4)^2 = (-3.4)^2 = 11.56$$
$$(4 - 5.4)^2 = (-1.4)^2 = 1.96$$
$$(8 - 5.4)^2 = (2.6)^2 = 6.76$$
$$(3 - 5.4)^2 = (-2.4)^2 = 5.76$$
$$(10 - 5.4)^2 = (4.6)^2 = 21.16$$
$$(17 - 5.4)^2 = (11.6)^2 = 134.56$$
$$(2 - 5.4)^2 = (-3.4)^2 = 11.56$$
$$(7 - 5.4)^2 = (1.6)^2 = 2.56$$

Next, add up the results from the squared differences:

$$29.16 + 19.36 + 11.56 + 1.96 + 6.76 + 5.76 + 21.16 + 134.56 + 11.56 + 2.56 = 244.4$$

Finally, plug the numbers into the formula for the sample standard deviation:

$$s = \sqrt{\frac{\sum(x - \bar{x})^2}{n-1}}$$

$$= \sqrt{\frac{244.4}{10-1}}$$

$$= \sqrt{27.156}$$

$$= 5.21$$

The question asks for the nearest year, so round to 5 years.

38. 8 years

The formula for the sample standard deviation of a data set is

$$s = \sqrt{\frac{\sum(x - \bar{x})^2}{n-1}}$$

where x is a single value, $\bar{x}$ is the mean of all the values, $\sum(x - \bar{x})^2$ represents the sum of the squared differences from the mean, and n is the sample size.

First, find the mean of the data set by adding together the data points and then dividing by the sample size (in this case, $n = 12$):

$$\bar{x} = \frac{12 + 10 + 16 + 22 + 24 + 18 + 30 + 32 + 19 + 20 + 35 + 26}{12}$$

$$= \frac{264}{12} = 22$$

Then, subtract the mean from each number in the data set and square the differences, $(x - \bar{x})^2$:

$$(12 - 22)^2 = (-10)^2 = 100$$

$$(10 - 22)^2 = (-12)^2 = 144$$

$$(16\sqrt{2}22)^2 = (-6)^2 = 36$$

$$(22 - 22)^2 = (0)^2 = 0$$

$$(24 - 22)^2 = (2)^2 = 4$$

$$(18 - 22)^2 = (4)^2 = 16$$

$$(30 - 22)^2 = (8)^2 = 64$$

$$(32 - 22)^2 = (10)^2 = 100$$

$$(19 - 22)^2 = (-3)^2 = 9$$

$$(20 - 22)^2 = (-2)^2 = 4$$

$$(35 - 22)^2 = (13)^2 = 169$$

$$(26 - 22)^2 = (4)^2 = 16$$

Next, add up the results from the squared differences:

$$100 + 144 + 36 + 0 + 4 + 16 + 64 + 100 + 9 + 4 + 169 + 16 = 662$$

Finally, plug the numbers into the formula for the sample standard deviation:

$$s = \sqrt{\frac{\sum(x - \bar{x})^2}{n - 1}}$$

$$= \sqrt{\frac{662}{12 - 1}}$$

$$= \sqrt{60.1818}$$

$$= 7.76$$

The question asks for the nearest year, so round to 8 years.

39. 8.7 points

The formula for the sample standard deviation of a data set is

$$s = \sqrt{\frac{\sum(x - \bar{x})^2}{n - 1}}$$

where x is a single value, $\bar{x}$ is the mean of all the values, $\sum(x - \bar{x})^2$ represents the sum of the squared differences from the mean, and n is the sample size.

First, find the mean of the data set. Although you don't have a list of all the individual values, you do know the test score for each student in the sample. For example, you know that three students scored 92 points, so if you listed every student's score individually, you'd see 92 three times, or (92)(3). To find the mean this way, multiply each exam score by the number of students who received that score, add the products together, and then divide by the number of students in the sample $(n = 20)$:

$$(98)(2) = 196$$
$$(95)(1) = 95$$
$$(92)(3) = 276$$
$$(88)(4) = 352$$
$$(87)(2) = 174$$
$$(85)(2) = 170$$
$$(81)(1) = 81$$
$$(78)(2) = 156$$
$$(73)(1) = 73$$
$$(72)(1) = 72$$
$$(65)(1) = 65$$

$$\bar{x} = \frac{196 + 95 + 276 + 352 + 174 + 170 + 81 + 156 + 73 + 72 + 65}{20}$$

$$= \frac{1,710}{20} = 85.5$$

Next, subtract the mean from each different exam score in the data set and square the differences, $(x - \bar{x})^2$. *Note:* There are 11 different exam scores here — 98, 95, 92, 88, 87, 85, 81, 78, 73, 72, and 65 — but 20 students. First, work with the 11 exam scores.

$$(98 - 85.5)^2 = (12.5)^2 = 156.25$$
$$(95 - 85.5)^2 = (9.5)^2 = 90.25$$
$$(92 - 85.5)^2 = (6.5)^2 = 42.25$$
$$(88 - 85.5)^2 = (2.5)^2 = 6.25$$
$$(87 - 85.5)^2 = (1.5)^2 = 2.25$$
$$(85 - 85.5)^2 = (-0.5)^2 = 0.25$$
$$(81 - 85.5)^2 = (-4.5)^2 = 20.25$$
$$(78 - 85.5)^2 = (-7.5)^2 = 56.25$$
$$(73 - 85.5)^2 = (-12.5)^2 = 156.25$$
$$(72 - 85.5)^2 = (-13.5)^2 = 182.25$$
$$(65 - 85.5)^2 = (-20.5)^2 = 420.25$$

Now, multiply each value by the number of students who got that score:

$$(156.25)(2) = 312.5$$
$$(90.25)(1) = 90.25$$
$$(42.25)(3) = 126.75$$
$$(6.25)(4) = 25$$
$$(2.25)(2) = 4.5$$
$$(0.25)(2) = 0.5$$
$$(20.25)(1) = 20.25$$
$$(56.25)(2) = 112.5$$
$$(156.25)(1) = 156.25$$
$$(182.25)(1) = 182.25$$
$$(420.25)(1) = 420.25$$

Then, add up those results:

$$312.5 + 90.25 + 126.75 + 25 + 4.5 + 0.5 + 20.25 + 112.5 + 156.25 +$$
$$182.25 + 420.25 = 1,451$$

Finally, plug the numbers into the formula for the sample standard deviation:

$$s = \sqrt{\frac{\sum (x - \bar{x})^2}{n - 1}}$$
$$= \sqrt{\frac{1,451}{20 - 1}}$$
$$= \sqrt{76.37}$$
$$= 8.74$$

The question asks for the nearest tenth of a point, so round to 8.7.

40. **0.0036 cm**

The formula for the sample standard deviation of a data set is

$$s = \sqrt{\frac{\sum(x - \bar{x})^2}{n-1}}$$

where x is a single value, $\bar{x}$ is the mean of all the values, $\sum(x - \bar{x})^2$ represents the sum of the squared differences from the mean, and n is the sample size.

First, find the mean of the data set by adding together the data points and then dividing by the sample size (in this case, $n = 10$):

$$\bar{x} = \frac{5.001 + 5.002 + 5.005 + 5.010 + 5.009 + 5.003 + 5.002 + 5.001 + 5.000}{10}$$

$$= \frac{50.033}{10} = 5.0033$$

Then, subtract the mean from each number in the data set and square the differences, $(x - \bar{x})^2$:

$$\left(5.001 - 5.0033\right)^2 = \left(-0.0023\right)^2 = 0.00000529$$

$$\left(5.002 - 5.0033\right)^2 = \left(-0.0013\right)^2 = 0.00000169$$

$$\left(5.005 - 5.0033\right)^2 = \left(0.0017\right)^2 = 0.00000289$$

$$\left(5.000 - 5.0033\right)^2 = \left(-0.0033\right)^2 = 0.00001089$$

$$\left(5.010 - 5.0033\right)^2 = \left(0.0067\right)^2 = 0.00004489$$

$$\left(5.009 - 5.0033\right)^2 = \left(0.0057\right)^2 = 0.00003249$$

$$\left(5.003 - 5.0033\right)^2 = \left(-0.0003\right)^2 = 0.00000009$$

$$\left(5.002 - 5.0033\right)^2 = \left(-0.0013\right)^2 = 0.00000169$$

$$\left(5.001 - 5.0033\right)^2 = \left(-0.0023\right)^2 = 0.00000529$$

$$\left(5.000 - 5.0033\right)^2 = \left(-0.0033\right)^2 = 0.00001089$$

Next, add up the results from the squared differences:

$$0.00000529 + 0.00000169 + 0.00000289 + 0.00001089 + 0.00004489 + 0.00003249$$
$$+ 0.00000009 + 0.00000169 + 0.00000529 + 0.00001089 = 0.0001161$$

Finally, plug the numbers into the formula for the sample standard deviation:

$$s = \sqrt{\frac{\sum(x - \bar{x})^2}{n-1}}$$

$$= \sqrt{\frac{0.0001161}{10-1}}$$

$$= \sqrt{0.0000129}$$

$$= 0.0036$$

The sample standard deviation for the jet engine turbine part is 0.0036 centimeters.

41. There is more variation in salaries in Magna Company than in Ace Corp.

The larger standard deviation in Magna Company shows a greater variation of salaries in both directions from the mean than Ace Corp. The standard deviation measures on average how spread out the data is (for example, the high and low salaries at each company).

42. B. measuring the variation in circuitry components when manufacturing computer chips

The quality of the vast majority of manufacturing processes depends on reducing variation to as little as possible. If a manufacturing process has a large standard deviation, it indicates a lack of predictability in the quality and usefulness of the end product.

43. Lake Town has a lower average temperature and less variability in temperatures than Sunshine City.

Lake Town has a much smaller standard deviation than Sunshine City, so its temperatures change (or vary) less. You don't know the actual range of temperatures for either city.

44. There will be no change in the standard deviation.

All the data points will shift up $2,000, and as a result, the mean will also increase by $2,000. But each individual salary's distance (or deviation) from the mean will be the same, so the standard deviation will stay the same.

45. The sample variance is 2.3 ounces². The standard deviation is 1.5 ounces.

You find the sample variance with the following formula:

$$s^2 = \frac{\sum(x - \bar{x})^2}{n-1}$$

where x is a single value, $\bar{x}$ is the mean of all the values, $\sum(x - \bar{x})^2$ represents the sum of the squared difference scores, and n is the sample size.

First, find the mean by adding together the data points and dividing by the sample size (in this case, $n = 5$):

$$\bar{x} = \frac{7+6+5+6+9}{5}$$
$$= \frac{33}{5} = 6.6$$

Then, subtract the mean from each data point and square the differences, $(x - \bar{x})^2$:

$$\left(7 - 6.6\right)^2 = \left(0.4\right)^2 = 0.16$$
$$\left(6 - 6.6\right)^2 = \left(-0.6\right)^2 = 0.36$$
$$\left(5 - 6.6\right)^2 = \left(-1.6\right)^2 = 2.56$$
$$\left(6 - 6.6\right)^2 = \left(-0.6\right)^2 = 0.36$$
$$\left(9 - 6.6\right)^2 = \left(2.4\right)^2 = 5.76$$

Next, plug the numbers into the formula for the sample variance:

$$s^2 = \frac{\sum(x-\bar{x})^2}{n-1}$$
$$= \frac{0.16+0.36+2.56+0.36+5.76}{5-1}$$
$$= \frac{9.2}{4} = 2.3$$

The sample variance is 2.3 ounces². But these units don't make sense because there's no such thing as "square ounces." However, the standard deviation is the square root of the variance, so it can then be expressed in the original units: $s = 1.5$ ounces (rounded). For this reason, standard deviation is preferred over the variance when it comes to measuring and interpreting variability in a data set.

46.   **The sample variance is 15 minutes². The standard deviation is 4 minutes.**

You find the sample variance with the following formula:

$$s^2 = \frac{\sum(x-\bar{x})^2}{n-1}$$

where x is a single value, $\bar{x}$ is the mean of all the values, $\sum(x-\bar{x})^2$ represents the sum of the squared difference scores, and n is the sample size.

First, find the mean by adding together the data points and dividing by the sample size (in this case, $n = 5$):

$$\bar{x} = \frac{15+16+18+10+9}{5}$$
$$= \frac{68}{5} = 13.6$$

Then, subtract the mean from each data point and square the differences, $(x-\bar{x})^2$:

$$(15-13.6)^2 = (1.4)^2 = 1.96$$
$$(16-13.6)^2 = (2.4)^2 = 5.76$$
$$(18-13.6)^2 = (4.4)^2 = 19.36$$
$$(10-13.6)^2 = (-3.6)^2 = 12.96$$
$$(9-13.6)^2 = (-4.6)^2 = 21.16$$

Next, plug the numbers into the formula for the sample variance:

$$s^2 = \frac{\sum(x-\bar{x})^2}{n-1}$$
$$= \frac{1.96+5.76+19.36+12.96+21.16}{5-1}$$
$$= \frac{61.2}{4} = 15.3$$

The sample variance is 15.3 minutes². But these units don't make sense because there's no such thing as "square minutes." However, the standard deviation is the square root of the variance, so it can then be expressed in the original units: $s = 3.91$ minutes (rounded up to 4). For this reason, standard deviation is preferred over the variance when it comes to measuring and interpreting variability in a data set.

47. The standard deviation is 13 kilometers/hour.

You find the standard deviation with the following formula:

$$s = \sqrt{\frac{\sum(x - \bar{x})^2}{n-1}}$$

where x is a single value, $\bar{x}$ is the mean of all the values, $\sum(x - \bar{x})^2$ represents the sum of the squared difference scores, and n is the sample size.

First, find the mean by adding together the data points and dividing by the sample size (in this case, $n = 5$):

$$\bar{x} = \frac{10 + 15 + 35 + 40 + 30}{5}$$

$$= \frac{130}{5} = 26$$

Then, subtract the mean from each data point and square the differences, $(x - \bar{x})^2$:

$$(10 - 26)^2 = (-16)^2 = 256$$
$$(15 - 26)^2 = (-11)^2 = 121$$
$$(35 - 26)^2 = (9)^2 = 81$$
$$(40 - 26) = (14)^2 = 196$$
$$(30 - 26) = (4)^2 = 16$$

Next, plug the numbers into the formula for the standard deviation:

$$s = \sqrt{\frac{\sum(x - \bar{x})^2}{n-1}}$$

$$= \sqrt{\frac{256 + 121 + 81 + 196 + 16}{5-1}}$$

$$= \sqrt{\frac{670}{4}} = 12.942$$

Rounded to a whole number, the standard deviation is 13 kilometers/hour.

48. D. Choices (A) and (B) (Data Set 1; Data Set 2)

The original data set contains the numbers 1, 2, 3, 4, 5. Data Set 1 just shifts those numbers up by five units to get 6, 7, 8, 9, 10. Standard deviation represents typical (or average) distance from the mean, and although the mean in Data Set 1 changes from 3 to 8, the distances from each point to that new mean stay the same as they were for the original data set, so the average distance from the mean is the same.

Data Set 2 contains the numbers −2, −1, 0, 1, 2. These numbers shift the original data set's values down by three units. For example, $1 - 3 = -2$, $2 - 3 = -1$, and so forth. Therefore, the standard deviation doesn't change from the original data set.

Data Set 3 divides all the numbers in the original data set by 10, making them closer to the mean, on average, than the original data set. Therefore, the standard deviation is smaller.

49. approximately 68%

The empirical rule states that in a normal (bell-shaped) distribution, approximately 68% of values are within one standard deviation of the mean.

50. about 95%

The empirical rule states that in a normal (bell-shaped) distribution, approximately 95% of values are within two standard deviations of the mean.

51. about 57 to 71 years

The empirical rule states that in a normal distribution, 95% of values are within two standard deviations of the mean. "Within two standard deviations" means two standard deviations below the mean *and* two standard deviations above the mean. In this case, the mean is 64 years, and the standard deviation is 3.5 years. So, two standard deviations is $(3.5)(2) = 7$ years.

To find the lower end of the range, subtract two standard deviations from the mean: $64 - 7$ years $= 57$ years. And then to find the upper end of the range, add two standard deviations to the mean: $64 + 7$ years $= 71$ years.

So, about 95% of people who retire do so between the ages of about 57 to 71 years.

52. about 16%

The empirical rule states that approximately 68% of values are within one standard deviation of the mean. "Within one standard deviation" means one standard deviation below the mean *and* one standard deviation above the mean.

In this case, the mean is $48,000, so about 50% of the engineers made less than $48,000. In a normal distribution, half of the values are above the mean and half are below the mean. The standard deviation is $7,000.

To find the lower end of the range within one standard deviation of the mean, subtract the standard deviation from the mean: $48,000 - $7,000 = $41,000; to find the upper end of the range, add the standard deviation to the mean: $48,000 + $7,000 = $55,000.

Because the normal distribution is symmetrical, 34% of the values will be between $41,000 and $48,000, and 34% will be between $48,000 and $55,000.

Therefore, $50\% + 34\% = 84\%$ of the data is $55,000 and below, which leaves 16% of the data above $55,000.

53. if a population has a normal distribution

You can use the empirical rule only if the distribution of the population is normal. Note that the rule says that *if* the distribution is normal, *then* approximately 68% of the values lie within one standard deviation of the mean, not the other way around. Many distributions have 68% of the values within one standard deviation of the mean that don't look like a normal distribution.

54. the mean and the standard deviation of the population

The empirical rule describes the distribution of the data in a population in terms of the mean and the standard deviation. For example, the first part of the empirical rule says that about 68% of the values lie within one standard deviation of the mean, so all you need to know is the mean and the standard deviation to use the rule.

55. 30

If you assume that the 600 lenses tested come from a population with a normal distribution (which they do), you can apply the empirical rule (also known as the 68–95–99.7 rule).

Using the empirical rule, approximately 95% of the data lies within two standard deviations of the mean, and 5% of the data lies outside this range. Because the lenses that are more than two standard deviations from the mean are rejected, you can expect about 5% of the 600 lenses, or $(0.05)(600) = 30$ lenses to be rejected.

56. cannot be determined with the information given

You could use the empirical rule (also known as the 68–95–99.7 rule) if the shape of the distribution of fish lengths was normal; however, this distribution is said to be "very much skewed left," so you can't use this rule. With the information given, you can't answer the question.

57. percentile

A percentile splits the data into two parts, the percentage below the value and the percentage above the value. In other words, a percentile measures where an individual data value stands compared to the rest of the data values. For example, the 90th percentile is the value where 90% of the values lie below it and 10% of the values lie above it.

58. median

The 50th percentile is the value where 50% of the data fall below it and 50% fall above it. This is the same as the definition of the median.

59. It means that 15% of the scores were better than your score.

Percentile is the relative standing in a set of data from the lowest values to highest values. If your score is in the 85th percentile, it means that 85% of the scores are below your score and 15% are above your score.

60. Bob's score is above 70.

A score in the 90th percentile means that 90% of the scores were lower. With 60 scores and an approximately normal distribution, the 90th percentile will certainly be above the mean, here given as 70.

61. **It means that 30% of the students scored above you and that you correctly answered 82% of the test questions.**

The 70th percentile means that 70% of the scores were below your score, and 30% were above your score. Your actual score was 82%, which means that you answered 82% of the test questions correctly.

62. 65%

The 50th percentile doesn't mean a score of 50%; it's the median (or middle number) of the data set. The middle number is 65%, so that is the 50th percentile.

63. 95

A percentage score is different than a percentile. In this case, a percentage score is the percent of questions answered correctly; a percentile expresses the relative standing of a score in terms of the other scores.

Use the procedure for calculating a percentile. To calculate the kth percentile (where k is any number between 0 and 100), follow these steps:

1. Put all the numbers in the data set in order from smallest to largest:

    80, 80, 82, 84, 85, 86, 88, 90, 91, 92, 92, 94, 96, 98, 100

2. Multiply k percent by the total number of values, n:

    $(0.80)(15) = 12$

3. Because your result in Step 2 is a whole number, count the numbers in the data set from left to right until you reach the number you found in Step 2 (in this case, the 12th number). The kth percentile is the average of that corresponding value in the data set and the value that directly follows it.

    Find the average of the 12th and 13th numbers in the data set:

    $$\frac{94+96}{2} = \frac{190}{2} = 95$$

    The 80th percentile is 95.

64. **Bill, Mary, Jose, Lisa, then Paul**

If your score is at the kth percentile, that means k percent of the students scored less than you did, and the rest scored better than you did. For example, someone scoring at the 95th percentile knows that 95% of the other students scored lower than they did and 5% scored higher.

When talking about exam scores, the person that scores at the highest percentile scored better than everyone else on the list. So, in ranking from highest to lowest in terms of where their grades stand compared to each other, you have Bill first, followed by Mary, then Jose, then Lisa, and then Paul.

65. B. the median

The median of a data set is the middle value after you've put the data in order from smallest to largest (or the average of the two middle values if your data set contains an even number of values). Because the median concerns only the very middle of the data set, adding an outlier won't affect its value much (if any). It adds only one more value to one end or the other of the sorted data set.

The mean is based on the sum of all the data values, which includes the outlier, so the mean will be affected by adding an outlier. The standard deviation involves the mean in its calculation; hence it's also affected by outliers. The range is perhaps the most affected by an outlier, because it's the distance between the minimum and maximum values, so adding an outlier makes either the minimum value smaller or the maximum value larger. Either way, the distance between the minimum and maximum increases.

66. E. None of the above.

It's strange but true that all the scenarios are possible. You can use one data set as an example where all four scenarios occur at the same time: 5, 5, 5, 5, 5, 5, 5. In this case, the minimum and maximum are both 5, and the median (middle value) is 5. The median cuts the data set in half, creating an upper half and a lower half of the data set. To find the 1st quartile, take the median of the lower half of the data set, which gives you 5 in this case; to find the 3rd quartile, take the median of the upper half of the data set (also 5). The range is the distance from the minimum to the maximum, which is $5 - 5 = 0$. The IQR is the distance from the 1st to the 3rd quartile, which is $5 - 5 = 0$. Hence, the range and IQR are the same.

67. 1st quartile $= 79$, median $= 90$, 3rd quartile $= 96$

The 1st quartile is the 25th percentile, the median is the 50th percentile, and the 3rd quartile is the 75th percentile.

To find the values for these numbers, use the procedure for calculating a percentile. To calculate the kth percentile (where k is any number between 1 and 100), follow these steps:

1. Put all the numbers in the data set in order from smallest to largest:

    72, 74, 75, 77, 79, 82, 83, 87, 88, 90, 91, 91, 91, 92, 96, 97, 97, 98, 100

2. Multiply k percent times the total number of values, n.

    For the 1st quartile (or 25%): (0.25)(19) = 4.75.

    For the median (or 50%): (0.50)(19) = 9.5.

    For the 3rd quartile (or 75%): (0.75)(19) = 14.25.

3. Because the results in Step 2 aren't whole numbers, round up to the nearest whole number, and then count the numbers in the data set from left to right (from the smallest to the largest number) until you reach the value of the rounded number. The corresponding value in the data set is the kth percentile.

    For the 1st quartile, round 4.75 up to 5 and then find the fifth number in the data set: 79.

For the median, round 9.5 up to 10 and then find the tenth number in the data set: 90.

For the 3rd quartile, round 14.25 up to 15 and then find the 15th number in the data set: 96.

68. The five-number summary can't be found.

The five-number summary of a data set includes the minimum value, the 1st quartile, the median, the 3rd quartile, and the maximum value. You're not given the minimum value or the maximum value here, so you can't fill out the five-number summary.

Note that even though you're given the range, which is the distance between the maximum and minimum values, you can't determine the actual values of the minimum and maximum.

69. The median can't be greater in value than the 3rd quartile.

The five numbers in the five-number summary are the minimum (smallest) number in the data set, the 25th percentile (also known as the 1st quartile, or Q_1), the median (50th percentile), the 75th percentile (also known as the 3rd quartile, or Q_3), and the maximum (largest) number in the data set.

Because the median is at the 50th percentile, its value must be between the value of the 1st and 3rd quartiles, inclusive. In this case, the 1st quartile is 50, and the 3rd quartile is 80, so a median of 85 isn't possible.

70. B. 15, 15, 15

Many data sets containing three numbers can have a mean of 15. However, if you force the standard deviation to be 0, you have only one choice: 15, 15, 15. A standard deviation of 0 means the average distance from the data values to the mean is 0. In other words, the data values don't deviate from the mean at all, and hence they have to be the same value.

71. The median can't be less than Q_1.

Q_1 represents the 25th percentile, indicating that 25% of the salaries lie below that value. The median represents the 50th percentile, indicating that 50% of the salaries lie below that value. That means that the median has to be at least as large as Q_1.

Regarding the variance, the fact that it's large isn't alarming; variance represents the approximate average squared distance from the data values to the mean and isn't related to the other statistics directly.

The median doesn't have to be halfway between Q_1 and Q_3; it does have to lie somewhere between them (inclusive), but it can lie anywhere in that range.

The mean and median don't have to be equal, and the other descriptive statistics shown don't require a certain relationship between them.

72. E. All of the above.

A data set is divided into four parts, each containing 25% of the data: (1) the minimum value to the 1st quartile, (2) the 1st quartile to the median, (3) the median to the

3rd quartile, and (4) the 3rd quartile to the maximum value. Each statement represents a distance that covers two adjacent parts out of the four, which gives a total percentage of $25\%(2) = 50\%$ in every case.

73. **E. None of the above.**

The mean and median can have any type of relationship depending on the data set. The range, being the distance between the minimum and maximum values in the data set, must be greater than or equal to the IQR, which is the distance between the 1st and 3rd quartiles — that is, the range of the middle 50% of the data. The variance is the square of the standard deviation and is larger than the standard deviation only if the standard deviation is greater than 1. (If you pick a standard deviation less than 1, say 0.50, then the variance equals 0.50 squared, which is 0.25, a smaller value.)

74. **the variance of the weights**

You calculate the variance by taking the distances from the mean and squaring each of them and then taking (an approximate) average. During the squaring process, the units become squared also. In this case, the units for the variance are square pounds, which doesn't make sense, so you typically don't report units for variance.

The standard deviation is the square root of the variance, which takes you from square units (square pounds) back to the original units (pounds).

The other statistics, such as the mean, median, and range, all stay in the original units because nothing is being done during their calculations to affect the units of the data.

75. **none of the above**

Measuring spread (variability) in a data set involves measuring distances of various types. The range is the distance from the minimum value to the maximum value, the IQR is the distance from the 1st quartile to the 3rd quartile, the standard deviation is the (approximate) average distance from the data points to the mean, and the variance is the (approximate) average squared distance from the data points to the mean. Different measures of spread are more appropriate than others in different situations, but they all involve something to do with distances.

76. **C. 2, 2, 3, 4, 4**

To begin, the standard deviation must be 1, so the (approximate) average from the data values to the mean must be 1. The data set 2, 2, 3, 4, 4 meets both criteria — it has a mean of 3, and its standard deviation (average distance from the data values to the mean) equals 1. (You find the standard deviation by subtracting the mean from each value in the set, squaring the results, adding them together, dividing by $n - 1$, where n is the number of values, and then taking the square root.)

77. **IQR and median**

The distribution is skewed right, meaning that there are a large number of low values concentrated over a relatively small range, and a long tail to the right, indicating a large number of high values spread over a wide range. When a distribution isn't approximately normal, the spread and center are best measured with the IQR and median, respectively. (The mean and standard deviation are sensitive to the outliers in the lower end of the test scores.)

78. standard deviation and mean

The distribution is approximately normal because the data is approximately symmetrical. The best measures of spread and center are the standard deviation and mean, respectively.

79. IQR and median

At least one outlier exists at the upper end of the salary range — $2 million — and the standard deviation is large. The mean is pulled up to $78,000 by the high outlier(s), whereas the median shows that half the company salaries are below $45,000, and half are above.

All these factors show that the distribution isn't symmetric and that the IQR and median are the best measures of spread and center, respectively.

80. IQR and median

The mean is much lower than the median, which tells you that the data is skewed (either the data is shifted off-center or at least one outlier exists). Because the median is highest, the data is skewed to the left and isn't symmetric.

The best measures for center and spread for skewed data are the median and the interquartile range (IQR) because they're less affected by the skewness than the mean and standard deviation are, and they represent the "typical" center and spread for most of the data.

81. Business

The Business College has 25% of the student enrollment, which is the largest proportion of any college.

82. D. Choices (B) and (C) (a separate pie chart for each college showing what percentage are enrolled and what percentage aren't; a bar graph where each bar represents a college and the height shows what percentage of students are enrolled)

To get the information across correctly, you want to show the results from each college separately by using separate pie charts or one bar graph (note that the heights of all the bars in a bar graph don't need to sum to 100%).

83. 39%

To find the percentage of students enrolled in these two colleges, simply add the percentages together: $23\% + 16\% = 39\%$.

84. 86%

To find the percentage of students *not* enrolled in the College of Engineering, subtract the percentage of students in the College of Engineering from the whole: $100\% - 14\% = 86\%$.

85. **impossible to tell without further information**

This pie chart contains information only about percentages, not the total number of students nor the number in each category.

86. **5,500**

To find how many students are enrolled in the College of Arts & Sciences, you multiply the total number of students by the percentage enrolled in that college, which is 22%: $(25,000)(0.22) = 5,500$.

87. **They distort the proportions of the slices.**

Three-dimensional pie charts distort the proportions of the slices by making those in the front look larger than they should.

88. **Attend a university.**

University is the tallest bar in the graph and therefore represents the greatest number of students of any category.

89. **Go into the military.**

Military is the shortest bar in the graph and therefore represents the fewest number of students.

90. **140**

Judging by the height of the bars, about 120 students plan to attend a university, and about 20 plan to take a gap year. So, $120 + 20 = 140$ students.

91. **322**

You can find the number of students in each category from the height of the bars and then add them together: $120 + 82 + 18 + 82 + 20 = 322$.

92. **25%**

You can find the number of students in each category from the height of the bars. The total number of students is $322(120 + 82 + 18 + 82 + 20 = 322)$. The number going to a community college is 82.

To find the percentage of students going to a community college, divide the number of students in that category by the total number of students: $82/322 = 0.2546$, or about 25%.

93. **63%**

Judging from the height of the bars, there are 322 students total $(120 + 82 + 18 + 82 + 20 = 322)$, and 120 plan to attend a university. To find the percentage of students not planning to attend a university, subtract the number that do plan to attend from the total number of students $(322 - 120 = 202)$, and then divide by the total: $202 / 322 = 0.627$, or about 63%.

94. **The y-axis has been stretched far beyond the range of the data.**

The largest value is 120 but the y-axis runs to 300, making it difficult to compare the height of the bars.

95. **histogram**

Because the data is continuous with many possible values, a histogram would be the best choice for displaying this type of data.

96. **Yes, each bar corresponds to a specific interval of values on the real number line.**

For example, a histogram bar covering the range 5 to 10 on the x-axis would represent all values in the data set in the range of 5 through 10.

97. **because the bars represent the distribution of continuous data**

The bars in a histogram represent cutoff values of continuous data that takes on real numbers. Bars for categorical data stand for groups like male and female, which can be shown in either order.

98. **the BMI scores**

The x-axis represents the values of the variable: BMI scores, a continuous variable that, in this case, runs from 18 to 31.

99. **the range of BMI scores for each group in the bar**

The BMI scores were divided into groups on the number line. The width of a bar represents the range of the BMI scores in that group.

100. **the number of adults with a BMI score in a particular range**

The y-axis of a histogram represents how many individuals are in each group, either as a count (frequency) or as a percentage (relative frequency). In this case, the y-axis represents the number of adults (frequency) with a BMI score in the given range.

101. approximately normal

The shape of this distribution is approximately normal because it has bell-shaped characteristics.

102. 18 to 31

You can see from the *x*-axis that the lowest bar has a lower bound of 18 and the highest bar has an upper bound of 31, so no data is outside that range.

103. the seventh bar

The data is approximately symmetric (similar looking on each side of the mean) and is likely centered in the range of 24 to 25. This corresponds to the seventh bar on the histogram.

104. 21

You can determine the number of people in each BMI range by the height of the bar. In this case, 8 are in the 22 to 23 range, and 13 are in the 23 to 24 range. Add these numbers to get your answer: $8 + 13 = 21$.

105. 15%

You can judge the number of people in each BMI range by the height of the bar. In this case, 9 are in the 28 to 29 range, 4 are in the 29 to 30 range, and 2 are in the 30 to 31 range. Add these numbers to get your answer: $9 + 4 + 2 = 15$. Find the proportion by dividing by the total sample size (101) to get 0.15, and then multiply by 100 to get the percentage (15%).

106. right-skewed

The data is right-skewed because it has a large number of values in the lower income range and a long tail extending to the right, representing a few cases with high incomes.

107. median

The median is the best measure of the center when the data is highly skewed, and it's less affected by extreme values than the mean. Because the mean uses the value of every number in its calculation, the few high values pull the mean upward, while the median stays in the middle and is a better representative of the "center" of the data.

108. The mean will be higher.

In a right-skewed distribution, the mean will be greater than the median because it has a few high values compared to the rest, pulling the average up. The median is just the middle number when the data is ordered, so the few high values can't pull up the median.

109. 0

The lowest possible value is 0 because the values in the first bar (looking at the *x*-axis) go from 0 to 5,000. *Note:* You can't tell whether 0 is actually in the data set, so you know only that the lowest *possible* value is 0.

110. 85,000

Each bar represents a range of 5,000 (looking at the *x*-axis). The lower bound of the highest bar is 80,000, so its upper bound is 85,000.

111. 35

The height of a bar represents the number of people in that range. Each bar covers a range of $5,000 in income, so the first two bars represent incomes less than $10,000. The first bar has a height of about 11 and the second has a height of about 24. Together, that's $35(11+24=35)$.

112. $10,000 to $15,000

The sample includes 110 adults, and if you sort their incomes from lowest to highest, the median is the number in the middle (between the 55th and 56th numbers). To find the bar that contains the median, count the heights of the bars until you reach 55 and 56. The third bar contains the median and ranges from $10,000 to $15,000.

113. Section 1 is approximately normal; Section 2 is approximately uniform.

Section 1 is clearly close to normal because it has an approximate bell shape. Section 2 is close to uniform because the heights of the bars are roughly equal all the way across.

114. They are the same.

The range of values lets you know where the highest and lowest values are. The grades are shown on the *x*-axis of each graph. Section 1's grades go from 70 to 90, and Section 2's grades go from 70 to 90, so they are the same.

115. They will be similar.

In both cases, the data appear to be fairly symmetric, which means that if you draw a line right down the middle of each graph, the shape of the data looks about the same on each side. For symmetric data, no skewness or outliers exist, so the average and the middle value (median) are similar.

116. They will be similar.

In both cases, the data appear to be fairly symmetric, which means that if you draw a line right down the middle of each graph, the shape of the data looks about the same on each side. For symmetric data, no skewness or outliers exist, so the average and the middle value (median) are similar.

117. **78.75 to 80**

Because the sample size is 100, the median will be between the 50th and 51st data value when the data is sorted from lowest to highest. To find the bar that contains the median, count the heights of the bars until you reach 50 and 51. The bar containing the median has the range 78.75 to 80.

118. **77.5 to 82.5**

Because the sample size is 100, the median will be between the 50th and 51st data value when the data is sorted from lowest to highest. To find the bar that contains the median, count the heights of the bars until you reach or pass 50 and 51. The bar containing the 50th data value has the range 77.5 to 80. The bar containing the 51st data value has the range 80 to 82.5. Thus the median is approximately 80 (the value that borders both intervals).

119. **Section 2, because a flat histogram has more variability than a bell-shaped histogram of a similar range.**

Standard deviation is the average distance the data is from the mean. When the data is flat, it has a large average distance from the mean, overall, but if the data has a bell shape (normal), much more data is close to the mean, and the standard deviation is lower.

120. **the median**

NBA salaries are highly skewed to the right — a few players make very high salaries, while most earn considerably less. The median, which is the salary in the middle when the salaries are arranged in order, is less affected by extreme values than the mean.

121. **2.5**

The range of data is from 1.5 to 4.0, which is $4.0 - 1.5 = 2.5$.

122. **3.0**

The thick line within the box indicates the median (or middle number) for the data.

123. **1.125**

The interquartile range (IQR) is the distance between the 1st and 3rd quartiles (Q_1 and Q_3). In this case, IQR $= 3.5 - 2.375 = 1.125$.

124. **the GPA values**

The numerical axis is a scale showing the GPAs of individual students ranging from 1.5 to 4.0.

125. **cannot tell**

A box plot includes five values: the minimum value, the 25th percentile (Q_1), the median, the 75th percentile (Q_3), and the maximum value. The value of the mean isn't included on a box plot.

126. **skewed left**

You can't tell the exact distribution of data from a box plot. But because the median is located above the center of the box and the lower tail is longer than the upper tail, this data is skewed left.

127. **50%**

The actual box part of a box plot includes the middle 50% of the data, so the remaining 50% of the total must be outside the box.

128. **50%**

The definition of a median is that half the data in a distribution is below it and half is above it. In a box plot, the median is indicated by the location of the line inside the box part of the box plot.

129. **E. Choices (A), (B), and (C) (the total sample size; the number of students in each college; the mean of each data set)**

The sample size isn't accessible from a box plot. You know that 25% of the data lies within each section, but you don't know the total sample size. You also don't know the mean; you see the median (the line inside the box), but the mean isn't included on a box plot.

130. **College 1**

The median is indicated by the line within the actual box part of the box plot. Comparing the medians, you can see College 1's median has a greater value than College 2's.

131. **College 2**

The interquartile range (IQR) is the distance between the 3rd and 1st quartiles and represents the length of the box. If you compare the IQR of the two box plots, the IQR for College 2 is larger than the IQR for College 1.

132. **Impossible to tell without further information.**

Just because one box plot has a longer box than another one doesn't mean it has more data in it. It just means that the data inside the box (the middle 50% of the data) is more spread out for that group. Each section marked off on a box plot represents 25% of the data; but you don't know how many values are in each section without knowing the total sample size.

133. **The two data sets have the same percentage of GPAs above their medians.**

The median is the place in the data set that divides the data in half: 50% above and 50% below. So, both data sets have 50% of their GPAs above their medians.

134. **City 2, City 1, City 3**

The bar in the center of the box represents the median of each distribution; City 2 is the highest, followed by City 1 and City 3.

135. **City 2**

To find the number of homes that sold for more than $72,000, look at the numerical axis, where you can see that City 2 has three-quarters of the data lying past 72 (because Q_1 for City 2 is greater than 72). The other cities don't.

136. **City 2**

The lower edge of the box, which represents Q_1, is above 72 for City 2, whereas the medians for City 1 and City 3 are below 72. Therefore, if all three cities had the same number of home sales in 2012, City 2 must have had the most above $72,000.

137. **City 1**

The range is the maximum value minus the minimum value of a data set (shown in the very top line and the very bottom line of the box plot).

> City 1 range: $80 - 62 = 18$
>
> City 2 range: $125 - 43 = 82$
>
> City 3 range: $80 - 38 = 42$
>
> City 1 has the smallest range, 18.

138. **D. Choices (A) and (B) (More than half of the homes in City 1 sold for more than $50,000; more than half of the homes in City 2 sold for more than $75,000.)**

As indicated by the median lines within the boxes for each city, more than half of the homes in City 1 sold for more than $50,000, and more than half of the homes in City 2 sold for more than $75,000.

139. E. Choices (A) and (C) (About 25% of homes in City 1 sold for $75,000 or more; about 25% of homes in City 2 sold for $98,000 or more.)

The cutoff for the upper 25% of the values in a box plot is indicated by the largest number in the box. For City 1, about 25% of homes sold for $75,000 or more, and about 25% of homes in City 2 sold for $100,000 or more.

140. decreasing

Although not every year shows a decrease from the previous year, the clear overall pattern of the dropout rate is one of a steady decrease.

141. 3.8%

Find where the value for 2005 on the x-axis intersects with the value for the dropout rate on the y-axis; it's at 3.8%.

142. drop of 2%

The dropout rate in 2001 was about 5%, and in 2011, it was about 3%. So, the dropout rate from 2001 to 2011 is $3\% - 5\% = -2\%$.

A change of −2% is the same as a drop of 2 percentage points.

143. rise of 0.2%

The dropout rate in 2003 was about 4.0%, and in 2004, it was 4.2%. So, the dropout rate from 2003 to 2004 changed by $4.2\% - 4.0\% = 0.2\%$. This is the same as an increase (rise) of 0.2 percentage points.

144. The y- axis scale has been stretched (the range of y-values goes well outside the data).

The change in dropout rates is made to look less significant in this chart because the y-axis scale runs from 0 to 14, while the data runs from 3 to 5.

145. because you need to take into account the number of students enrolled each year, not just the number who dropped out

Suppose that one year, out of 1,000 students in the system, 10 of them dropped out, so $10/1,000 = 0.01$, or 1% of the students dropped out. However, if the number of students in the system was only 500, but the number of dropouts was the same, then $10/500 = 0.02$, or 2% of the students dropped out. Dividing by the total number of students calculates a dropout rate, which allows for a fair comparison.

146. Some points have been omitted from the scale of the x- axis.

Data for the years 2002 and 2009 have been omitted, but there's no indication of this omission on the chart. (That is, no broken line calls attention to the missing years.)

147. normal, uniform, bimodal

Income 1 has the bell shape of a normal distribution. Income 2 has a roughly flat shape. Income 3 has two peaks in it and is bimodal.

148. Income 3, Income 2, Income 1

Because you measure variability in terms of average distance from the mean, graphs with values more concentrated around the mean (such as the bell shape) have less variability than graphs with values not concentrated around the mean (like the uniform). So, of the three graphs, Income 1 has the lowest variability, and Income 3 has the most. (Income 3 is relatively flat but has peaks at the extremes; these peaks increase the variability.)

149. D. the number of books purchased by a student in a year

A discrete random variable is one that can take on only values that are integers (positive and negative whole numbers and 0). The values of a discrete random variable can have a finite stopping point, like –1, 0, and 1, or they can go to infinity (for example, 1, 2, 3, 4, . . .).

A continuous random variable takes on all possible values within an interval on the real number line (such as all real numbers between –2 and 2, written as [–2, 2]).

A number of books takes on only positive integer values, such as 0, 1, or 2, and thus is a discrete random variable.

150. B. the annual rainfall in a city

A discrete random variable is one that can take on only values that are integers (positive and negative whole numbers and 0). The values of a discrete random variable can have a finite stopping point, like –1, 0, and 1, or they can go to infinity (for example, 1, 2, 3, 4, . . .).

A continuous random variable takes on all possible values within an interval on the real number line (such as all real numbers between –2 and 2, written as [–2, 2]). The amount of rain that falls on a city in a year can take on any non-negative value on the real number line, such as 11.45 inches or 37.9 inches, and therefore, is continuous rather than discrete.

151. C. the number of cars registered in a state

A discrete random variable is one that can take on only values that are integers (positive and negative whole numbers and 0). The values of a discrete random variable can have a finite stopping point, like –1, 0, and 1, or they can go to infinity (for example, 1, 2, 3, 4, . . .).

A continuous random variable takes on all possible values within an interval on the real number line (such as all real numbers between –2 and 2, written as [–2, 2]). The number of cars registered in a state must be a non-negative integer, such as 0, 1, or 2, and thus is a discrete random variable.

152.

C. the proportion of the American population that believes in ghosts

A discrete random variable is one that can take on only values that are integers (positive and negative whole numbers and 0). The values of a discrete random variable can have a finite stopping point, like –1, 0, and 1, or they can go to infinity (for example, 1, 2, 3, 4, . . .).

A continuous random variable takes on all possible values within an interval on the real number line (such as all real numbers between –2 and 2, written as [–2, 2]).

A proportion can take on any real number between 0 and 1 on the real number line and is therefore continuous.

153.

E. the amount of gasoline used in the United States in 2012

A discrete random variable is one that can take on only values that are integers (positive and negative whole numbers and 0). The values of a discrete random variable can have a finite stopping point, like –1, 0, and 1, or they can go to infinity (for example, 1, 2, 3, 4, . . .).

A continuous random variable takes on all possible values within an interval on the real number line (such as all real numbers between –2 and 2, written as [–2, 2]).

The amount of gasoline can take on any value on the real number line that's greater than or equal to 0 and is therefore continuous.

154.

D. the number of bird species observed in an area

A discrete random variable is one that can take on only values that are integers (positive and negative whole numbers and 0). The values of a discrete random variable can have a finite stopping point, like –1, 0, and 1, or they can go to infinity (for example, 1, 2, 3, 4, . . .).

A continuous random variable takes on all possible values within an interval on the real number line (such as all real numbers between –2 and 2, written as [–2, 2]).

The number of bird species takes on non-negative integer values, such as 0, 1, or 2, and thus is a discrete random variable.

Note: The amount of money spent by a household for food during a year is considered continuous because of all the possible values it can take on (even though the amounts would be rounded to dollars and cents).

155.

0.10

From the table, you see that 0.10 or 10% of the adults in the city are employed part-time. Using notation, this means that $P(part\text{-}time) = 0.10$.

156.

0.82

Because total probability is always equal to 1, the probability that someone isn't retired is 1 minus the probability that the person is retired (which, according to the table, is 0.18 in this case). So, the probability that the adult isn't retired is $1 - 0.18 = 0.82$, or 82%. Using notation, this means that $P(not\ retired) = 0.82$.

157. 0.75

Because the categories don't overlap, the probability that someone is working either part-time or full-time is the sum of their individual probabilities. You can see from the table that the probability for part-time employment is 0.10 and for full-time employment, 0.65. Add these two probabilities to get your answer: $0.10 + 0.65 = 0.75$, or 75%. Using notation, this means that $P(part\text{-}time \ or \ full\text{-}time) = 0.75$.

158. 3.4

In this case, X represents the number of classes. The possible values of X are 4 and 3, denoted x_1 and x_2, respectively; their proportions (probabilities) are equal to 0.40 and 0.60 (denoted p_1 and p_2, respectively).

To find the average number of classes, or the mean of X (denoted by μ_x), multiply each value, x_i, by its probability, p_i, and then add the products:

$$
\begin{aligned}
\mu_x &= \sum x_i p_i \\
&= (4)(0.40) + (3)(0.60) \\
&= 1.6 + 1.8 \\
&= 3.4
\end{aligned}
$$

159. 18.75

In this case, X represents the age of a student. The possible values of X are 18, 19, and 20, denoted x_1, x_2, and x_3, respectively; their proportions (probabilities) are equal to 0.50, 0.25, and 0.25 (denoted p_1, p_2, and p_3, respectively).

To find the mean of X, or the average age of the students in the class (denoted by μ_x), msultiply each value, x_i, by its probability, p_i, and then add the products:

$$
\begin{aligned}
\mu_x &= \sum x_i p_i \\
&= (18)(0.50) + (19)(0.25) + (20)(0.25) \\
&= 9 + 4.75 + 5 \\
&= 18.75
\end{aligned}
$$

160. 0.10

The total probability must equal 1, so you can subtract the sum of the known values in the table from 1 to find the missing value: $1 - (0.25 + 0.60 + 0.05) = 1 - 0.90 = 0.10$.

161. 0.95

In this case, X represents the number of automobiles. The possible values of X are 0, 1, 2, and 3, denoted x_1, x_2, x_3, and x_4, respectively; their proportions (probabilities) are equal to 0.25, 0.60, 0.10, and 0.05 (denoted p_1, p_2, p_3, and p_4, respectively).

To find the mean of X (denoted by μ_x), multiply each value, x_i, by its probability, p_i, and then add the products:

$$\mu_x = \sum x_i p_i$$
$$= (0)(0.25) + (1)(0.60) + (2)(0.10) + (3)(0.05)$$
$$= 0 + 0.60 + 0.20 + 0.15$$
$$= 0.95$$

162. **1.20**

In this case, X represents the number of automobiles owned. You want the mean of X, designated as μ_x.

If every family not currently owning a car bought a car, then the total proportion of families owning one car would increase from 0.25 to $0.25 + 0.60 = 0.85$, and the proportion of families with no car would now be 0. You still have 0.10, or 10%, of the families owning 2 cars and 0.05, or 5%, of the families owning 3 cars. So, the values for X are 0, 1, 2, and 3, (denoted x_1, x_2, x_3, and x_4, respectively) and their proportions (probabilities) are 0, 0.85, 0.10, and 0.05 (denoted p_1, p_2, p_3, and p_4, respectively).

To find the mean of X, multiply each value, x_i, by its probability, p_i, and then add the products:

$$\mu_x = \sum x_i p_i$$
$$= (0)(0) + (1)(0.85) + (2)(0.10) + (3)(0.05)$$
$$= 0 + 0.85 + 0.20 + 0.15$$
$$= 1.20$$

163. **1.00**

In this case, X represents the number of automobiles owned. You want the mean of X, designated as μ_x.

If every family currently owning three cars bought a fourth car, the values of X would be 0, 1, 2, 3, and 4. The proportion of families owning three cars would be 0, and the total proportion of families owning four cars would be 0.05. You still have 0.25 of the families owning 0 cars, 0.60 of the families owning 1 car, and 0.10 of the families owning 2 cars.

So, the values of X are 0, 1, 2, 3, and 4 (denoted x_1, x_2, x_3, x_4, and x_5, respectively), and their proportions (probabilities) are 0.25, 0.60, 0.10, 0, and 0.05 (denoted p_1, p_2, p_3, p_4, and p_5, respectively).

To find the mean of X, multiply each value, x_i, by its probability, p_i, and then add the products:

$$\mu_x = \sum x_i p_i$$
$$= (0)(0.25) + (1)(0.60) + (2)(0.10) + (3)(0) + (4)(0.05)$$
$$= 0 + 0.60 + 0.20 + 0 + 0.20$$
$$= 1.00$$

164. 1.73

The standard deviation is the square root of the variance, so if the variance of X is 3, the standard deviation of X is $s\sqrt{3} = 1.73$ (rounded to two decimal places).

165. 0.42

The variance is the square of the standard deviation, so if the standard deviation of X is 0.65, the variance of X is $(0.65)^2 = 0.4225$. Rounded to two decimal places, the answer is 0.42.

166. 0.80

In this case, X represents the number of siblings a student has. The question asks for the mean of X, designated as μ_x.

The possible values of X are 0, 1, and 2, denoted x_1, x_2, and x_3, respectively; their proportions (probabilities) are equal to 0.34, 0.52, and 0.14 (denoted p_1, p_2, and p_3, respectively).

To find the mean of X, multiply each value, x_i, by its probability, p_i, and then add the products:

$$\mu_x = \sum x_i p_i$$
$$= (0)(0.34) + (1)(0.52) + (2)(0.14)$$
$$= 0 + 0.52 + 0.28$$
$$= 0.80$$

167. 0.44

To find the variance of X, denoted by σ_x^2, you take the first value of X, call it x_1, subtract the mean of X (denoted by μ_x), square that result, and then multiply it by the probability for x_1 (denoted p_1). Do the same thing for every other possible value of X, and then sum up all the results.

In this case, X represents the number of siblings. The values of X are 0, 1, and 2, denoted x_1, x_2, and x_3, respectively. Their probabilities are 0.34, 0.52, and 0.14, respectively.

You need to first find the mean of X because it's part of the formula to calculate the variance. Multiply each value, x_i, by its probability, p_i, and then add the products:

$$\mu_x = \sum x_i p_i$$
$$= (0)(0.34) + (1)(0.52) + (2)(0.14)$$
$$= 0 + 0.52 + 0.28$$
$$= 0.80$$

Now, plug this value into the formula for finding the variance:

$$\sigma_x^2 = \sum (x_i - \mu_x)^2 p_i$$
$$= (0-0.8)^2(0.34) + (1-0.8)^2(0.52) + (2-0.8)^2(0.14)$$
$$= 0.2176 + 0.0208 + 0.2016$$
$$= 0.4400$$

168. 0.66

To find the standard deviation of X, (denoted by σ_x), you first find the variance of X (denoted by σ_x^2), and then take the square root of that result.

To find the variance of X, you take the first value of X, call it x_1, subtract the mean of X (denoted by μ_x), and square the result. Then, you multiply that result by the probability for x_1, denoted p_1. Do this for each possible value of X, and then add up all the results.

In this case, X represents the number of siblings. The values of X are 0, 1, and 2, denoted x_1, x_2, and x_3, respectively. Their probabilities are 0.34, 0.52, and 0.14, respectively.

You need to first find the mean of X because it's part of the formula to calculate the variance. Multiply each value, x_i, by its probability, p_i, and then add the products:

$$\mu_x = \sum x_i p_i$$
$$= (0)(0.34) + (1)(0.52) + (2)(0.14)$$
$$= 0 + 0.52 + 0.28$$
$$= 0.80$$

Now, plug this value into the formula to calculate the variance of X:

$$\sigma_x^2 = \sum (x_i - \mu_x)^2 p_i$$
$$= (0-0.8)^2(0.34) + (1-0.8)^2(0.52) + (2-0.8)^2(0.14)$$
$$= 0.2176 + 0.0208 + 0.2016$$
$$= 0.4400$$

Finally, find the standard deviation of X. The standard deviation is the square root of the variance, or $\sqrt{0.44} = 0.66$.

169. The variance would be four times as large, and the standard deviation would be twice as large.

If you double all the values of X, their average distance from the mean (and hence the standard deviation) doubles as well. And because the variance of X is the square of the standard deviation, the variance of X becomes larger by a factor of $2^2 = 4$.

170. 1.45

In this case, X represents the number of books required. The question asks for the mean of X, designated as μ_x.

From the table, the possible values of X are 0, 1, 2, 3, and 4, denoted x_1, x_2, x_3, x_4, and x_5, respectively; their proportions (probabilities) are 0.30, 0.25, 0.25, 0.10, and 0.10, respectively.

To find the mean of X, multiply each value of X by its probability, and then add the products:

$$\mu_x = \sum x_i p_i$$
$$= (0)(0.30) + (1)(0.25) + (2)(0.25) + (3)(0.10) + (4)(0.10)$$
$$= 0 + 0.25 + 0.50 + 0.30 + 0.40$$
$$= 1.45$$

171. 1.65

To find the variance of X, you take the first value of X, call it x_1, subtract the mean of X (denoted by μ_x), and square the result. Then, you multiply that result by the probability for x_1, denoted p_1. Do this for each possible value of X, and then add up all the results.

In this case, X represents the number of books required. The values of X are 0, 1, 2, 3, and 4, denoted x_1, x_2, x_3, x_4, and x_5, respectively; their probabilities are 0.30, 0.25, 0.25, 0.10, and 0.10, respectively.

You need to first find the mean of X because it's part of the formula to calculate the variance. Multiply each value of X by its probability, and then add up the results:

$$\mu_x = \sum x_i p_i$$
$$= (0)(0.30) + (1)(0.25) + (2)(0.25) + (3)(0.10) + (4)(0.10)$$
$$= 0 + 0.25 + 0.50 + 0.30 + 0.40$$
$$= 1.45$$

Now, plug this value into the formula to calculate the variance of X:

$$\sigma_x^2 = \sum (x_i - \mu_x)^2 p_i$$
$$= (0 - 1.45)^2 (0.30) + (1 - 1.45)^2 (0.25) + (2 - 1.45)^2 (0.25)$$
$$+ (3 - 1.45)^2 (0.10) + (4 - 1.45)^2 (0.10)$$
$$= 0.63075 + 0.050625 + 0.075625 + 0.24025 + 0.65025$$
$$= 1.6475$$

Rounded to two decimal places, the answer is 1.65.

172. 1.28

To find the standard deviation of X, you first find the variance, and take the square root of the result.

To find the variance of X (denoted σ_x^2), you take the first value of X, call it x_1, subtract the mean of X (denoted by μ_x), and square the result. Then, you multiply that result by the probability for x_1, denoted p_1. Do this for each possible value of X, and then sum up all the results.

In this case, X represents the number of books required. The values of X are 0, 1, 2, 3, and 4, denoted x_1, x_2, x_3, x_4, and x_5, respectively; their probabilities are 0.30, 0.25, 0.25, 0.10, and 0.10, respectively.

You need to first find the mean of X because it's part of the formula to calculate the variance. To find the mean of X, multiply each value of X by its probability, and then add up the results.

$$\mu_x = \sum x_i p_i$$
$$= (0)(0.30) + (1)(0.25) + (2)(0.25) + (3)(0.10) + (4)(0.10)$$
$$= 0 + 0.25 + 0.50 + 0.30 + 0.40$$
$$= 1.45$$

Now, plug this value into the formula to find the variance of X:

$$\sigma_x^2 = \sum (x_i - \mu_x)^2 p_i$$
$$= (0 - 1.45)^2 (0.30) + (1 - 1.45)^2 (0.25) + (2 - 1.45)^2 (0.25)$$
$$+ (3 - 1.45)^2 (0.10) + (4 - 1.45)^2 (0.10)$$
$$= 0.63075 + 0.050625 + 0.075625 + 0.24025 + 0.65025$$
$$= 1.6475$$

The standard deviation, σ, is the square root of the variance:

$$\sigma = \sqrt{1.6475} = 1.28.$$

173. **Both would increase.**

With the changes described, fewer students have the number of books near the mean (middle), and some students are getting a higher number of books. The mean number of books overall would increase a bit, but the number of books students have will be more spread out (compared to the mean) than they were before. That means that the variance increases and so does the standard deviation.

174. **E. all of the above**

For a random variable to be binomial, it must have a fixed number of trials, with exactly two outcomes possible on each trial, a constant probability of success on all trials, and each trial must be independent.

175. **the total number of heads**

For a random variable to be binomial, it must have a fixed number of trials (n), with exactly two outcomes possible on each trial, a constant probability of success on all trials, and each trial must be independent. Then you define X as the number of "successes" (what you're interested in counting).

In this case, you have 25 trials (coin flips), with exactly two possible outcomes on each flip: heads or tails. According to this problem, a success is a heads, the flips are independent, and the probability of a heads on each trial is the same for each flip (0.5), so the random variable X is binomial. Here, X represents the total number of heads.

176. because each trial has more than two possible outcomes

In this case, you have six possible outcomes on each trial, but a binomial trial may have only two possible outcomes: success or failure. Here, X represents the outcome of one roll of the die (1, 2, 3, 4, 5, or 6), not the total number of die with a certain outcome (such as the total number of 6s rolled).

177. because the number of trials isn't fixed

For a binomial experiment, the number of trials must be specified in advance. In this case, although you know in the end that you need 30 employees who say they graduated from high school, you don't know how many employees you'll have to ask before finding 30 who graduated from high school. Because the total number of trials, n, is unknown, X isn't binomial.

178. because the trials aren't independent

If one sibling has the mutation, a higher chance exists that the other one will, too, so the results for each person aren't independent. Instead of recruiting 30 pairs of siblings for this test, you should recruit 60 people at random.

179. 7.2

The mean of a binomial random variable X is represented by the symbol μ. A binomial distribution has a special formula for the mean, which is $\mu = np$. Here, $n = 18$ and $p = 0.4$, so $\mu = (18)(0.4) = 7.2$.

180. 8.75

The mean of a binomial random variable X is represented by the symbol μ. A binomial distribution has a special formula for the mean, which is $\mu = np$. Here, $n = 25$ and $p = 0.35$, so $\mu = (25)(0.35) = 8.75$.

181. 2.08

The standard deviation of X is represented by σ and represents the square root of the variance. If X has a binomial distribution, the formula for the standard deviation is $\sigma = \sqrt{np(1-p)}$, where n is the number of trials and p is the probability of success on each trial. For this situation, $n = 18$ and $p = 0.4$, so

$$\sigma = \sqrt{18(0.4)(1-0.4)}$$
$$= \sqrt{4.32} = 2.08$$

182. 5.69

The variance is represented by σ^2 and represents the typical squared distance from the mean for all values of X.

For a binomial distribution, the variance has its own formula: $\sigma^2 = np(1-p)$. In this case, $n = 25$ and $p = 0.35$, so

$$\begin{aligned}
\sigma^2 &= 25(0.35)(1-0.35) \\
&= 25(0.35)(0.65) \\
&= 5.6875
\end{aligned}$$

Rounded to two decimal places, the answer is 5.69.

183. 130

The mean of a random variable X is denoted μ. For a binomial distribution, the mean has a special formula: $\mu = np$. In this case, $p = 0.14$ and μ is 18.2, so you need to find n. Plug the known values into the formula for the mean, so $18.2 = n(0.14)$, and then divide both sides by 0.14 to get $n = 18.2 / 0.14 = 130$.

184. 8

The value of $\binom{n}{x}$, called "n choose x" tells you the number of ways you can get X successes in n trials. The more ways to get those X successes, the higher the probability becomes. The formula for calculating "n choose x" is

$$\binom{n}{x} = \frac{n!}{x!(n-x)!}$$

The $n!$ stands for "n factorial." To calculate $n!$, you do a string of multiplications, starting from n and going down to 1. For example $5! = (5)(4)(3)(2)(1) = 120$; $2! = (2)(1) = 2$; $1! = 1$; and by convention, $0! = 1$.

To find $\binom{n}{x}$ in this problem, where $n = 8$ and $x = 1$, plug the numbers into the formula:

$$\begin{aligned}
\binom{8}{1} &= \frac{8!}{1!(8-1)!} \\
&= \frac{(8)(7)(6)(5)(4)(3)(2)(1)}{(1)[(7)(6)(5)(4)(3)(2)(1)]} \\
&= \frac{40,320}{5,040} = 8
\end{aligned}$$

185. 0.0164

The formula for calculating a probability for a binomial distribution is

$$P(X=x)=\binom{n}{x}p^x(1-p)^{n-x}$$

Here, $\binom{n}{x}=\dfrac{n!}{x!(n-x)!}$ and $n!$ means $n(n-1)(n-2)...(3)(2)(1)$. For example

$5!=(5)(4)(3)(2)(1)=120$; $2!=(2)(1)=2$; $1!=1$; and by convention, $0!=1$.

To find the probability of exactly one success in eight trials, you need $P(X=1)$, where $n=8$ (remember that $p=0.55$ here):

$$P(X=1)=\binom{8}{1}(0.55)^1(1-0.55)^{8-1}$$

$$=\frac{8!}{1!(8-1)!}(0.55)^1(1-0.55)^{8-1}$$

$$=\frac{(8)(7)(6)(5)(4)(3)(2)(1)}{(1)[(7)(6)(5)(4)(3)(2)(1)]}(0.55)(0.45)^7$$

$$=(8)(0.55)(0.00373669453125)$$

$$=0.0164414559375$$

Rounded to four decimal places, the answer is 0.0164.

186. 0.0703

The formula for calculating a probability for a binomial distribution is

$$P(X=x)=\binom{n}{x}p^x(1-p)^{n-x}$$

Here, $\binom{n}{x}=\dfrac{n!}{x!(n-x)!}$ and $n!$ means $n(n-1)(n-2)...(3)(2)(1)$. For example

$5!=(5)(4)(3)(2)(1)=120$; $2!=(2)(1)=2$; $1!=1$; and by convention, $0!=1$.

To find the probability of exactly two successes in eight trials, you want $P(X=2)$, where $n=8$ (remember that $p=0.55$ here):

$$P(X=2)=\binom{8}{2}(0.55)^2(1-0.55)^{8-2}$$

$$=\frac{8!}{2!(8-2)!}(0.55)^2(1-0.55)^{8-2}$$

$$=\frac{(8)(7)(6)(5)(4)(3)(2)(1)}{[(2)(1)][(6)(5)(4)(3)(2)(1)]}(0.55)^2(0.45)^6$$

$$=(28)(0.3025)(0.008303765625)$$

$$=0.07033289484375$$

Rounded to four decimal places, the answer is 0.0703.

187. 1

The formula for calculating "n choose x" is

$$\binom{n}{x} = \frac{n!}{x!(n-x)!}$$

The $n!$ stands for "n factorial." To calculate $n!$, you do a string of multiplications, starting from n and going down to 1. For example $5! = (5)(4)(3)(2)(1) = 120$; $2! = (2)(1) = 2$; $1! = 1$; and by convention, $0! = 1$.

So, find $\binom{n}{x}$ in this problem, plugging in the numbers for $n = 8$ and $x = 0$:

$$\binom{8}{0} = \frac{8!}{0!(8-0)!}$$

$$= \frac{(8)(7)(6)(5)(4)(3)(2)(1)}{(1)[(8)(7)(6)(5)(4)(3)(2)(1)]}$$

$$= 1$$

Incidentally, when $x = 0$, $\binom{n}{x}$ is always 1.

188. 0.9983

The formula for calculating a probability for a binomial distribution is

$$P(X = x) = \binom{n}{x} p^x (1-p)^{n-x}$$

Here, $\binom{n}{x} = \frac{n!}{x!(n-x)!}$ and $n!$ means $n(n-1)(n-2)\ldots(3)(2)(1)$. For example $5! = (5)(4)(3)(2)(1) = 120$; $2! = (2)(1) = 2$; $1! = 1$; and by convention, $0! = 1$.
In this case, X is the number of successes in n trials. You want $P(X \geq 1)$ because "at least one" means the same as "one or more." The easiest way to answer this question is to take 1 minus $P(X = 0)$, because that's the opposite of $P(X \geq 1)$ and easier to find.

$$P(X = 0) = \binom{8}{0}(0.55)^0(1-0.55)^{8-0}$$

$$= \frac{8!}{0!(8-0)!}(0.55)^0(0.45)^8$$

$$= \frac{(8)(7)(6)(5)(4)(3)(2)(1)}{(1)[(8)(7)(6)(5)(4)(3)(2)(1)]}(0.55)^0(0.45)^8$$

$$= (1)(1)(0.0016815125390625)$$

$$= 0.0016815125390625$$

Rounded to four decimal places, this answer is 0.017. Now, plug the value of $P(X = 0)$ in the formula to find $P(X > 0)$:

$$P(X > 0) = 1 - P(X = 0)$$

$$= 1 - 0.0017$$

$$= 0.9983$$

189. 0.0439

The formula for calculating a probability for a binomial distribution is

$$P(X=x) = \binom{n}{x} p^x (1-p)^{n-x}$$

Here, $\binom{n}{x} = \dfrac{n!}{x!(n-x)!}$ and $n!$ means $n(n-1)(n-2)...(3)(2)(1)$. For example $5! = (5)(4)(3)(2)(1) = 120$; $2! = (2)(1) = 2$; $1! = 1$; and by convention, $0! = 1$.

To find the probability of exactly eight successes in ten trials, you want $P(X=8)$, where $n=10$ (remember that $p=0.50$ here):

$$P(X=8) = \binom{10}{8}(0.50)^8(1-0.50)^{10-8}$$

$$= \frac{10!}{8!(10-8)!}(0.50)^8(1-0.50)^{10-8}$$

$$= \frac{(10)(9)(8)(7)(6)(5)(4)(3)(2)(1)}{[(8)(7)(6)(5)(4)(3)(2)(1)][(2)(1)]}(0.50)^8(0.50)^2$$

$$= (45)(0.00390625)(0.25)$$

$$= 0.0439453125$$

Rounded to four decimal places, the answer is 0.0439.

190. 0.3125

The formula for calculating a probability for a binomial distribution is

$$P(X=x) = \binom{n}{x} p^x (1-p)^{n-x}$$

Here, $\binom{n}{x} = \dfrac{n!}{x!(n-x)!}$ and $n!$ means $n(n-1)(n-2)...(3)(2)(1)$. For example $5! = (5)(4)(3)(2)(1) = 120$; $2! = (2)(1) = 2$; $1! = 1$; and by convention, $0! = 1$.

In this case, $n=5$ trials, $x=3$ successes, and $p=0.5$, the probability of success on each trial. You want $P(X=3)$:

$$P(X=3) = \binom{5}{3}(0.50)^3(1-0.50)^{5-3}$$

$$= \frac{5!}{3!(5-3)!}(0.50)^3(1-0.50)^{5-3}$$

$$= \frac{(5)(4)(3)(2)(1)}{[(3)(2)(1)][(2)(1)]}(0.50)^3(0.50)^2$$

$$= (10)(0.125)(0.25)$$

$$= 0.3125$$

191.   0.1563

The formula for calculating a probability for a binomial distribution is

$$P(X = x) = \binom{n}{x} p^x (1-p)^{n-x}$$

Here, $\binom{n}{x} = \dfrac{n!}{x!(n-x)!}$ and $n!$ means $n(n-1)(n-2)\ldots(3)(2)(1)$. For example $5! = (5)(4)(3)(2)(1) = 120$; $2! = (2)(1) = 2$; $1! = 1$; and by convention, $0! = 1$. In this case, $n = 5$ trials, $x = 4$ successes, and $p = 0.50$. You want $P(X = 4)$:

$$P(X = 4) = \binom{5}{4}(0.50)^4 (1-0.50)^{5-4}$$

$$= \frac{5!}{4!(5-4)!}(0.50)^4 (1-0.50)^{5-4}$$

$$= \frac{(5)(4)(3)(2)(1)}{[(4)(3)(2)(1)](1)}(0.50)^4 (0.50)^1$$

$$= (5)(0.0625)(0.50)$$

$$= 0.15625$$

Rounded to four decimal places, the answer is 0.1563.

192.   1

The formula for calculating "n choose x" is

$$\binom{n}{x} = \frac{n!}{x!(n-x)!}$$

The $n!$ stands for "n factorial." To calculate $n!$, you do a string of multiplications, starting from n and going down to 1. For example $5! = (5)(4)(3)(2)(1) = 120$; $2! = (2)(1) = 2$; $1! = 1$; and by convention, $0! = 1$.

So, find $\binom{n}{x}$ in this problem, plugging in the numbers for $n = 5$ and $x = 5$:

$$\binom{5}{5} = \frac{5!}{5!(5-5)!}$$

$$= \frac{(5)(4)(3)(2)(1)}{[(5)(4)(3)(2)(1)](1)}$$

$$= 1$$

Incidentally, $\binom{n}{x}$ is always 1 when n and x are the same number.

193.   0.4688

The formula for calculating a probability for a binomial distribution is

$$P(X = x) = \binom{n}{x} p^x (1-p)^{n-x}$$

Here, $\binom{n}{x} = \dfrac{n!}{x!(n-x)!}$ and $n!$ means $n(n-1)(n-2)\ldots(3)(2)(1)$. For example

$5! = (5)(4)(3)(2)(1) = 120$; $2! = (2)(1) = 2$; $1! = 1$; and by convention, $0! = 1$.

In this problem, $n = 5$ trials, $x = 3$ or 4 successes, and $p = 0.50$, the probability of success on each trial. You want $P(X = 3 \text{ or } 4) = P(X = 3) + P(X = 4)$. First, find the probability of each outcome separately:

$$P(X = 3) = \binom{5}{3}(0.50)^3(1 - 0.50)^{5-3}$$

$$= \frac{5!}{3!(5-3)!}(0.50)^3(1 - 0.50)^{5-3}$$

$$= \frac{(5)(4)(3)(2)(1)}{[(3)(2)(1)][(2)(1)]}(0.50)^3(0.50)^2$$

$$= (10)(0.125)(0.25)$$

$$= 0.3125$$

$$P(X = 4) = \binom{5}{4}(0.50)^4(1 - 0.50)^{5-4}$$

$$= \frac{5!}{4!(5-4)!}(0.50)^4(1 - 0.50)^{5-4}$$

$$= \frac{(5)(4)(3)(2)(1)}{[(4)(3)(2)(1)](1)}(0.50)^4(0.50)^1$$

$$= (5)(0.0625)(0.50)$$

$$= 0.15625$$

Then, add the results together:

$$P(X = 3 \text{ or } 4) = P(X = 3) + P(X = 4)$$

$$= 0.3125 + 0.15625$$

$$= 0.46875$$

Rounded to four decimal places, the answer is 0.4688.

194.   0.5001

The formula for calculating a probability for a binomial distribution is

$$P(X = x) = \binom{n}{x}p^x(1 - p)^{n-x}$$

Here, $\binom{n}{x} = \dfrac{n!}{x!(n-x)!}$ and $n!$ means $n(n-1)(n-2)\ldots(3)(2)(1)$. For example

$5! = (5)(4)(3)(2)(1) = 120$; $2! = (2)(1) = 2$; $1! = 1$; and by convention, $0! = 1$.

In this case, $n = 5$ trials, $x = 3$, 4, or 5 successes, and $p = 0.50$, the probability of success on each trial. *Note:* The probability of *at least three* successes means the probability of three, four, or five successes. In other words, you need to find $P(X = 3 \text{ or } 4 \text{ or } 5) = P(X = 3) + P(X = 4) + P(X = 5)$. First, find the probability of each outcome separately:

$$P(X=3) = \binom{5}{3}(0.50)^3(1-0.50)^{5-3}$$

$$= \frac{5!}{3!(5-3)!}(0.50)^3(1-0.50)^{5-3}$$

$$= \frac{(5)(4)(3)(2)(1)}{[(3)(2)(1)][(2)(1)]}(0.50)^3(0.50)^2$$

$$= (10)(0.125)(0.25)$$

$$= 0.3125$$

$$P(X=4) = \binom{5}{4}(0.50)^4(1-0.50)^{5-4}$$

$$= \frac{5!}{4!(5-4)!}(0.50)^4(1-0.50)^{5-4}$$

$$= \frac{(5)(4)(3)(2)(1)}{[(4)(3)(2)(1)](1)}(0.50)^4(0.50)^1$$

$$= (5)(0.0625)(0.50)$$

$$= 0.15625$$

$$P(X=5) = \binom{5}{5}(0.50)^5(1-0.50)^{5-5}$$

$$= \frac{5!}{5!(5-0)!}(0.50)^5(1-0.50)^{5-5}$$

$$= \frac{(5)(4)(3)(2)(1)}{[(5)(4)(3)(2)(1)][(5)(4)(3)(2)(1)]}(0.50)^5(0.50)^0$$

$$= 0.0313$$

Then add the results together:

$$P(X=3 \text{ or } 4 \text{ or } 5) = P(X=3) + P(X=4) + P(X=5)$$

$$= 0.3125 + 0.15625 + 0.0313$$

$$= 0.50005$$

Rounded to four decimal places, the answer is 0.5001.

195. 0.4999

The formula for calculating a probability for a binomial distribution is

$$P(X=x) = \binom{n}{x}p^x(1-p)^{n-x}$$

Here, $\binom{n}{x} = \frac{n!}{x!(n-x)!}$ and $n!$ means $n(n-1)(n-2)...(3)(2)(1)$. For example

$5! = (5)(4)(3)(2)(1) = 120$; $2! = (2)(1) = 2$; $1! = 1$; and by convention, $0! = 1$.
In this case, $n = 5$ trials, $x =$ no more than 2 successes, and $p = 0.50$, the probability of success on each trial. To find the probability of no more than two successes, you can either find $P(X \le 2)$ or find the probability of at least three successes, $P(X \ge 3)$ and subtract the sum of these probabilities from 1. For example, you find $P(X=3)$, $P(X=4)$, and $P(X=5)$:

$$P(X=3) = \binom{5}{3}(0.50)^3(1-0.50)^{5-3}$$

$$= \frac{5!}{3!(5-3)!}(0.50)^3(1-0.50)^{5-3}$$

$$= \frac{(5)(4)(3)(2)(1)}{[(3)(2)(1)][(2)(1)]}(0.50)^3(0.50)^2$$

$$= (10)(0.125)(0.25)$$

$$= 0.3125$$

$$P(X=4) = \binom{5}{4}(0.50)^4(1-0.50)^{5-4}$$

$$= \frac{5!}{4!(5-4)!}(0.50)^4(1-0.50)^{5-4}$$

$$= \frac{(5)(4)(3)(2)(1)}{[(4)(3)(2)(1)](1)}(0.50)^4(0.50)^1$$

$$= (5)(0.0625)(0.50)$$

$$= 0.15625$$

$$P(X=5) = \binom{5}{5}(0.50)^5(1-0.50)^{5-5}$$

$$= \frac{5!}{5!(5-0)!}(0.50)^5(1-0.50)^{5-5}$$

$$= \frac{(5)(4)(3)(2)(1)}{[(5)(4)(3)(2)(1)][(5)(4)(3)(2)(1)]}(0.50)^5(0.50)^0$$

$$= 0.0313$$

And then you add the results:

$$P(X \geq 3) = P(X=3) + P(X=4) + P(X=5)$$

$$= 0.3125 + 0.15625 + 0.0313$$

$$= 0.50005$$

You can round this answer to four decimal places: 0.5001. Finally, you subtract the probability of $P(X \geq 3)$ from $1 : 1 - 0.5001 = 0.4999$.

---

196. 0.012

The binomial table (Table A-3 in the appendix) has a series of mini-tables inside of it, one for each selected value of n. To find $P(X=6)$, where $n=15$ and $p=0.7$, locate the mini-table for $n=15$, find the row for $x=6$, and follow across to where it intersects with the column for $p=0.7$. This value is 0.012.

---

197. 0.219

The binomial table (Table A-3 in the appendix) has a series of mini-tables inside of it, one for each selected value of n. To find $P(X=11)$, where $n=15$ and $p=0.7$, locate the mini-table for $n=15$, find the row for $x=11$, and follow across to where it intersects with the column for $p=0.7$. This value is 0.219.

198. 0.995

In this case, 15 is the greatest possible value of X (because there are only 15 total trials), so to find $P(X < 15)$, you can first find $P(X = 15)$ and subtract this result from 1 to get what you need. (This makes the calculations much easier.)

The binomial table (Table A-3 in the appendix) has a series of mini-tables inside of it, one for each selected value of n. To find $P(X = 15)$, where $n = 15$ and $p = 0.7$, locate the mini-table for $n = 15$, find the row for $x = 15$, and follow across to where it intersects with the column for $p = 0.7$. This value is 0.005.

Now, subtract that from 1 so you have $P(X \geq 15) = 1 - 0.005 = 0.995$.

199. 0.015

You want the probability between 4 and 7, but you don't want to include 4 and 7. So, you want only the probabilities for $X = 5$ and $X = 6$. You know that $n = 15$ and $p = 0.7$, which is the probability of success on each trial.

To find each of these probabilities, use the binomial table (Table A-3 in the appendix), which has a series of mini-tables inside of it, one for each selected value of n. To find $P(X = 5)$, where $n = 15$ and $p = 0.7$, locate the mini-table for $n = 15$, find the row for $x = 5$, and follow across to where it intersects with the column for $p = 0.7$. This value is 0.003. Now do the same for $P(X = 6)$ to get 0.012. Then add these probabilities together:

$$P(4 < X < 7) = P(X = 5) + P(X = 6)$$
$$= 0.003 + 0.012$$
$$= 0.015$$

200. 0.051

Here, you want to find the probability equal to 4 and 7 and everything in between. In other words, you want the probabilities for $X = 4$, $X = 5$, $X = 6$, and $X = 7$. You know that $n = 15$ and $p = 0.7$, which is the probability of success on each trial.

To find each of these probabilities, use the binomial table (Table A-3 in the appendix), which has a series of mini-tables inside of it, one for each selected value of n. To find $P(X = 4)$, where $n = 15$ and $p = 0.7$, locate the mini-table for $n = 15$, find the row for $x = 4$, and follow across to where it intersects with the column for $p = 0.7$. This value is 0.001. Now do the same for the other probabilities: $P(X = 5) = 0.003; P(X = 6) = 0.012$; and $P(X = 7) = 0.035$. Finally, add these probabilities together:

$$P(4 \leq X \leq 7) = P(X = 4) + P(X = 5) + P(X = 6) + P(X = 7)$$
$$= 0.001 + 0.003 + 0.012 + 0.035$$
$$= 0.051$$

201. 0.221

The binomial table (Table A-3 in the appendix) has a series of mini-tables inside of it, one for each selected value of n. To find $P(X = 5)$, where $n = 11$ and $p = 0.4$, locate the mini-table for $n = 11$, find the row for $x = 5$, and follow across to where it intersects with the column for $p = 0.4$. This value is 0.221.

202. 0.996

To find the probability that X is greater than 0, find the probability that X is equal to 0, and then subtract that probability from 1. This makes the calculations much easier.

The binomial table (Table A-3 in the appendix) has a series of mini-tables inside of it, one for each selected value of n. To find $P(X = 0)$, where $n = 11$ and $p = 0.4$, locate the mini-table for $n = 11$, find the row for $x = 0$ and follow across to where it intersects with the column for $p = 0.4$. This value is 0.004. Now subtract that from 1:

$$P(X > 0) = 1 - P(X = 0)$$
$$= 1 - 0.004$$
$$= 0.996$$

203. 0.120

To find the probability that X is less than or equal to 2, you first need to find the probability of each possible value of X less than 2. In other words, you find the values for $P(X = 0)$, $P(X = 1)$, and $P(X = 2)$.

To find each of these probabilities, use the binomial table (Table A-3 in the appendix), which has a series of mini-tables inside of it, one for each selected value of n. To find $P(X = 0)$, where $n = 11$ and $p = 0.4$, locate the mini-table for $n = 11$, find the row for $x = 0$, and follow across to where it intersects with the column for $p = 0.4$. This value is 0.004. Now do the same for the other probabilities: $P(X = 1) = 0.027$ and $P(X = 2) = 0.089$. Finally, add these probabilities together:

$$P(X \le 2) = P(X = 0) + P(X = 1) + P(X = 2)$$
$$= 0.004 + 0.027 + 0.089$$
$$= 0.120$$

204. 0.001

To find the probability that X is greater than 9, first find the probability that X is equal to 10 or 11 (in this case, 11 is the greatest possible value of x because there are only 11 total trials).

To find each of these probabilities, use the binomial table (Table A-3 in the appendix), which has a series of mini-tables inside of it, one for each selected value of n. To find $P(X = 10)$, where $n = 11$ and $p = 0.4$, locate the mini-table for $n = 11$, find the row for $x = 10$, and follow across to where it intersects with the column for $p = 0.4$. This value is 0.001. Now do the same for $P(X = 11)$ which gives you 0.000. (*Note:* $P(X = 11)$ isn't exactly 0.000 here; it's just a smaller probability than can be expressed in the four decimal places used in this table.) Finally, add the two probabilities together:

$$P(X > 9) = P(X = 10) + P(X = 11)$$
$$= 0.001 + 0.000$$
$$= 0.001$$

205. 0.634

Here, you want to find the probability equal to 3 and 5 and everything in between. In other words, you want the probabilities for $X = 3$, $X = 4$, and $X = 5$. You know that $n = 11$ and $p = 0.4$, which is the probability of success on each trial.

To find each of these probabilities, use the binomial table (Table A-3 in the appendix), which has a series of mini-tables inside of it, one for each selected value of n. To find $P(X = 3)$, where $n = 11$ and $p = 0.4$, locate the mini-table for $n = 11$, find the row for $x = 3$, and follow across to where it intersects with the column for $p = 0.4$. This value is 0.177. Now do the same for the other probabilities: $P(X = 4) = 0.236$ and $P(X = 5) = 0.221$. Finally, add these probabilities together:

$$P(3 \leq X \leq 5) = P(X = 3) + P(X = 4) + P(X = 5)$$
$$= 0.177 + 0.236 + 0.221$$
$$= 0.634$$

206. $n = 30, p = 0.4$

Two conditions must be met to use the normal approximation to the binomial: Both np and $n(1 - p)$ must be at least 10. Using the choices you're given, the only one that works is $n = 30$ and $p = 0.4$: $np = 30(0.4) = 12$, and $n(1 - p) = 30(1 - 0.4) = 30(0.6) = 18$.

207. 20

Two conditions must be met to use the normal approximation to the binomial: Both np and $n(1 - p)$ must be at least 10. So, you need the smallest value of n that meets both of these conditions, knowing that $p = 0.5$ here.

First, take $np \geq 10$, or $n(0.5) \geq 10$. To get n by itself, divide both sides by 0.5, which gives you $n \geq 20$. Then, take $n(1 - p) \geq 10$, or $n(1 - 0.5) \geq 10$. Again, you get $n \geq 20$. The minimum sample size (n) that meets both of these requirements is 20.

Note: Sometimes p is very small or very large, which changes the values of np and $n(1 - p)$, so you must always check and meet both conditions every time.

208. 34

To use the normal approximation to the binomial, both np and $n(1 - p)$ must be at least 10. Here the value of p is 0.30 ("success" = green marble).

First, take $np \geq 10$, or $n(0.3) \geq 10$. To get n by itself, divide both sides by 0.3, which gives you $n \geq 33.33$ (round up to 34 to make sure the condition is met). Then, take $n(1 - p) \geq 10$, or $n(1 - 0.3) \geq 10$. That gives you $n \geq 14.29$ (round up to 15 to make sure the condition is met).

For the first condition, you need $n \geq 34$; for the second condition, you need $n \geq 15$. To meet both conditions, you need the larger n, which is 34.

209. $P(X \geq 50)$

You can rephrase the probability of getting at least 50 heads out of 80 flips (that is, X is at least 50) as the probability that X is greater than or equal to 50 (because 50 is the lower limit of possible values): $P(X \geq 50)$.

210. $P(X \leq 30)$

You can rephrase the probability of getting no more than 30 heads out of 80 flips (that is, X is no more than 30) as the probability that X is less than or equal to 30 (because 30 is the upper limit of possible values): $P(X \leq 30)$.

211. 40

For a binomial distribution, the mean, μ, is equal to np. In this case, $n = 80$ trials (flips of the coin) and $p = 0.50$ (chance of a heads/success on each flip). Therefore, $\mu = (80)(0.50) = 40$.

212. 4.47

For a binomial distribution, the standard deviation, σ, is equal to $\sqrt{np(1-p)}$. In this case, you have $n = 80$ trials (flips of the coin) and $p = 0.5$ (chance of a heads/success on each flip). Therefore,

$$\sigma = \sqrt{80(0.50)(1-0.50)}$$
$$= \sqrt{20} = 4.47$$

213. 1.12

To find the z-value for an x-value, subtract the population mean from x, and divide by the population standard deviation:

$$z = \frac{x - \mu}{\sigma}$$

The question is what do you use for μ and σ? Because X has a binomial distribution, you use the mean and standard deviation of the binomial distribution. The mean of a binomial distribution is $\mu = np$. In this case, $n = 80$ and $p = 0.50$, so $\mu = (80)(0.50) = 40$. And the formula for the standard deviation of a binomial distribution is $\sigma = \sqrt{np(1-p)}$. So, you get

$$\sigma = \sqrt{80(0.50)(1-0.50)}$$
$$= \sqrt{20} = 4.47$$

Now, plug these numbers into the formula for the z-value, where $x = 45$:

$$z = \frac{x - \mu}{\sigma}$$
$$= \frac{(45 - 40)}{4.47} = 1.12$$

214. **0.1314**

Here, you want the probability that X is at least 45, or $P(X \geq 45)$. In this case, $n = 80$ trials (flips of the coin) and $p = 0.50$ (chance of a heads/success on each flip).

Because n is so large, you may want to use the normal approximation to the binomial to solve this problem. But first you need to determine whether the two conditions are met — that is, both np and $n(1-p)$ must be at least 10. So, plug in the numbers: $np = (80)(0.50) = 40$, and $n(1-p) = 80(1-0.50) = (80)(0.50) = 40$. Both conditions are at least 10, so you can move forward.

The first step in doing the normal approximation to find a binomial probability is to find the z-value. To find the z-value for an x-value, subtract the population mean from x, and divide by the population standard deviation:

$$z = \frac{x - \mu}{\sigma}$$

Because X has a binomial distribution, you use the mean and standard deviation of the binomial distribution. The mean of a binomial distribution is $\mu = np$. In this case, $n = 80$ and $p = 0.50$, so $\mu = (80)(0.50) = 40$. The formula for the standard deviation of a binomial distribution is $\sigma = \sqrt{np(1-p)}$. So, you get

$$\sigma = \sqrt{80(0.50)(1-0.50)}$$
$$= \sqrt{20} = 4.47$$

Now, plug these numbers into the formula for the z-value, where $x = 45$:

$$z = \frac{x - \mu}{\sigma}$$
$$= \frac{(45 - 40)}{4.47} = 1.12$$

Use a Z-table, such as Table A-1 in the appendix, to find $P(Z \leq 1.12)$, and then subtract that from 1 to find $P(Z \geq 1.12)$. Locate the row for $z = 1.1$ and follow it across to where it intersects with the column for 0.02, which gives you 0.8686. This also corresponds to $P(Z \leq 1.12) = P(X \leq 45)$. Then subtract from 1 to get $P(X \geq 45): 1 - 0.8686 = 0.1314$.

215. $\mu = 45, \sigma = 4.97$

For a binomial distribution, the mean is $\mu = np$, and the standard deviation is $\sigma = \sqrt{np(1-p)}$. In this case, $n = 100$ trials and $p = 0.45$, the probability of success on each trial. Therefore, the mean of X is $\mu = (100)(0.45) = 45$, and the standard deviation is

$$\sigma = \sqrt{np(1-p)}$$
$$= \sqrt{100(0.45)(1-0.45)}$$
$$= \sqrt{24.75} = 4.97$$

216. **−1.00**

To find the z-value for an x-value, subtract the population mean from x, and divide by the population standard deviation:

$$z = \frac{x - \mu}{\sigma}$$

For a binomial distribution, the mean is $\mu = np$, and the standard deviation is $\sigma = \sqrt{np(1-p)}$. In this case, $n = 100$ and $p = 0.45$, so $\mu = (100)(0.45) = 45$, and the standard deviation is

$$\sigma = \sqrt{np(1-p)}$$
$$= \sqrt{100(0.45)(1-0.45)}$$
$$= \sqrt{24.75} = 4.97$$

Now, plug these numbers into the formula for the z-value, where $x = 40$:

$$z = \frac{x - \mu}{\sigma}$$
$$= \frac{(40 - 45)}{4.97} = -1.00$$

217. 0.1587

Because n is so large, you may want to use the normal approximation to the binomial to solve this problem. But first you need to determine whether the two conditions are met — that is, both np and $n(1-p)$ must be at least 10. In this case, $np = 100(0.45) = 45$, and $n(1-p) = 100(1-0.45) = (100)(0.55) = 55$. Both conditions are at least 10, so you can move forward.

To find a "less than" probability for an x-value from a normal distribution, you convert the x-value to a z-value and then find the corresponding probability for that z-value by using a Z-table, such as Table A-1 in the appendix.

To find the z-value for an x-value, subtract the population mean from x, and divide by the population standard deviation:

$$z = \frac{x - \mu}{\sigma}$$

For a binomial distribution, the mean is $\mu = np$, and the standard deviation is $\sigma = \sqrt{np(1-p)}$. In this case, $n = 100$ and $p = 0.45$, so $\mu = (100)(0.45) = 45$, and the standard deviation is

$$\sigma = \sqrt{np(1-p)}$$
$$= \sqrt{100(0.45)(1-0.45)}$$
$$= \sqrt{24.75} = 4.97$$

Now, plug these numbers into the formula for the z-value, where $x = 40$:

$$z = \frac{x - \mu}{\sigma}$$
$$= \frac{(40 - 45)}{4.97} = -1.00$$

Then find $P(Z \leq -1.00)$, using Table A-1. Look in the row for $z = -1.0$ and follow across to where it intersects with the column for 0.00, which gives you 0.1587. Remember this is only an approximation because you started with a binomial and were able to use the normal approximation to find the probability.

218.   0.8413

Because n is so large, you may want to use the normal approximation to the binomial to solve this problem. But first you need to determine whether the two conditions are met — that is, both np and $n(1-p)$ must be at least 10. In this case, $np = 100(0.45) = 45$, and $n(1-p) = 100(1-0.45) = (100)(0.55) = 55$. Both conditions are at least 10, so you can move forward.

To find a "greater than" probability for an x-value from a normal distribution, you convert the x-value to a z-value and find the corresponding probability for that z-value by using a Z-table, such as Table A-1 in the appendix; then, you subtract that result from 1 (because Table A-1 gives you "less than" probabilities only).

To find the z-value for an x-value, subtract the population mean from x, and divide by the population standard deviation:

$$z = \frac{x - \mu}{\sigma}$$

For a binomial distribution, the mean is $\mu = np$, and the standard deviation is $\sigma = \sqrt{np(1-p)}$. In this case, $n = 100$ and $p = 0.45$, so $\mu = (100)(0.45) = 45$, and the standard deviation is

$$\sigma = \sqrt{np(1-p)}$$
$$= \sqrt{100(0.45)(1-0.45)}$$
$$= \sqrt{24.75} = 4.97$$

Now, plug these numbers into the formula for the z-value, where $x = 45$:

$$z = \frac{x - \mu}{\sigma}$$
$$= \frac{(40 - 45)}{4.97} = -1.00$$

Then, find $P(Z \le -1.00)$, using Table A-1. Look in the row for $z = -1.0$ and follow across to where it intersects with the column for 0.00, which gives you 0.1587. To get $P(Z \le -1.00)$, subtract that value from $1 : 1 - 0.1587 = 0.8413$. Remember this is only an approximation because you started with a binomial random variable and were able to use the normal approximation to find the probability.

219.   0

Even though X is binomial and discrete, the Z-distribution is continuous, so the probability at a single value is 0. Probability for a continuous random variable is represented by the area under a curve. There's no area under the curve at a single point. You would have to use continuity correction to solve $P(X = 40)$ using a normal approximation.

220.   2.21

To find the z-value for an x-value, subtract the population mean from x, and divide by the population standard deviation:

$$z = \frac{x - \mu}{\sigma}$$

For a binomial distribution, the mean is $\mu = np$, and the standard deviation is $\sigma = \sqrt{np(1-p)}$. In this case, $n = 100$ and $p = 0.45$, so $\mu = (100)(0.45) = 45$, and the standard deviation is

$$\sigma = \sqrt{np(1-p)}$$
$$= \sqrt{100(0.45)(1-0.45)}$$
$$= \sqrt{24.75} = 4.97$$

Now, plug these numbers into the formula for the z-value, where $x = 56$:

$$z = \frac{x - \mu}{\sigma}$$
$$= \frac{(56 - 45)}{4.97} = 2.21$$

221.   0.0136

Because n is so large, you may want to use the normal approximation to the binomial to solve this problem. But first you need to determine whether the two conditions are met — that is, both np and $n(1-p)$ must be at least 10. In this case, $np = 100(0.45) = 45$, and $n(1-p) = 100(1-0.45) = (100)(0.55) = 55$. Both conditions are at least 10, so you can move forward.

To find a "greater than" probability for an x-value from a normal distribution, you convert the x-value to a z-value and then find the corresponding probability for that z-value by using a Z-table, such as Table A-1 in the appendix. You then subtract that result from 1 (because Table A-1 gives you "less than" probabilities only).

To find the z-value for an x-value, subtract the population mean from x, and divide by the population standard deviation:

$$z = \frac{x - \mu}{\sigma}$$

For a binomial distribution, the mean is $\mu = np$ and the standard deviation is $\sigma = \sqrt{np(1-p)}$. In this case, $n = 100$ and $p = 0.45$, so $\mu = (100)(0.45) = 45$, and the standard deviation is

$$\sigma = \sqrt{np(1-p)}$$
$$= \sqrt{100(0.45)(1-0.45)}$$
$$= \sqrt{24.75} = 4.97$$

Now, plug these numbers into the formula for the z-value, where $x = 56$:

$$z = \frac{x - \mu}{\sigma}$$
$$= \frac{(56 - 45)}{4.97} = 2.21$$

Then, find $P(Z \leq 2.21)$, using Table A-1. Look in the row for $z = 2.2$ and follow across to where it intersects with the column for 0.01, which gives you 0.9864. To get $P(Z \geq 2.21)$, subtract that value from 1: $1 - 0.9864 = 0.0136$. This corresponds to $P(X \geq 56)$.

Remember this is only an approximation because you started with a binomial random variable and were able to use the normal approximation to find the probability.

222. **0.9864**

Because n is so large, you may want to use the normal approximation to the binomial to solve this problem. But first you need to determine whether the two conditions are met — that is, both np and $n(1-p)$ must be at least 10. In this case, $np = 100(0.45) = 45$, and $n(1-p) = 100(1-0.45) = (100)(0.55) = 55$. Both conditions are at least 10, so you can move forward.

To find a "less than" probability for an x-value from a normal distribution, you convert the x-value to a z-value and then find the corresponding probability for that z-value by using a Z-table, such as Table A-1 in the appendix.

To find the z-value for an x-value, subtract the population mean from x, and divide by the population standard deviation:

$$z = \frac{x - \mu}{\sigma}$$

For a binomial distribution, the mean is $\mu = np$, and the standard deviation is $\sigma = \sqrt{np(1-p)}$. In this case, $n = 100$ and $p = 0.45$, so $\mu = (100)(0.45) = 45$, and the standard deviation is

$$\sigma = \sqrt{np(1-p)}$$
$$= \sqrt{100(0.45)(1-0.45)}$$
$$= \sqrt{24.75} = 4.97$$

Now, plug these numbers into the formula for the z-value, where $x = 56$:

$$z = \frac{x - \mu}{\sigma}$$
$$= \frac{(56 - 45)}{4.97} = 2.21$$

Then find $P(Z \leq 2.21)$, using Table A-1. Look in the row for $z = 2.2$ and follow across to where it intersects with the column for 0.01, which gives you 0.9864. Remember this is only an approximation because you started with a binomial and were able to use the normal approximation to find the probability.

223. **0.0023**

Because n is so large, you may want to use the normal approximation to the binomial to solve this problem. But first you need to determine whether the two conditions are met — that is, both np and $n(1-p)$ must be at least 10. In this case, $np = 100(0.45) = 45$, and $n(1-p) = 100(1-0.45) = (100)(0.55) = 55$. Both conditions are at least 10, so you can move forward.

To find the probability of "being between" two x-values from a normal distribution, convert each x-value to a z-value, and find the corresponding probability for each z-value by using a Z-table, such as Table A-1 in the appendix. Then subtract the lowest probability from the highest probability.

To find the z-value for an x-value, subtract the population mean from x, and divide by the population standard deviation:

$$z = \frac{x - \mu}{\sigma}$$

For a binomial distribution, the mean is $\mu = np$, and the standard deviation is $\sigma = \sqrt{np(1-p)}$. In this case, $n = 100$ and $p = 0.45$, so $\mu = (100)(0.45) = 45$, and the standard deviation is

$$\begin{aligned} \sigma &= \sqrt{np(1-p)} \\ &= \sqrt{100(0.45)(1-0.45)} \\ &= \sqrt{24.75} = 4.97 \end{aligned}$$

Now, plug these numbers into the formula for the z-value, where $x = 56$:

$$\begin{aligned} z &= \frac{x - \mu}{\sigma} \\ &= \frac{(56 - 45)}{4.97} = 2.21 \end{aligned}$$

Then find $P(Z \le 2.21)$, using Table A-1. Look in the row for $z = 2.2$ and follow across to where it intersects with the column for 0.01, which gives you 0.9864.

Follow these steps for $x = 60$:

$$\begin{aligned} z &= \frac{x - \mu}{\sigma} \\ &= \frac{(60 - 45)}{4.97} = 3.02 \end{aligned}$$

Find $P(Z \le 3.02)$ in Table A-1, which is 0.9987.

Finally, to find the probability of being between the two z-values, subtract the lowest probability from the highest probability to get $0.9987 - 0.9864 = 0.0023$. Remember this is only an approximation because you started with a binomial and were able to use the normal approximation to find the probability because the necessary conditions were met.

224. All of the above.

The properties of the normal distribution are that it's symmetrical, mean and median are the same, the most common values are near the mean and less common values are farther from it, and the standard deviation marks the distance from the mean to the inflection point.

225. 68%

The empirical rule (also known as the 68-95-99.7 rule) says that about 68% of the values in a normal distribution are within one standard deviation of the mean.

226. 95%

The empirical rule (also known as the 68-95-99.7 rule) says that about 95% of the values in a normal distribution are within two standard deviations of the mean.

227. **99.7%**

The empirical rule (also known as the 68–95–99.7 rule) says that about 99.7% of the values in a normal distribution are within three standard deviations of the mean.

228. **the mean and the standard deviation**

You can re-create any normal distribution if you know two parameters: the mean and the standard deviation. The mean is the center of the bell-shaped picture, and the standard deviation is the distance from the mean to the inflection point (the place where the concavity of the curve changes on the graph).

229. **C.** $\mu = 5, \sigma = 1.75$

The larger the standard deviation (σ), the greater the spread for a normal distribution. The value of the mean (μ) doesn't affect the spread of the normal distribution; it just shows you where the center is.

230. **6.2**

You want to find a value of X where 34% of the values lie between the mean (5) and x (and x is in the right side of the mean). First, note that the normal distribution has a total probability of 100%, and each half takes up 50%. You'll use the 50% idea to do this problem.

Because this is a normal distribution, according to the empirical rule, about 68% of values are within one standard deviation from the mean on either side. That means that about 34% are within one standard deviation above the mean.

If x is one standard deviation above its mean, x equals the mean (5) plus 1 times the standard deviation (1.2): $x = 5 + 1(1.2) = 6.2$.

231. **7.4**

You want to find a value of X where 2.5% of the values are greater than x. First, note that the normal distribution has a total probability of 100%, and each half takes up 50%. You'll use this 50% idea to do this problem.

Because this is a normal distribution, according to the empirical rule, about 95% of values are within two standard deviations from the mean on either side. That means that about 47.5% are within two standard deviations above the mean, and beyond that point, you have about $(50 - 47.5) = 2.5\%$ of the values. (Remember that the total percentage above the mean equals 50%.) The value of X where this occurs is the one that's two standard deviations above its mean.

If x is two standard deviations above its mean, x equals the mean (5) plus 2 times the standard deviation (1.2): $x = 5 + 2(1.2) = 7.4$.

232. **3.8**

You want to find a value of X where 16% of the values are less than x. First, note that the normal distribution has a total probability of 100%, and each half takes up 50%. You'll use this 50% idea to do this problem.

Because this is a normal distribution, according to the empirical rule (or the 68-95-99.7 rule), about 68% of values are within one standard deviation from the mean on either side. That means that about 34% are within one standard deviation below the mean, and 16% of the values lie below that value. (Remember the total percentage below the mean equals 50%, so you have $50 - 34 = 16\%$.) The value of X where all this occurs is one standard deviation below its mean.

If x is one standard deviation below its mean, x equals the mean (5) minus 1 times the standard deviation (1.2): $x = 5 - 1(1.2) = 3.8$.

233. 1.5

The empirical rule (or the 68-95-99.7 rule) says that in a normal distribution, about 99.7% of the values lie within three standard deviations above and below the mean (8), which includes the numbers between $8 - 3\sigma$ and $8 + 3\sigma$.

If the upper and lower values of this range are supposed to be 3.5 and 12.5, you know that $3.5 = 8 - 3\sigma$ and $12.5 = 8 + 3\sigma$. Solving for σ in the first equation, you get $3\sigma = 4.5$, so

$$\sigma = \frac{4.5}{3} = 1.5$$

The same answer works in the second equation.

234. $\mu = 0, \sigma = 1$

The Z-distribution, also called the standard normal distribution, has a mean (μ) of 0 and a standard deviation (σ) of 1.

235. 1.2

To calculate the z-score for a value of X, subtract the population mean from x and then divide by the standard deviation:

$$z = \frac{x - \mu}{\sigma}$$
$$= \frac{21.2 - 17}{3.5}$$
$$= 1.2$$

236. −1.0

To calculate the z-score for a value of X, subtract the population mean from x and then divide by the standard deviation:

$$z = \frac{x - \mu}{\sigma}$$
$$= \frac{13.5 - 17}{3.5}$$
$$= -1.0$$

237. 2.5

To calculate the z-score for a value of X, subtract the population mean from x and then divide by the standard deviation:

$$z = \frac{x - \mu}{\sigma}$$
$$= \frac{25.75 - 17}{3.5}$$
$$= 2.5$$

238. **15.6**

The question gives you a z-score and asks for its corresponding x-value. The z-formula contains both x and z, so as long as you know one of them you can always find the other:

$$z = \frac{x - \mu}{\sigma}$$

You know that $z = -0.4$, $\mu = 17$, and $\sigma = 3.5$, so you just plug these numbers into the z-formula and then solve for x:

$$-0.4 = \frac{x - 17}{3.5}$$
$$x = 17 - 0.4(3.5)$$
$$= 15.6$$

239. **24.7**

The question gives you a z-score and asks for its corresponding x-value. The z-formula contains both x and z, so as long as you know one of them you can always find the other:

$$z = \frac{x - \mu}{\sigma}$$

You know that $z = 2.2$, $\mu = 17$, and $\sigma = 3.5$, so you just plug these numbers into the z-formula and then solve for x:

$$2.2 = \frac{x - 17}{3.5}$$
$$x = 17 + 2.2(3.5)$$
$$= 24.7$$

240. **Form A = 85, Form B = 86**

The exam scores from the two forms have different means and different standard deviations, so you can convert both of them to Z-distributions so they're on the same scale and then do the work from there.

The formula that changes an x-value to a z-value is

$$z = \frac{x - \mu}{\sigma}$$

For Form A, you want the x-value corresponding to a z-score of 1.5. Form A has a mean of 70 and standard deviation of 10, so the x-value is

$$1.5 = \frac{x-70}{10}$$
$$x = 70 + 1.5(10)$$
$$= 85$$

Similarly for Form B, with a mean of 74 and standard deviation of 8, you have

$$1.5 = \frac{x-74}{8}$$
$$x = 74 + 1.5(8)$$
$$= 86$$

241. **Form A = 50, Form B = 58**

The exam scores from the two forms have different means and different standard deviations, so you can convert both of them to Z-distributions so they're on the same scale and then do the work from there.

The formula that changes an x-value to a z-value is

$$z = \frac{x - \mu}{\sigma}$$

For Form A, you want the x-value corresponding to a z-score of -2.0. Form A has a mean of 70 and standard deviation of 10, so the x-value is

$$-2.0 = \frac{x-70}{10}$$
$$x = 70 - 2.0(10)$$
$$= 50$$

Similarly for Form B, with a mean of 74 and standard deviation of 8, you have

$$-2.0 = \frac{x-74}{8}$$
$$x = 74 - 2.0(8)$$
$$= 58$$

242. **82**

To find and/or use x-values on two different normal distributions, you use the z-formula to get everything on the same scale and then work from there. For this question, first find the z-score that goes with a score of 80 on Form A, and then find the score on Form B that corresponds to that same z-score.

To find the z-score for Form A, use the z-formula with an x-score of 80, a mean of 70, and a standard deviation of 10:

$$z = \frac{x - \mu}{\sigma}$$
$$= \frac{80 - 70}{10}$$
$$= 1$$

So, a score of 80 on Form A has a z-score of 1, meaning its score is one standard deviation above its mean. To find the corresponding score on Form B, add 1 standard deviation (8) to its mean (74): $74 + (1)(8) = 82$.

243. 86

To find and/or use x-values on two different normal distributions, you use the z-formula to get everything on the same scale and then work from there. For this question, first find the z-score that goes with a score of 85 on Form A, and then find the score on Form B that corresponds to that same z-score.

To find the z-score for Form A, use the z-formula with an x-score of 85, a mean of 70, and a standard deviation of 10:

$$
\begin{aligned}
z &= \frac{x - \mu}{\sigma} \\
&= \frac{85 - 70}{10} \\
&= 1.5
\end{aligned}
$$

So, a score of 85 on Form A has a z-score of 1.5, meaning its score is one and a half standard deviations above its mean. To find the corresponding score on Form B, add 1.5 standard deviations (8) to its mean (74): $74 + (1.5)(8) = 86$.

244. 75

To find and/or use x-values on two different normal distributions, you use the z-formula to get everything on the same scale and then work from there. For this question, first find the z-score that goes with a score of 78 on Form B, and then find the score on Form A that corresponds to that same z-score.

To find the z-score for Form B, use the z-formula with an x-score of 78, a mean of 74, and a standard deviation of 8:

$$
\begin{aligned}
z &= \frac{x - \mu}{\sigma} \\
&= \frac{78 - 74}{8} \\
&= 0.5
\end{aligned}
$$

So, a score of 78 on Form B has a z-score of 0.5, meaning its score is half a standard deviation above its mean. To find the corresponding score on Form A, add 0.5 standard deviations (10) to its mean (70): $70 + (0.5)(10) = 75$.

245. 62.5

To find and/or use x-values on two different normal distributions, you use the z-formula to get everything on the same scale and then work from there. For this question, first find the z-score that goes with a score of 68 on Form B, and then find the score on Form A that corresponds to that same z-score.

To find the z-score for Form B, use the z-formula with an x-score of 68, a mean of 74, and a standard deviation of 8:

$$z = \frac{x - \mu}{\sigma}$$
$$= \frac{68 - 74}{8}$$
$$= -0.75$$

So, a score of 68 on Form B is 0.75 standard deviations below its mean. To find the corresponding score on Form A, add -0.75 of its standard deviations (10) to its mean (70): $70 + (-0.75)(10) = 62.5$.

246. $P(Z \leq 2)$

Looking at the graph, you see that the shaded area represents the probability of all z-values of 2 or less. The probability notation for this is $P(Z \leq 2)$.

247. $P(0 \leq Z \leq 2)$

Looking at the graph, you see that the shaded area represents the probability that Z is between 0 and 2, expressed as $P(0 \leq Z \leq 2)$.

248. $P(Z \geq -2)$

Looking at the graph, you see that the shaded area represents the probability of a z-value of -2 or higher, expressed as $P(Z \geq -2)$.

249. 0.9332

To find $P(Z \leq 1.5)$, using the Z-table (Table A-1 in the appendix), find where the row for 1.5 intersects with the column for 0.00; this value is 0.9332. The Z-table shows only "less than" probabilities so it gives you exactly what you need for this question. *Note:* No probability is exactly at one single point, so $P(Z \leq 1.5) = P(Z < 1.5)$.

250. 0.0668

You want $P(Z \geq 1.5)$, so use the Z-table (Table A-1 in the appendix) to find where the row for 1.5 intersects with the column for 0.00, which is 0.9332. Because the Z-table gives you only "less than" probabilities, subtract $P(Z < 1.5)$ from 1 (remember that the total probability for the normal distribution is 1.00, or 100%):

$$P(Z \geq 1.5) = 1 - P(Z < 1.5)$$
$$= 1 - 0.9332 = 0.0668$$

251. 0.7734

You want $P(Z \geq -0.75)$, so use the Z-table (Table A-1 in the appendix) to find where the row for -0.7 intersects with the column for 0.05, which is 0.2266. Because the Z-table gives you only "less than" probabilities, subtract $P(Z < -0.75)$ from 1 (remember that the total probability for the normal distribution is 1.00, or 100%):

$$P(Z \geq -0.75) = 1 - P(Z < -0.75)$$
$$= 1 - 0.2266 = 0.7734$$

252. 0.5328

To find the probability that Z is between two values, use the Z-table (Table A-1 in the appendix) to find the probabilities corresponding to each z-value, and then find the difference between the probabilities.

Here, you want the probability that Z is between –0.5 and 1.0. First, use the Z-table to find the value where the row for –0.5 intersects with the column for 0.00, which is 0.3085. Then, find the value where the row for 1.0 intersects with the column for 0.00, which is 0.8413. Because the Z-table gives you only "less than" probabilities, find the difference between the probability less than 1.0 (written as $P[Z \leq 1.0]$) and the probability less than –0.5 (written as $P[Z \leq -0.5]$):

$$P(-0.5 \leq Z \leq 1.0) = P(Z \leq 1.0) - P(Z \leq -0.50)$$
$$= 0.8413 - 0.3085 = 0.5328$$

253. 0.6826

To find the probability that Z is between two values, use the Z-table (Table A-1 in the appendix) to find the probabilities corresponding to each z-value, and then find the difference between the probabilities.

Here, you want the probability that Z is between –1.0 and 1.0. First, use the Z-table to find the value where the row for –1.0 intersects with 0.00, which is 0.1587. Then, find the value where the row for 1.0 intersects with the column for 0.00, which is 0.8413. Because the Z-table gives you only "less than" probabilities, find the difference between probability less than 1.0 (written as $P[Z \leq 1.0]$) and the probability less than –1.0 (written as $P[Z \leq -1.0]$):

$$P(-1.0 \leq Z \leq 1.0) = P(Z \leq 1.0) - P(Z \leq -1.0)$$
$$= 0.8413 - 0.1587 = 0.6826$$

254. 1.5

To find a z-score for a particular value of X, subtract the population mean from x, and then divide by the population standard deviation:

$$z = \frac{x - \mu}{\sigma}$$
$$= \frac{13 - 10}{2}$$
$$= 1.5$$

255. 0.0668

To find a "greater than" probability for an x-value, you first convert the x-value to a z-score and then find the corresponding probability for that z-score by using a z-table, such as Table A-1 in the appendix. Then, you subtract that result from 1 (because Table A-1 gives you "less than" probabilities only).

To find the z-score for an x-value, subtract the population mean from x, and then divide by the population standard deviation:

$$z = \frac{x - \mu}{\sigma}$$

Here, $x = 13$ centimeters in diameter, and you want $P(X \geq 13)$, the mean, μ, is 10, and the standard deviation, σ, is 2. Plug these numbers into the z-formula to convert to a z-score:

$$z = \frac{13 - 10}{2} = 1.5$$

Using Table A-1, find where the row for 1.5 and the column for 0.00 intersect; you get $P(Z < 1.5) = 0.9332$. Now, subtract this value from 1 to get $P(Z > 1.5) = 1 - 0.9332 = 0.0668$.

256. 0.9332

To find the probability of a value "no greater than 13" means that the value must be "less than or equal to 13." So, first, convert 13 into a z-score, and then use the Z-table (Table A-1 in the appendix) to find the probability (because Table A-1 provides "less than" probabilities only).

To find the z-score for an x-value, subtract the population mean from x, and then divide by the population standard deviation:

$$z = \frac{x - \mu}{\sigma}$$

Here, $x = 13$ centimeters in diameter, and you want $P(X \leq 13)$, the mean, μ, is 10, and the standard deviation, σ, is 2. Plug these numbers into the z-formula to convert to a z-score:

$$z = \frac{13 - 10}{2} = 1.5$$

Using Table A-1, find where the row for 1.5 and the column for 0.00 intersect; you get $P(Z < 1.5) = 0.9332$. Now, subtract this value from 1 to get $P(Z > 1.5) = 1 - 0.9332 = 0.0668$.

257. 0.4332

To find the probability that X is between two values, change both values to z-scores and then use the Z-table (Table A-1 in the appendix) to find the probabilities corresponding to each z-value; finally, find the difference between the probabilities.

Here, you want the probability that X is between 10 and 13. To find the z-score for an x-value, subtract the population mean from x, and then divide by the population standard deviation:

$$z = \frac{x - \mu}{\sigma}$$

Change $x = 10$ to a z-score with a mean of 10 and a standard deviation of 2:

$$z = \frac{10 - 10}{2} = 0$$

Then, do the same for $x = 13$:

$$z = \frac{13 - 10}{2} = 1.5$$

Now, find the probabilities that Z is between 0 and 1.5. First, use the Z-table to find the value where the row for 0.0 intersects with the column for 0.00, which is 0.5000. Then, find the value where the row for 1.5 intersects with the column for 0.00, which is 0.9332. Because the Z-table gives you only "less than" probabilities, find the difference between the probability less than 1.5 (written as $P[Z \leq 1.5]$) and the probability less than 0 (written as $P[Z \leq 0]$):

$$P(Z \leq 1.5) - P(Z \leq 0) = 0.9332 - 0.5000 = 0.4332$$

258.   **0.4322**

To find the probability that X is between two values, change both values to z-scores and then use the Z-table (Table A-1 in the appendix) to find the probabilities corresponding to each z-value; finally, find the difference between the probabilities.

Here, you want the probability that X is between 7 and 10. To find the z-score for an x-value, subtract the population mean from x, and then divide by the population standard deviation:

$$z = \frac{x - \mu}{\sigma}$$

Change $x = 7$ to a z-score with a mean of 10 and a standard deviation of 2:

$$z = \frac{7 - 10}{2} = -1.5$$

Then, do the same for $x = 10$:

$$z = \frac{10 - 10}{2} = 0$$

Now, find the probabilities that Z is between −1.5 and 0. First, use the Z-table to find the value where the row for −1.5 intersects with the column for 0.00, which is 0.0668. Then, find the value where the row for 0 intersects with the column for 0.00, which is 0.5000. Because the Z-table gives you only "less than" probabilities, find the difference between the probability less than 0 (written as $P[Z \leq 0]$) and the probability less than −1.5 (written as $P[Z \leq -1.5]$):

$$P(Z \leq 0) - P(Z \leq -1.5) = 0.5000 - 0.0668 = 0.4332$$

259.   **−1.25**

To find a z-score for a value of X, subtract the population mean (μ) from x, and then divide by the population standard deviation (σ):

$$z = \frac{x - \mu}{\sigma}$$

$$= \frac{135 - 160}{20} = -1.25$$

260. 0.5

To find a z-score for a value of X, subtract the population mean (μ) from x, and then divide by the population standard deviation (σ):

$$z = \frac{x - \mu}{\sigma}$$
$$= \frac{170 - 160}{20} = 0.5$$

261. −2.25

To find a z-score for a value of X, subtract the population mean (μ) from x, and then divide by the population standard deviation (σ):

$$z = \frac{x - \mu}{\sigma}$$
$$= \frac{115 - 160}{20} = -2.25$$

262. 3

To find a z-score for a value of X, subtract the population mean (μ) from x, and then divide by the population standard deviation (σ):

$$z = \frac{x - \mu}{\sigma}$$
$$= \frac{220 - 160}{20} = 3$$

263. 2.25

To find a z-score for a value of X, subtract the population mean (μ) from x, and then divide by the population standard deviation (σ):

$$z = \frac{x - \mu}{\sigma}$$
$$= \frac{205 - 160}{20} = 2.25$$

264. 0.0013

To find a "greater than" probability for an x-value, first convert the x-value to a z-score and then find the corresponding probability for that z-score by using a Z-table, such as Table A-1 in the appendix; finally, you subtract that result from 1 (because Table A-1 gives you "less than" probabilities only).

To find the z-score for an x-value, subtract the population mean from x, and then divide by the population standard deviation:

$$z = \frac{x - \mu}{\sigma}$$

Here, x is 220 pounds, and you want $P(X > 220)$; the mean, μ, is 160, and the standard deviation, σ, is 20. Plug these numbers into the z-formula to convert to a z-score:

$$z = \frac{220 - 160}{20} = 3$$

Using Table A-1, find where the row for 3.0 and the column for 0.00 intersect; you get $P(Z \leq 3.0) = 0.9987$. Now, subtract this value from 1 to get $P(Z > 3) = 1 - 0.9987 = 0.0013$.

265. **0.9987**

To find a "less than" probability for a value of x from a normal distribution, first convert the value into a z-score, and then use the Z-table (Table A-1 in the appendix) to find the probability.

To find the z-score for an x-value, subtract the population mean from x, and then divide by the population standard deviation:

$$z = \frac{x - \mu}{\sigma}$$

Here, x is 220 pounds, the mean, μ, is 160, and the standard deviation, σ, is 20. Plug these numbers into the z-formula to convert to a z-score:

$$z = \frac{220 - 160}{20} = 3$$

Using Table A-1, find where the row for 3.0 and the column for 0.00 intersect; you get $P(Z \leq 3.0) = 0.9987$.

266. **0.3944**

To find the probability that X is between two values, change both values to z-scores and then use the Z-table (Table A-1 in the appendix) to find the probabilities corresponding to each z-value; finally, find the difference between the probabilities.

Here, you want the probability that X is between 135 and 160. To find the z-score for an x-value, subtract the population mean from x, and then divide by the population standard deviation:

$$z = \frac{x - \mu}{\sigma}$$

Change $x = 135$ to a z-score with a mean of 160 and a standard deviation of 20:

$$z = \frac{135 - 160}{20} = -1.25$$

Then, do the same for $x = 160$:

$$z = \frac{160 - 160}{20} = 0$$

Now, find the probabilities that Z is between -1.25 and 0. First, use the Z-table to find the value where the row for -1.2 intersects with the column for 0.05, which is 0.1056. Then, find the value where the row for 0.0 intersects with the column for 0.00, which is 0.5000. Because the Z-table gives you only "less than" probabilities, find the difference between the probability less than 0 (written as $P[Z \le 0]$) and the probability less than −1.25 (written as $P[Z \le -1.25]$):

$$P(135 \le X \le 160) = P(-1.25 \le Z \le 0)$$
$$= P(Z \le 0) - P(Z \le -1.25)$$
$$= 0.5000 - 0.1056 = 0.3944$$

267. 0.0109

To find the probability that X is between two values, change both values to z-scores and then use the Z-table (Table A-1 in the appendix) to find the probabilities corresponding to each z-value; finally, find the difference between the probabilities.

Here, you want the probability that X is between 205 and 220, written as $P(205 \le X \le 220)$. To find the z-score for an x-value, subtract the population mean from x, and then divide by the population standard deviation:

$$z = \frac{x - \mu}{\sigma}$$

Change $x = 205$ to a z-score with a mean of 160 and a standard deviation of 20:

$$z = \frac{205 - 160}{20} = 2.25$$

Then, do the same for $x = 220$:

$$z = \frac{220 - 160}{20} = 3$$

Now, find the probabilities that Z is between 2.25 and 3. First, use the Z-table to find the value where the row for 2.2 intersects with the column for 0.05, which is 0.9878. Then, find the value where the row for 3.0 intersects with the column for 0.00, which is 0.9987. Because the Z-table gives you only "less than" probabilities, find the difference between the probability less than 3.0 (written as $P[Z \le 3.0]$) and the probability less than 2.25 (written as $P[Z \le 2.25]$). In essence, you're starting with everything below 3.0 and taking off what you don't want, which is everything below 2.25:

$$P(205 \le X \le 220) = P(2.25 \le Z \le 3.0)$$
$$= P(Z \le 3.0) - P(Z \le 2.25)$$
$$= 0.9987 - 0.9878 = 0.0109$$

268. 0.0934

To find the probability that X is between two values, change both values to z-scores and then use the Z-table (Table A-1 in the appendix) to find the probabilities corresponding to each z-value; finally, find the difference between the probabilities.

Here, you want the probability that X is between 115 and 135, written as $P(115 \le X \le 135)$. To find the z-score for an x-value, subtract the population mean from x, and then divide by the population standard deviation:

$$z = \frac{x - \mu}{\sigma}$$

Change $x = 115$ to a z-score with a mean of 160 and a standard deviation of 20:

$$z = \frac{115 - 160}{20} = -2.25$$

Then, do the same for $x = 135$:

$$z = \frac{135 - 160}{20} = -1.25$$

Now, find the probabilities that Z is between −2.25 and −1.25. First, use the Z-table to find the value where the row for −2.2 intersects with the column for 0.05, which is 0.0122. Then, find the value where the row for −1.2 intersects with the column for 0.05, which is 0.1056. Because the Z-table gives you only "less than" probabilities, find the difference between the probability less than −1.25 (written as $P[Z < -1.25]$) and the probability less than −2.25 (written as $P[Z < -2.25]$). In essence, you're starting with everything below −1.25 and taking off what you don't want, which is everything below −2.25:

$$P(115 \leq X \leq 135) = P(-2.25 \leq Z \leq -1.25)$$
$$= P(Z \leq -1.25) - P(Z \leq -2.25)$$
$$= 0.1056 - 0.0122 = 0.0934$$

269. **0.5859**

To find the probability that X is between two values, change both values to z-scores and then use the Z-table (Table A-1 in the appendix) to find the probabilities corresponding to each z-value; finally, find the difference between the probabilities.

Here, you want the probability that X is between 135 and 170, written as $P(135 \leq X \leq 170)$. To find the z-score for an x-value, subtract the population mean from x, and then divide by the population standard deviation:

$$z = \frac{x - \mu}{\sigma}$$

Change $x = 135$ to a z-score with a mean of 160 and a standard deviation of 20:

$$z = \frac{135 - 160}{20} = -1.25$$

Then, do the same for $x = 170$:

$$z = \frac{170 - 160}{20} = 0.50$$

Now, find the probabilities that Z is between −1.25 and −0.50. First, use the Z-table to find the value where the row for −1.2 intersects with the column for 0.05, which is 0.1056. Then, find the value where the row for 0.5 intersects with the column for 0.00, which is 0.6915. Because the Z-table gives you only "less than" probabilities, find the difference between the probability less than 0.50 (written as $P[Z < 0.50]$) and the probability less than −1.25 (written as $P[Z < -1.25]$). In essence, you're starting with everything below 0.50 and taking off what you don't want, which is everything below −1.25:

$$P(135 \leq X \leq 170) = P(-1.25 \leq Z \leq 0.50)$$
$$= P(Z \leq 0.50) - P(Z \leq -1.25)$$
$$= 0.6915 - 0.1056 = 0.5859$$

270.   0.3072

To find the probability that X is between two values, change both values to z-scores and then use the Z-table (Table A-1 in the appendix) to find the probabilities corresponding to each z-value; finally, find the difference between the probabilities.

Here, you want the probability that X is between 170 and 220, written as $P(170 \leq X \leq 220)$. To find the z-score for an x-value, subtract the population mean from x, and then divide by the population standard deviation:

$$z = \frac{x - \mu}{\sigma}$$

Change $x = 170$ to a z-score with a mean of 160 and a standard deviation of 20:

$$z = \frac{170 - 160}{20} = 0.5$$

Then, do the same for $x = 220$:

$$z = \frac{220 - 160}{20} = 3$$

Now, find the probabilities that Z is between 0.5 and 3.0. First, use the Z-table to find the value where the row for 0.5 intersects with the column for 0.00, which is 0.6915. Then, find the value where the row for 3.0 intersects with the column for 0.00, which is 0.9987. Because the Z-table gives you only "less than" probabilities, find the difference between the probability less than 3.0 (written as $P[Z \leq 3.0]$) and the probability less than 0.50 (written as $P[Z \leq 0.50]$). In essence, you're starting with everything below 3.0 and taking off what you don't want, which is everything below 0.50:

$$P(170 \leq X \leq 220) = P(0.50 \leq Z \leq 3.0)$$
$$= P(Z \leq 3.0) - P(Z \leq 0.50)$$
$$= 0.9987 - 0.6915 = 0.3072$$

271.   27

In this case, using intuition is very helpful. If you have a normal distribution for the population, then half of the values lie below the mean (because it's symmetrical and the total percentage is 100%). Here, the mean is 27, so 50%, or half, of the population of adults has a BMI lower than 27.

272.   23.65

You want to find the value of X (BMI) where 25% of the population lies below it. In other words, you want to find the 25th percentile of X. First, you need to find the 25th percentile for Z (using the Z-table, or Table A-1 in the appendix) and then change the z-value to an x-value by using the z-formula:

$$z = \frac{x - \mu}{\sigma}$$

To find the 25th percentile for Z (or the cutoff point where 25% of the population lies below it), look at the Z-table and find the probability that's closest to 0.25. (*Remember:* The probabilities for the Z-table are the values *inside* the table. The numbers on the outsides that tell which row/column you're in are actual z-values, not probabilities.) Searching Table A-1, you see that the closest probability to 0.25 is 0.2514.

Next, find what z-score this probability corresponds to. After you've located 0.2514 inside the table, find its corresponding row (–0.6) and column (0.07). Put these numbers together and you get the z-score of –0.67. This is the 25th percentile for Z. In other words, 25% of the z-values lie below –0.67.

To find the corresponding BMI that marks the 25th percentile, use the z-formula and solve for x. You know that $z = -0.67$, $\mu = 27$, and $\sigma = 5$:

$$z = \frac{x - \mu}{\sigma}$$
$$-0.67 = \frac{x - 27}{5}$$
$$x = 27 - 0.67(5)$$
$$= 23.65$$

So, 25% of the population has a BMI lower than 23.65.

---

273. **18.80**

You want to find the value of X (BMI) where 5% of the population lies below it. In other words, you want to find the 5th percentile of X. First, you need to find the 5th percentile for Z (using the Z-table, or Table A-1 in the appendix) and then change the z-value to an x-value by using the z-formula:

$$z = \frac{x - \mu}{\sigma}$$

To find the 5th percentile for Z (or the cutoff point where 5% of the population lies below it), look at the Z-table and find the probability that's closest to 0.05. (*Remember:* The probabilities for the Z-table are the values *inside* the table. The numbers on the outsides that tell which row/column you're in are actual z-values, not probabilities.) Searching Table A-1, you see that the closest probability to 0.05 is either 0.0495 or 0.0505 (use 0.0505 in this case).

Next, find what z-score this probability corresponds to. After you've located 0.0505 inside the table, find its corresponding row (–1.6) and column (0.04). Put these numbers together and you get the z-score of –1.64. This is the 5th percentile for Z. In other words, 5% of the z-values lie below –1.64.

To find the corresponding BMI that marks the 5th percentile, use the z-formula and solve for x. You know that $z = -1.64$, $\mu = 27$, and $\sigma = 5$:

$$z = \frac{x - \mu}{\sigma}$$
$$-1.64 = \frac{x - 27}{5}$$
$$x = 27 - 1.64(5)$$
$$= 18.80$$

So, 5% of the population has a BMI lower than 18.80.

274. 20.60

You want to find the value of X (BMI) where 10% of the population lies below it. In other words, you want to find the 10th percentile of X. First, you need to find the 10th percentile for Z (using the Z-table, or Table A-1 in the appendix) and then change the z-value to an x-value by using the z-formula:

$$z = \frac{x - \mu}{\sigma}$$

To find the 10th percentile for Z (or the cutoff point where 10% of the population lies below it), look at the Z-table and find the probability that's closest to 0.10. (*Remember:* The probabilities for the Z-table are the values *inside* the table. The numbers on the outsides that tell which row/column you're in are actual z-values, not probabilities.) Searching Table A-1, you see that the closest probability to 0.10 is 0.1003.

Next, find what z-score this probability corresponds to. After you've located 0.1003 inside the table, find its corresponding row (–1.2) and column (0.08). Put these numbers together and you get the z-score of –1.28. This is the 10th percentile for Z. In other words, 10% of the z-values lie below –1.28.

To find the corresponding BMI that marks the 10th percentile, use the z-formula and solve for x. You know that $z = -1.28$, $\mu = 27$, and $\sigma = 5$:

$$z = \frac{x - \mu}{\sigma}$$
$$-1.28 = \frac{x - 27}{5}$$
$$x = 27 + (-1.28)(5)$$
$$= 20.60$$

So, 10% of the population has a BMI lower than 20.60.

275. 33.40

You want to find the value of X (BMI) where 10% of the population lies above it. Because you need to use the Z-table to solve this problem and because the Z-table shows only "less than" probabilities, work this problem as if you wanted the cutoff for the lower 90%. In other words, you want to find the 90th percentile of X (don't worry; you'll get the same answer). First, you need to find the 90th percentile for Z (using the Z-table, or Table A-1 in the appendix) and then change the z-value to an x-value by using the z-formula:

$$z = \frac{x - \mu}{\sigma}$$

To find the 90th percentile for Z, look at the Z-table and find the probability that's closest to 0.90. (*Remember:* The probabilities for the Z-table are the values *inside* the table. The numbers on the outsides that tell which row/column you're in are actual z-values, not probabilities.) Searching Table A-1, you see that the closest probability to 0.90 is 0.8997.

Next, find what z-score this probability corresponds to. After you've located 0.8997 inside the table, find its corresponding row (1.2) and column (0.08). Put these numbers together and you get the z-score of 1.28. This is the 90th percentile for Z. In other words, 90% of the z-values lie below 1.28 (and 10% are above it).

To find the corresponding BMI that marks the 90th percentile, use the z-formula and solve for x. You know that $z = 1.28$, $\mu = 27$, and $\sigma = 5$:

$$z = \frac{x - \mu}{\sigma}$$
$$1.28 = \frac{x - 27}{5}$$
$$x = 27 + (1.28)(5)$$
$$= 33.40$$

So, the BMI marking the upper 10% for this population is 33.40.

276. 35.25

You want to find the value of X (BMI) where 5% of the population lies above it. Because you need to use the Z-table to solve this problem and because the Z-table shows only "less than" probabilities, work this problem as if you wanted the cutoff for the lower 95%. In other words, you want to find the 95th percentile of X (don't worry; you'll get the same answer). First, you need to find the 95th percentile for Z (using the Z-table, or Table A-1 in the appendix) and then change the z-value to an x-value by using the z-formula:

$$z = \frac{x - \mu}{\sigma}$$

To find the 95th percentile for Z, look at the Z-table and find the probability that's closest to 0.95. (*Remember:* The probabilities for the Z-table are the values *inside* the table. The numbers on the outsides that tell which row/column you're in are actual z-values, not probabilities.) In Table A-1, use the probability 0.9505.

Next, find what z-score this probability corresponds to. After you've located 0.9505 inside the table, find its corresponding row (1.6) and column (0.05). Put these numbers together and you get the z-score of 1.65. This is the 95th percentile for Z. In other words, 95% of the z-values lie below 1.65 (and 5% are above it).

To find the corresponding BMI that marks the 95th percentile, use the z-formula and solve for x. You know that $z = 1.65$, $\mu = 27$, and $\sigma = 5$:

$$z = \frac{x - \mu}{\sigma}$$
$$1.65 = \frac{x - 27}{5}$$
$$x = 27 + 1.65(5)$$
$$= 35.25$$

So, the BMI marking the upper 5% for this population is 35.25.

277. 29.60

You want to find the value of X (BMI) where 30% of the population lies above it. Because you need to use the Z-table to solve this problem and because the Z-table shows only "less than" probabilities, work this problem as if you wanted the cutoff for the lower 70%. In other words, you want to find the 70th percentile of X (don't worry;

you'll get the same answer). First, find the 70th percentile for Z (using the Z-table, or Table A-1 in the appendix) and then change the z-value to an x-value by using the z-formula:

$$z = \frac{x - \mu}{\sigma}$$

To find the 70th percentile for Z, look at the Z-table and find the probability that's closest to 0.70. (*Remember:* The probabilities for the Z-table are the values *inside* the table. The numbers on the outsides that tell which row/column you're in are actual z-values, not probabilities.) Searching Table A-1, you see that the closest probability to 0.70 is 0.6985.

Next, find what z-score this probability corresponds to. After you've located 0.6985 inside the table, find its corresponding row (0.5) and column (0.02). Put these numbers together and you get the z-score of 0.52. This is the 70th percentile for Z. In other words, 70% of the z-values lie below 0.52 (and 30% are above it).

To find the corresponding BMI that marks the 70th percentile, use the z-formula and solve for x. You know that $z = 0.52$, $\mu = 27$, and $\sigma = 5$:

$$z = \frac{x - \mu}{\sigma}$$
$$0.52 = \frac{x - 27}{5}$$
$$x = 27 + (0.52)(5)$$
$$= 29.60$$

So, the BMI marking the upper 30% for this population is 29.60.

278. 23.65, 30.35

The 1st quartile (Q_1) is the value with 25% of the distribution below it, and the 3rd quartile (Q_3) is the value with 75% of the values below it. Using the Z-table (Table A-1 in the appendix), you can find that the value closest to 0.25 is 0.2514, corresponding to a z-score of –0.67 (the value where the row for –0.6 and the column for 0.07 intersect).

To find the 1st quartile (Q_1) of X (a BMI score) corresponding to $z = -0.67$, use the z-formula and solve for x:

$$z = \frac{x - \mu}{\sigma}$$
$$-0.67 = \frac{x - 27}{5}$$
$$x = 27 + (-0.67)(5)$$
$$= 23.65$$

To find the 3rd quartile (Q_3) for X, follow the same procedure: First, find the value in the Z-table that's closest to 0.75, which is 0.67 (note the symmetry in the values). Then, use the z-formula to solve for x:

$$z = \frac{x - \mu}{\sigma}$$

$$0.67 = \frac{x - 27}{5}$$

$$x = 27 + (0.67)(5)$$

$$= 30.35$$

---

279. **69.96**

You want to find the value of X (exam score) where 20% of the population lies below it. In other words, you want to find the 20th percentile of X. First, find the 20th percentile for Z (using the Z-table, or Table A-1 in the appendix) and then change the z-value to an x-value by using the z-formula:

$$z = \frac{x - \mu}{\sigma}$$

To find the 20th percentile for Z (or the cutoff point where 20% of the population lies below it), look at the Z-table and find the probability that's closest to 0.20. (*Remember:* The probabilities for the Z-table are the values *inside* the table. The numbers on the outsides that tell which row/column you're in are actual z-values, not probabilities.) Searching Table A-1, you see that the closest probability to 0.20 is 0.2005.

Next, find what z-score this probability corresponds to. After you've located 0.2005 inside the table, find its corresponding row (–0.8) and column (0.04). Put these numbers together and you get the z-score of –0.84. This is the 20th percentile for Z. In other words, 20% of the z-values lie below –0.84.

To find the corresponding exam score that marks the 20th percentile, use the z-formula and solve for x. You know that $z = -0.84$, $\mu = 75$, and $\sigma = 6$:

$$z = \frac{x - \mu}{\sigma}$$

$$-0.84 = \frac{x - 75}{6}$$

$$x = 75 + (-0.84)(6)$$

$$= 69.96$$

So, 20% of the students scored below 69.96.

---

280. **65.16**

You want to find the value of X (exam score) where 5% of the population lies below it. In other words, you want to find the 5th percentile of X. First, you need to find the 5th percentile for Z (using the Z-table, or Table A-1 in the appendix) and then change the z-value to an x-value by using the z-formula:

$$z = \frac{x - \mu}{\sigma}$$

To find the 5th percentile for Z (or the cutoff point where 5% of the population lies below it), look at the Z-table and find the probability that's closest to 0.05. (*Remember:*

The probabilities for the Z-table are the values *inside* the table. The numbers on the outsides that tell which row/column you're in are actual z-values, not probabilities.) Searching Table A-1, you see that the closest probability to 0.05 is either 0.0495 or 0.0505 (use 0.0505 in this case).

Next, find what z-score this probability corresponds to. After you've located 0.0505 inside the table, find its corresponding row (–1.6) and column (0.04). Put these numbers together and you get the z-score of –1.64. This is the 5th percentile for Z. In other words, 5% of the z-values lie below –1.64.

To find the corresponding exam score that marks the 5th percentile, use the z-formula and solve for x. You know that $z = -1.64$, $\mu = 75$, and $\sigma = 6$:

$$z = \frac{x - \mu}{\sigma}$$
$$-1.64 = \frac{x - 75}{6}$$
$$x = 75 - 1.64(6)$$
$$= 65.16$$

So, 5% of the students scored below 65.16.

281. 82.68

You want to find the value of X (exam score) where 10% of the population lies above it. Because you need to use the Z-table to solve this problem and because the Z-table shows only "less than" probabilities, work this problem as if you wanted the cutoff for the lower 90%. In other words, you want to find the 90th percentile of X (don't worry; you'll get the same answer). First, you need to find the 90th percentile for Z (using the Z-table, or Table A-1 in the appendix) and then change the z-value to an x-value by using the z-formula:

$$z = \frac{x - \mu}{\sigma}$$

To find the 90th percentile for Z, look at the Z-table and find the probability that's closest to 0.90. (*Remember:* The probabilities for the Z-table are the values *inside* the table. The numbers on the outsides that tell which row/column you're in are actual z-values, not probabilities.) Searching Table A-1, you see that the closest probability to 0.90 is 0.8997.

Next, find what z-score this probability corresponds to. After you've located 0.8997 inside the table, find its corresponding row (1.2) and column (0.08). Put these numbers together and you get the z-score of 1.28. This is the 90th percentile for Z. In other words, 90% of the z-values lie below 1.28 (and 10% lie above it).

To find the corresponding exam score that marks the 90th percentile, use the z-formula and solve for x. You know that $z = 1.28$, $\mu = 75$, and $\sigma = 6$:

$$z = \frac{x - \mu}{\sigma}$$
$$1.28 = \frac{x - 75}{6}$$
$$x = 75 + 1.28(6)$$
$$= 82.68$$

So, 10% of the students scored above 82.68.

282. 88.98

You want to find the value of X (exam score) where 1% of the population lies above it. Because you need to use the Z-table to solve this problem and because the Z-table shows only "less than" probabilities, work this problem as if you wanted the cutoff for the lower 99%. In other words, you want to find the 99th percentile of X (don't worry; you'll get the same answer). First, you need to find the 99th percentile for Z (using the Z-table, or Table A-1 in the appendix) and then change the z-value to an x-value by using the z-formula:

$$z = \frac{x - \mu}{\sigma}$$

To find the 99th percentile for Z, look at the Z-table and find the probability that's closest to 0.99. (*Remember:* The probabilities for the Z-table are the values *inside* the table. The numbers on the outsides that tell which row/column you're in are actual z-values, not probabilities.) Searching Table A-1, you see that the closest probability to 0.99 is 0.9901.

Next, find what z-score this probability corresponds to. After you've located 0.9901 inside the table, find its corresponding row (2.3) and column (0.03). Put these numbers together and you get the z-score of 2.33. This is the 99th percentile for Z. In other words, 99% of the z-values lie below 2.33 and 1% lie above it.

To find the corresponding exam score that marks the 99th percentile, use the z-formula and solve for x. You know that $z = 2.33$, $\mu = 75$, and $\sigma = 6$:

$$z = \frac{x - \mu}{\sigma}$$
$$2.33 = \frac{x - 75}{6}$$
$$x = 75 + 2.33(6)$$
$$= 88.98$$

So, 1% of the students scored above 88.98.

283. 86.76

You want to find the value of X (exam score) where 2.5% of the population lies above it. Because you need to use the Z-table to solve this problem and because the Z-table shows only "less than" probabilities, work this problem as if you wanted the cutoff for the lower 97.5%. In other words, you want to find the 97.5th percentile of X (don't worry; you'll get the same answer). First, you need to find the 97.5th percentile for Z (using the Z-table, or Table A-1 in the appendix) and then change the z-value to an x-value by using the z-formula:

$$z = \frac{x - \mu}{\sigma}$$

To find the 97.5th percentile for Z, look at the Z-table and find the probability that's closest to 0.975. (*Remember:* The probabilities for the Z-table are the values *inside* the table. The numbers on the outsides that tell which row/column you're in are actual z-values, not probabilities.) Searching Table A-1, you see that the closest probability to 0.975 is exactly 0.9750.

Next, find what z-score this probability corresponds to. After you've located 0.9750 inside the table, find its corresponding row (1.9) and column (0.06). Put these numbers together and you get the z-score of 1.96. This is the 97.5th percentile for Z. In other words, 97.5% of the z-values lie below 1.96 (and 2.5% lie above it).

To find the corresponding exam score that marks the 97.5th percentile, use the z-formula and solve for x. You know that $z = 1.96$, $\mu = 75$, and $\sigma = 6$:

$$z = \frac{x - \mu}{\sigma}$$
$$1.96 = \frac{x - 75}{6}$$
$$x = 75 + (1.96)(6)$$
$$= 86.76$$

So, 2.5% of the students scored above 86.76.

284. 84.84

You want to find the value of X (exam score) where 5% of the population lies above it. Because you need to use the Z-table to solve this problem and because the Z-table shows only "less than" probabilities, work this problem as if you wanted the cutoff for the lower 95%. In other words, you want to find the 95th percentile of X (don't worry; you'll get the same answer). First, you need to find the 95th percentile for Z (using the Z-table, or Table A-1 in the appendix) and then change the z-value to an x-value by using the z-formula:

$$z = \frac{x - \mu}{\sigma}$$

To find the 95th percentile for Z, look at the Z-table and find the probability that's closest to 0.95. (*Remember:* The probabilities for the Z-table are the values *inside* the table. The numbers on the outsides that tell which row/column you're in are actual z-values, not probabilities.) In Table A-1, use the probability 0.9495.

Next, find what z-score this probability corresponds to. After you've located 0.9495 inside the table, find its corresponding row (1.6) and column (0.04). Put these numbers together and you get the z-score of 1.64. This is the 95th percentile for Z. In other words, 95% of the z-values lie below 1.64 (and 5% are above it).

To find the corresponding exam score that marks the 95th percentile, use the z-formula and solve for x. You know that $z = 1.64$, $\mu = 75$, and $\sigma = 6$:

$$z = \frac{x - \mu}{\sigma}$$
$$1.64 = \frac{x - 75}{6}$$
$$x = 75 + (1.64)(6)$$
$$= 84.84$$

So, 5% of the students scored above 84.84.

285. 310.8

The fastest (and best) times are at the lower end of the distribution. Using the Z-table (Table A-1 in the appendix), find the value where only 5% of the times are below it.

The closest table value to 0.05 is 0.0505, which corresponds to a z-value of -1.64.

To find the time corresponding to a particular z-score, use the z-formula to solve for x:

$$z = \frac{x - \mu}{\sigma}$$
$$-1.64 = \frac{x - 360}{30}$$
$$x = 360 + (-1.64)(30)$$
$$= 310.8$$

This means a time of 310.8 seconds is the cutoff for the fastest 5% of the times.

286. 360

In this case, using intuition is very helpful. If you have a normal distribution for the population, then half of the values lie below the mean (because it's symmetrical and the total percentage is 100%). Here, the mean is 360, so 50%, or half, of the military recruits have a time of 360 seconds.

287. 398.4

The slowest (and worst) times are at the upper end of the distribution. Using the Z-table (Table A-1 in the appendix), find the value where 90% of the times are below it. The closest table value to 0.90 is 0.8997, corresponding to a z-value of 1.28.

To find the time corresponding to a particular z-score, use the z-formula to solve for x:

$$z = \frac{x - \mu}{\sigma}$$
$$1.28 = \frac{x - 360}{30}$$
$$x = 360 + (1.28)(30)$$
$$= 398.4$$

So, the slowest 10% of the recruits had a time of 398.4 seconds or more.

288. 321.6

The fastest (and best) times are at the lower end of the distribution. Using the Z-table (Table A-1 in the appendix), find the value where only 10% of the times are below it. The closest table value to 0.10 is 0.1003, which corresponds to a z-value of -1.28.

To find the time corresponding to a particular z-score, use the z-formula to solve for x:

$$z = \frac{x - \mu}{\sigma}$$
$$-1.28 = \frac{x - 360}{30}$$
$$x = 360 + (-1.28)(30)$$
$$= 321.6$$

So, the fastest 10% of the recruits had a time of 321.6 seconds or less.

289. **339.9**

The fastest (and best) times are at the lower end of the distribution. Using the Z-table (Table A-1 in the appendix), find the value where only 25% of the times are below it. The closest table value to 0.25 is 0.2514, which corresponds to a z-value of −0.67.

To find the time corresponding to a particular z-score, use the z-formula to solve for x:

$$z = \frac{x - \mu}{\sigma}$$
$$-0.67 = \frac{x - 360}{30}$$
$$x = 360 + (-0.67)(30)$$
$$= 339.9$$

So, the fastest 25% of the recruits had times of 339.9 seconds or less.

290. **380.1**

The slowest (and worst) times are at the upper end of the distribution. Using the Z-table (Table A-1 in the appendix), you first have to rewrite what you're looking for in terms of a "less than" probability; so you find the value where 75% of the times are below it. The closest table value to 0.75 is 0.7486, corresponding to a z-value of 0.67.

To find the time corresponding to a particular z-score, use the z-formula to solve for x:

$$z = \frac{x - \mu}{\sigma}$$
$$0.67 = \frac{x - 360}{30}$$
$$x = 360 + (0.67)(30)$$
$$= 380.1$$

So, the slowest 25% of the recruits had times of 380.1 seconds or more.

291. **E. Choices (A), (B), and (C) (The t-distribution has thicker tails than the Z-distribution; the t-distribution has a proportionately larger standard deviation than the Z-distribution; the t-distribution is bell-shaped but has a lower peak than the Z-distribution.)**

Compared to the Z-distribution, the t-distribution has thicker tails and a proportionately larger standard deviation. It's still bell-shaped, but it has a lower peak than the Z-distribution.

292. t_{29}

A t-distribution for a study with one population with a sample size of 30 has $n - 1 = 30 - 1 = 29$ degrees of freedom, so the correct distribution is t_{29}.

293. **25**

A t_{24} distribution has $n-1=24$ degrees of freedom, so the sample size, n, is 25.

294. **The peak of the Z-distribution would be higher.**

In general, the t-distribution is bell-shaped but is flatter and has a lower peak than the standard normal (Z-) distribution, particularly with smaller degrees of freedom for the t-distribution.

295. **The t-distribution would have thicker tails.**

The t-distribution is flatter, has a lower peak, and has thicker tails compared to the standard normal (Z-) distribution, particularly with smaller degrees of freedom for the t-distribution.

296. **100**

As the degrees of freedom increase, the t-distribution tends to look more like the Z-distribution. So, the t-distribution with the highest degrees of freedom most resembles the Z-distribution.

297. **the degrees of freedom**

A t-distribution is defined by its degrees of freedom, unlike the normal distribution, which is defined by its mean and standard deviation. The t-distribution always has a mean of 0 (like the Z-distribution), and the more degrees of freedom that a t-distribution has, the smaller its standard deviation gets (because the tails aren't as thin).

298. **the one-sample t-test**

You're testing the mean of one population, so the answer has to be a one-sample test. You can't use the one-sample Z-test without knowing the population standard deviation, so in this instance, you'd use the one-sample t-test.

299. **the paired t-test**

You use the paired t-test to study mean differences among paired subjects according to some variable — for example, the mean difference in weight before and after a weight loss program or the mean difference in weight loss in study participants who are matched according to similar characteristics.

300. t_{24}

This type of study is called a matched-pairs design. A matched-pairs design with 50 observations from the two samples combined has 25 pairs of data, so $n=25$, and the degrees of freedom is $n-1=25-1=24$. So, the t-distribution corresponding to this scenario is t_{24}.

301. $df = 17$

The study involving one population and a sample size of 18 has $n-1=18-1=17$ degrees of freedom.

302. $df = 21$

A matched-pairs design with 44 total observations has 22 pairs. The degrees of freedom is one less than the number of pairs: $n-1=22-1=21$.

303. $p = 0.025$

The column headings of Table A-2 display upper-tail ("greater than") probabilities for specified t-values, so you can read the value for an upper-tail probability of 0.025 directly from the column heading for 0.025.

304. $p = 0.005$

For a hypothesis test, α is the level of significance; if the p-value for the test is less than α, then H_0 (the null hypothesis) is rejected. (The p-value is the probability of being beyond your test statistic.)

The column headings of Table A-2 display upper-tail ("greater than") probabilities for specified t-values. For a two-tailed test (where H_a, or the alternative hypothesis, is "not equal to") with significance level α, you select the column for $\alpha/2$, which gives you the probability for each tail. So, in this case, you need the column for $\alpha/2 = 0.01/2 = 0.005$.

305. $p = 0.025$

For a hypothesis test, α is the level of significance; if the p-value for the test is less than α, then H_0 (the null hypothesis) is rejected. (The p-value is the probability of being beyond your test statistic.)

The column headings of Table A-2 display upper-tail ("greater than") probabilities for specified t-values. For a two-tailed test with significance level α, you select the column for $\alpha/2$, which gives you the probability for each tail. So, in this case, you need the column for $\alpha/2 = 0.05/2 = 0.025$.

306. 0.05

The column headings of Table A-2 display upper-tail ("greater than") probabilities for specified t-values. To find the upper-tail probability for $t_{10} \geq 1.81$, locate the row for $df = 10$ and follow it across until you find the t-value 1.81. The column heading ("greater than" probability) for this value is 0.05.

307. 0.01

The column headings in Table A-2 display upper-tail ("greater than") probabilities for specified t-values. To find the upper-tail probability for $t_{25} \geq 2.49$, locate the row for

$df = 25$ and follow it across until you find the t-value 2.49. The column heading ("greater than" probability) for this value is 0.01.

308. 0.10

The column headings in Table A-2 display upper-tail ("greater than") probabilities for specified t-values. To find the upper-tail probability for $t_{15} \geq 1.34$, locate the row for $df = 15$ and follow it across until you find the t-value 1.34. The column heading ("greater than" probability) for this value is 0.10.

309. 0.05 and 0.025

Using Table A-2, locate the row with 22 degrees of freedom and look for 1.80. However, this exact value doesn't lie in this row, so look for the values on either side of it: 1.717144 and 2.07387. The upper-tail probabilities in Table A-2 appear in the column headings; the column heading for 1.717144 is 0.05, and the column heading for 2.07387 is 0.025. Hence, the upper-tail probability for a t-value of 1.80 must lie between 0.05 and 0.025.

310. 0.025 and 0.01

Using Table A-2, locate the row with 14 degrees of freedom and look for 2.35. However, this exact value doesn't lie in this row, so look for the values on either side of it: 2.14479 and 2.62449. The upper-tail probabilities in Table A-2 appear in the column headings; the column heading for 2.14479 is 0.025, and the column heading for 2.62449 is 0.01. Hence, the upper-tail probability for a t-value of 2.35 must lie between 0.025 and 0.01.

311. 0.01

The t-distribution is symmetrical, so the probability of being in the upper tail ("greater than") with a positive value of t is the same as the probability of being in the lower tail (less than) with the corresponding negative value of t. Table A-2 gives you upper-tail probabilities for positive values of t, so locate the row for $df = 15$ and follow it across to the t-value of 2.60; you find that $P(t_{15} \geq 2.60)$ is 0.01 (the column heading). This means (by symmetry) that $P(t_{15} \leq -2.60)$ is also 0.01.

312. 0.025

The t-distribution is symmetrical, so the probability of being in the upper tail ("greater than") with a positive value of t is the same as the probability of being in the lower tail ("less than") with the corresponding negative value of t. Table A-2 gives you upper-tail probabilities for positive values of t, so locate the row for $df = 27$ and follow it across to the t-value of 2.05; you find that $P(t_{27} \geq 2.05)$ is 0.025 (the column heading). This means (by symmetry) that $P(t_{27} \leq -2.05)$ is also 0.025.

313. 0.05

The t-distribution is symmetrical, so the probability of being in the upper tail ("greater than") with a positive value of t is the same as the probability of being in the lower tail ("less than") with the corresponding negative value of t. Table A-2 gives you upper-tail probabilities for positive values of t, so locate the row for $df = 27$ and follow it across to the t-value of 2.05; you find that $P(t_{27} \geq 2.05)$ is 0.025 (the column heading). This means (by symmetry) that $P(t_{27} \leq -2.05)$ is also 0.025. To find $P(t_{27} \geq 2.05$ or $\leq -2.05)$, you sum the two individual probabilities:

$0.005 + 0.005 = 0.01$.

314. 0.01

The t-distribution is symmetrical, so the probability of being in the upper tail ("greater than") with a positive value of t is the same as the probability of being in the lower tail ("less than") with the corresponding negative value of t. Table A-2 gives you upper-tail probabilities for positive values of t, so locate the row for $df = 27$, and follow it across to the t-value of 3.25; you find that $P(t_9 \geq 3.25)$ is 0.005 (the column heading). This means (by symmetry) that $P(t_9 \leq -3.25)$ is also 0.005. To find $P(t_9 \geq 3.25$ or $\leq -3.25)$, you sum the two individual probabilities: $0.005 + 0.005 = 0.01$

315. 1.81

The 95th percentile of a distribution is the value that 95% of values are less than and 5% are greater than. The column headings in Table A-2 show "greater than" probabilities, so locate the row for $df = 10$, and follow it across to the t-value with a greater than probability of 0.05 (this is the value of t that intersects column 0.05 and row 10), which is 1.81. Because 5% of the values are greater than 1.81, you know that 95% are less than 1.81; so the 95th percentile is $t = 1.81$.

316. 0.26

The 60th percentile of a distribution is the value that 60% of values are less than and 40% are greater than. The column headings in Table A-2 show "greater than" probabilities, so locate the row for $df = 28$, and follow it across to the t-value with a greater than probability of 0.40 (this is the value of t that intersects column 0.40 and row 28), which is 0.26. Because 40% of the values are greater than 0.26, you know that 60% are less than 0.26; so the 60th percentile is $t = 0.26$.

317. −1.33

The 10th percentile of a distribution is the value that 10% of values are less than and 90% are greater than. The column headings in Table A-2 show "greater than" probabilities. Note that 0.90 isn't one of them, but 0.10 is there, and because the t-distribution is symmetrical, the t-value for the 10th percentile is the negative of the t-value for the 90th percentile.

So, to find the 10th percentile value, locate the 90th percentile value in Table A-2 and then take its negative value. First, locate the row for $df = 20$, and follow it across to the value that intersects with the column *0.10*, which gives you 1.33 (the 90th percentile). So, because the t-distribution is symmetrical, you know that 10% of values are less than −1.33. That means that the 10th percentile is $t = -1.33$.

318. −0.69

The 25th percentile of a distribution is the value that 25% of values are less than and 75% are greater than. The column headings in Table A-2 show "greater than" probabilities. Note that 0.75 isn't one of them, but 0.25 is there, and because the t-distribution is symmetrical, the t-value for 25th percentile is the negative of the t-value for the 75th percentile.

So, to find the 25th percentile value, locate the 75th percentile in Table A-2 and then take its negative value. First, locate the row for $df = 20$, and follow it across to the value that intersects with the column *0.25*, which gives you 0.69 (the 75th percentile). So, because the t-distribution is symmetrical, you know that 25% of values are less than −0.69. That means that the 25th percentile is $t = -0.69$.

319. −1.34

The 10th percentile of a distribution is the value that 10% of values are less than and 90% are greater than. The column headings in Table A-2 show "greater than" probabilities. Note that 0.90 isn't one of them, but 0.10 is there, and because the t-distribution is symmetrical, the t-value for the 10th percentile is the negative of the t-value for the 90th percentile.

So, to find the 10th percentile value, locate the 90th percentile value in Table A-2 and then take its negative value. First, locate the row for $df = 16$, and follow it across to the value that intersects with the column *0.10*, which gives you 1.34 (the 90th percentile). So, because the t-distribution is symmetrical, you know that 10% of values are less than −1.34. That means that the 10th percentile is $t = -1.34$.

320. −1.75

The 5th percentile of a distribution is the value that 5% of values are less than and 95% are greater than. The column headings in Table A-2 show "greater than" probabilities. Note that 0.95 isn't one of them, but 0.05 is there, and because the t-distribution is symmetrical, the t-value for the 5th percentile is the negative of the t-value for the 95th percentile.

So, to find the 5th percentile value, locate the 95th percentile value in Table A-2 and then take its negative value. First, locate the row for $df = 16$, and follow it across to the value that intersects with the column *0.05*, which gives you 1.75 (the 95th percentile). So, because the t-distribution is symmetrical, you know that 5% of values are less than −1.75. That means that the 5th percentile is $t = -1.75$.

321. D. Choices (A) and (B) (You have a small sample size; you don't know the population standard deviation)

You use the t-distribution rather than the Z-distribution to calculate confidence intervals when you have a small sample size and/or when you don't know the population standard deviation (so you use the sample standard deviation instead). In both cases, you pay a penalty, hence the wider (flatter) t-distribution.

322. **99%**

Here, you want the t-values for a 99% confidence interval, so look at the last row of Table A-2 (the row labeled CI), and find the value 99%. Using this column, you can find the t-values for a 99% confidence level by following it to where it intersects with the row for the degrees of freedom you want.

323. **0.005**

A 99% confidence interval means that 99%, or 0.99, of all the values lie inside the confidence interval, and 10%, or 0.01, of all the values lie on the outside, with 0.01 / 2=0.005 of the values above (greater than) the confidence interval and 0.005 of the values below (less than) the confidence interval.

The first row (column headings) of Table A-2 show upper-tail ("greater than") probabilities only. To find a t-value for a 99% confidence interval using the first row of Table A-2, look in the column for 0.01 / 2 = 0.005. Then go to the row corresponding to the degrees of freedom to find the t-value you need.

324. **2.13**

Here, you want a t-value for a 95% confidence interval, so look at the last row of Table A-2 (labeled CI), find the value 95%, and then intersect this column with the row for $df = 15$, which gives you 2.13.

325. **2.81**

In Table A-2, find where the row for $df = 23$ and the column for 99% CI (in the last row of the table) intersect. The value at this intersection is 2.81.

326. **1.70**

In Table A-2, find where the row for $df = 30$ and the column for 90% CI (in the last row of the table) intersect. The value at this intersection is 1.697261, which rounds to 1.70.

327. **95%**

In Table A-2, you find the value 2.09 where the row for $df = 19$ and the column for 95% CI (in the last row of the table) intersect.

328. **99%**

In Table A-2, you find the value 2.78 where the row for $df = 26$ and the column for 99% CI (in the last row of the table) intersect.

329. **80%**

In Table A-2, you find the value 1.36 where the row for $df = 12$ and the column for 80% CI (in the last row of the table) intersect.

330. 50

As the degrees of freedom increase, the t-distribution resembles the Z-distribution more closely. The peak in the bell shape rises higher and higher, and the tails become thinner and thinner, until the two distributions are practically indistinguishable. So, the t-distribution with the highest degrees of freedom most resembles the Z-distribution.

331. 10

The smaller the degrees of freedom, the less the t-distribution resembles the Z-distribution. The peak in the bell shape gets lower and lower, and the tails become thicker and thicker. So, the t-distribution with the lowest degrees of freedom least resembles the Z-distribution.

332. 10

The column in Table A-2 with a heading of 0.05 shows all the different t-values with right-tail ("greater than") probabilities of 0.05 for various degrees of freedom. As the degrees of freedom decrease (moving from bottom to top in the column), the t-values increase because the tails in the t-distributions with fewer degrees of freedom are thicker, and you have to move out farther on the t-distribution to get to the 5% mark. Moving farther out requires a higher t-value; so the t-distribution with the lowest degrees of freedom is the one with the largest t-value.

333. 50

The column in Table A-2 with a heading of 0.10 shows all the different t-values with right-tail ("greater than") probabilities of 0.10 for various degrees of freedom. As the degrees of freedom increase (moving from top to bottom in the column), the t-values decrease because the tails in the t-distributions with higher degrees of freedom are thinner, and you don't have to move out as far on the t-distribution to get to the 10% mark. That means you'll have a smaller t-value; so the t-distribution with the highest degrees of freedom is the one with the smallest t-value.

334. 80%

For any distribution (including the t-distribution with 40 degrees of freedom), the greater the confidence level is for a confidence interval, the wider the confidence interval is; and the lower the confidence level is, the narrower the confidence interval is. Therefore, the confidence interval with the lowest confidence level (in this case, 80%) will be the narrowest.

335. 99%

For any distribution (including the t-distribution with 50 degrees of freedom), the greater the confidence level is for a confidence interval, the wider the confidence interval is; and the lower the confidence level is, the narrower the confidence interval is. Therefore, the confidence interval with the highest confidence level (in this case, 99%) will be the widest.

336. −2.74

Because you don't know the population standard deviation and are testing the mean of one population, you compute the one-sample t-test, using the following formula for the test statistic:

$$t = \frac{\bar{x} - \mu_0}{s/\sqrt{n}}$$

In this case, the sample mean, $\bar{x}$, is 4.8; the target population mean, μ_0, is 5 (this value goes in the null hypothesis H_0, hence the subscript 0 in both expressions); the sample standard deviation, s, is 0.4; the sample size, n, is 30; and the degrees of freedom, $n-1$, is 29. Now, plug these numbers into the formula and solve:

$$t = \frac{4.8 - 5.0}{0.4/\sqrt{30}} = -2.74$$

337. between 0.01 and 0.005

In Table A-2, using the row for $df = 29$, the upper-tail ("greater than") probability for 2.46202 is 0.01 (found in the column heading), and the probability for 2.75639 is 0.005. Because the t-distribution is symmetrical, the lower-tail ("less than") probability for −2.46202 is also 0.01, and the probability for −2.75639 is also 0.005. The t-value of −2.74 lies between these two numbers, so the probability is between 0.01 and 0.005.

338. (4.65, 4.95)

The formula for the confidence interval for one population mean, using the t-distribution, is

$$\bar{x} \pm t_{n-1}\frac{s}{\sqrt{n}}$$

In this case, the sample mean, $\bar{x}$, is 4.8; the sample standard deviation, s, is 0.4; the sample size, n, is 30; and the degrees of freedom, $n-1$, is 29. That means $t_{n-1} = 2.05$ (from Table A-2).

Now, plug in the numbers:

$$\bar{x} \pm t_{n-1}\frac{s}{\sqrt{n}}$$
$$= 4.8 \pm 2.05\frac{0.4}{\sqrt{30}}$$
$$= 4.8 \pm 0.1497$$
$$= 4.6503 \text{ to } 4.9497$$

Rounded to two decimal places, the answer is 4.65 to 4.95.

339. (4.68, 4.92)

The formula for the confidence interval for one population mean, using the t-distribution, is

$$\bar{x} \pm t_{n-1}\frac{s}{\sqrt{n}}$$

In this case, the sample mean, $\bar{x}$, is 4.8; the sample standard deviation, s, is 0.4; the sample size, n, is 30; and the degrees of freedom, $n-1$, is 29. That means that $t_{n-1} = 1.70$ (from Table A-2).

Now, plug in the numbers:

$$\bar{x} \pm t_{n-1} \frac{s}{\sqrt{n}}$$
$$= 4.8 \pm 1.70 \frac{0.4}{\sqrt{30}}$$
$$= 4.8 \pm 0.1242$$
$$= 4.6758 \text{ to } 4.9242$$

Rounded to two decimal places, the answer is 4.68 to 4.92.

340. **(4.60, 5.00)**

The formula for the confidence interval for one population mean, using the t-distribution, is

$$\bar{x} \pm t_{n-1} \frac{s}{\sqrt{n}}$$

In this case, the sample mean, $\bar{x}$, is 4.8; the sample standard deviation, s, is 0.4; the sample size, n, is 30; and the degrees of freedom, $n-1$, is 29. That means that $t_{n-1} = 2.76$ (from Table A-2).

Now, plug in the numbers:

$$\bar{x} \pm t_{n-1} \frac{s}{\sqrt{n}}$$
$$= 4.8 \pm 2.76 \frac{0.4}{\sqrt{30}}$$
$$= 4.8 \pm 0.2016$$
$$= 4.5984 \text{ to } 5.0016$$

Rounded to two decimal places, the answer is 4.60 to 5.00.

341. **random variable**

A *random variable* is an assignment of numbers to the outcome of some random (or partially random) event. Many things can be random variables. For example, on a coin toss, you can assign 1 to heads and 0 to tails, and the outcome of the coin toss would be a random variable. You can also toss a coin five times and count the number of times it comes up heads, and that number would be a random variable. If you rolled two dice and added the numbers that came up on both, the total of the roll would be a random variable.

342. **a sampling distribution of the sample means**

A *sampling distribution* is a collection of all the means from all possible samples of the same size taken from a population. In this case, the population is the 10,000 test scores, each sample is 100 test scores, and each sample mean is the average of the 100 test scores.

343. $\mu_X = 3.11$

Because you found the average GPA of every student in the university, you used a population value, which needs a Greek letter. μ_X refers to the mean of all individual values in the population.

344. $\mu_{\bar{x}} = 3.5$

Here, you take all possible samples (of the same size), find all their possible means, and treat those as a population. Then, you find the mean of that entire population of sample means. The notation for this is $\mu_{\bar{x}} = 3.5$.

345. $\bar{x} = 3.5$

Because the value is the result of only a sample of dice rools, and not the full population of all possible rolls, you must use the sample mean notation, $\bar{x} = 3.5$.

346. **D. Each of the observations in the distribution must consist of a statistic that describes a collection of data points.**

A sampling distribution is a set of all possible values in a population, except the values themselves represent statistics, like sample means or sample standard deviations.

The critical element in each case is that data points going into your distribution each represent a summary statistic for a sample.

347. **E. a distribution showing the weight of each individual football fan entering a stadium on game day**

A sampling distribution is a population of data points where each data point represents a summary statistic from one sample of individuals. A population distribution is a population of data points where each data point represents an individual.

348. **a random variable denoting the outcome from a single roll of the die**

X is a random variable with possible values 1, 2, 3, 4, 5, and 6, denoting the outcome from a single roll of the die.

349. **a random variable denoting the average value when you roll the die n times (where n is some fixed number)**

$\bar{X}$ is a random variable representing any calculated average from a certain number of rolls of the die. You just don't know what its value is yet because you haven't rolled the die yet.

350. 2.6

$\bar{x}$ represents the sample mean; you find it by adding the numbers and dividing by n (the sample size). Use the following formula:

$$\bar{x} = \frac{\sum_{n}^{i=1} x_i}{n}$$
$$= \frac{3+4+2+3+1}{5}$$
$$= \frac{13}{5}$$
$$= 2.6$$

Here, each x_i represents a value in the data set — x_1 is the first number, x_2 is the second number, and so on, and then x_n is the nth, or last, number.

351. 4.2

$\bar{x}$ represents the sample mean; you find it by adding the numbers and dividing by n (the sample size). Use the following formula:

$$\bar{x} = \frac{\sum_{n}^{i=1} x_i}{n}$$
$$= \frac{3+4+6+3+5}{5}$$
$$= \frac{21}{5}$$
$$= 4.2$$

Here, each x_i represents a value in the data set — x_1 is the first number, x_2 is the second number, and so on, and then x_n is the nth, or last, number.

352. C. $\sigma_{\bar{x}} = \frac{\sigma_X}{\sqrt{n}}$

The formula for the standard error of a sample mean is

$$\sigma_{\bar{x}} = \frac{\sigma_X}{\sqrt{n}}$$

where σ_X is the population standard deviation and n is the sample size.

353. standard deviation; standard error

The standard deviation represents the variability in the entire population, or the variability of X, while the standard error represents the variability of the sample means, or the variability of $\bar{X}$.

354. B. be approximately the same; be smaller

You don't expect the sample mean to change with the sample size. However, a larger sample size is expected to result in a smaller standard error because the formula for standard error includes dividing by the sample size:

$$\sigma_{\bar{x}} = \frac{\sigma_X}{\sqrt{n}}$$

where σ_X is the population standard deviation and n is the sample size.

Dividing the same population standard deviation by the square root of a larger n results in a smaller standard error. Larger samples have a smaller standard error because their mean changes less from sample to sample.

355. 3.68

To calculate the standard error, use the following formula:

$$\sigma_X = \frac{\sigma_X}{\sqrt{n}}$$

where σ_X is the population standard deviation and n is the sample size.

Substitute the known values into the formula and solve:

$$\sigma_{\bar{x}} = \frac{26}{\sqrt{50}}$$
$$= \frac{26}{7.071}$$
$$= 3.677$$

This rounds to 3.68.

356. 3.36

To calculate the standard error, use the following formula:

$$\sigma_X = \frac{\sigma_X}{\sqrt{n}}$$

where σ_X is the population standard deviation and n is the sample size.

Substitute the known values into the formula and solve:

$$\sigma_{\bar{x}} = \frac{26}{\sqrt{60}}$$
$$= \frac{26}{7.746}$$
$$= 3.356$$

This rounds to 3.36.

357. 4.75

To calculate the standard error, use the following formula:

$$\sigma_{\bar{x}} = \frac{\sigma_X}{\sqrt{n}}$$

where σ_X is the population standard deviation and n is the sample size.

Substitute the known values into the formula and solve:

$$\sigma_{\bar{x}} = \frac{26}{\sqrt{30}}$$

$$= \frac{26}{5.477}$$

$$= 4.747$$

This rounds to 4.75.

358. $\bar{x}$

A small x with a bar over it indicates the average of a set of individual scores.

359. μ_X

This notation represents the population parameter for the mean of the random variable X.

360. $\sigma_{\bar{x}}$

The standard error of the mean is also known as the standard deviation (σ) of the sampling distribution of the sample mean ($\bar{X}$), hence the subscript.

361. σ_X

Population parameters are represented by Greek letters — in this case, by the lowercase letter sigma (σ) with X as a subscript to indicate that it's the standard deviation of individual scores.

362. σ_X

In this case, you're taking repeated samples of size $n = 5$, computing a sample average each time, and then computing the dispersion around the average of all sample averages. Such a procedure is conceptually a way of getting the standard error of the mean, which is denoted as σ_X.

363. $\sigma_{\bar{x}}$

The fish boat owner is looking at the standard deviation of the average weights for their catches over time. The owner is looking at the standard deviation of the sample means (thinking of each catch as a sample). Statistically speaking, this term is the standard error of the sample mean and is denoted by $\sigma_{\bar{x}}$.

364. x_i

In this case, you're considering the sale price of an individual house, so you're dealing with an individual data point. The subscript indicates which data point (or home in this case) you're referring to in the data set.

365.

D. a smaller sample size

The formula for calculating standard error is

$$\sigma_{\bar{x}} = \frac{\sigma_X}{\sqrt{n}}$$

where σ_X is the population standard deviation and n is the sample size.

As you can see, the population mean has no effect on the standard error. A smaller population standard deviation will produce a smaller standard error because the population standard deviation is the numerator of the standard error formula. Because sample size is the denominator of the formula, a smaller sample size will produce a larger standard error, while a larger sample size will produce a smaller standard error. Larger samples have a smaller standard error because their mean changes less from sample to sample.

366.

It would lower the standard error of the sample mean.

Use the formula for calculating the standard error of the sample mean:

$$\sigma_{\bar{x}} = \frac{\sigma_X}{\sqrt{n}}$$

where σ_X is the population standard deviation and n is the sample size.

Dividing the same population standard deviation by the square root of a larger n results in a smaller standard error. In other words, increasing the sample sizes reduces the amount of change (standard error) in the sample means.

367.

Population B has a smaller standard error because of the smaller population standard deviation.

Use the formula for calculating the standard error of the mean:

$$\sigma_{\bar{x}} = \frac{\sigma_X}{\sqrt{n}}$$

where σ_X is the population standard deviation and n is the sample size.

Dividing a smaller population standard deviation by the square root of the same n results in a smaller standard error. Samples drawn from a population that is less variable, as shown by a smaller standard deviation, are more likely to have means closer to the sample mean and hence a smaller standard error.

368.

Quadruple the sample size.

Use the formula for calculating the standard error of the mean:

$$\sigma_{\bar{x}} = \frac{\sigma_X}{\sqrt{n}}$$

where σ_X is the population standard deviation and n is the sample size.

To double $\sigma_{\bar{x}}$, you have to divide σ_X by half as much. Because the divisor is the square root of n, you must quadruple the sample size to get a divisor twice as large.

369. **6.6667**

Use the formula for calculating the standard error of the mean:

$$\sigma_{\bar{x}} = \frac{\sigma_X}{\sqrt{n}}$$

where σ_X is the population standard deviation and n is the sample size.

Substitute the known values into the formula and solve:

$$\sigma_{\bar{x}} = \frac{20}{\sqrt{9}}$$
$$= \frac{20}{3}$$
$$= 6.\overline{6}$$

This rounds to 6.6667

370. **5**

Use the formula for calculating the standard error of the mean:

$$\sigma_{\bar{x}} = \frac{\sigma_X}{\sqrt{n}}$$

where σ_X is the population standard deviation and n is the sample size.

Substitute the known values into the formula and solve:

$$\sigma_{\bar{x}} = \frac{20}{\sqrt{16}}$$
$$= \frac{20}{4}$$
$$= 5$$

371. **E. Choices (B) and (D) (a smaller sample size; a larger population standard deviation)**

Given the formula for the standard error of the mean

$$\sigma_{\bar{x}} = \frac{\sigma_X}{\sqrt{n}}$$

where σ_X is the population standard deviation and n is the sample size, increasing the numerator or decreasing the denominator will both result in a larger standard error. A more variable population will result in more variable sample means, and a smaller sample size will also result in more variable sample means, in both cases resulting in a larger sample error.

372. **25**

The formula for standard error can be rearranged to find the population standard deviation, given sample size and standard error. Multiply both sides by the square root of n, substitute the values given, and solve:

$$\sigma_{\bar{x}} = \frac{\sigma_x}{\sqrt{n}}$$
$$\sigma_{\bar{x}}(\sqrt{n}) = \sigma_x$$
$$5(\sqrt{25}) = \sigma_x$$
$$5(5) = 25$$

373. **centimeter**

Units for standard error are the same as for the original measurements.

374. **A. 0.4856**

A smaller standard error will give you a more precise estimate of the mean because the sample means will cluster more closely around the population mean.

375. **No specific requirement for sample size is needed.**

Because you know that the individual scores come from a normal distribution, the distribution of the sample means will also have a normal distribution, regardless of the sample size.

376. **D. Individual scores x_i are normally distributed.**

If the individual scores are normally distributed, then the sampling distribution of the sample means is normal. The magic of the central limit theorem is that as samples become sufficiently large (30 or more), the sampling distribution of the sample means becomes approximately normal.

377. **It is exactly normal.**

Because the individual data points are normally distributed, the sampling distribution of sample means is also normal, no matter what the size of each sample is. (You don't need the central limit theorem and the $n \geq 30$ requirement if you start with a normal distribution.)

378. **C. right-skewed, 60**

If the population's distribution is normal, the sampling distribution of the sample means is also normal, so the central limit theorem is required only for non-normal (that is, right-skewed) populations.

379. **normally distributed**

When data points are drawn from a normal population of data points, the sampling distribution of the sample mean is normal.

380. **It is exactly normal for any sample size.**

When a sample is drawn from a normal population, the sampling distribution of the sample means is normal.

381. **the shape of a normal distribution**

Repeatedly sampling from a population of scores and then forming a histogram of the means of the samples creates a sampling distribution. When individual scores are normally distributed, the sampling distribution of the sample means is also normal, which makes a bell shape.

382. **It would be expected to be normally distributed.**

Although the sample size is small (four), the distribution of sample means from a population with an underlying normal distribution is also expected to be normal.

383. **It is exactly normal.**

When all samples of a fixed size, even small ones, are drawn from a normally distributed population, the sampling distribution of sample means is normal.

384. **a precise normal distribution**

When samples are drawn from a normally distributed population, the sampling distribution of sample means is normal.

385. **normal**

The sampling distribution of the sample means is expected to be normal, although the underlying distribution isn't normal, because the sample size is sufficiently large (35) that the central limit theorem applies.

386. **E. All of the above (Population A, Population B, Population C, Population D)**

According to the central limit theorem, with reasonably large samples ($n \geq 30$), the sampling distribution of sample means is expected to be normally distributed, no matter what shape the underlying distribution of individual observations has (most situations work well if the sample size is at least 30).

387. **Population D**

A sample size of 20 is too small to expect that the sampling distribution of the sample means will be approximately normal, unless the population has a normal distribution.

388. **normal for all four populations**

Because the sample size is fairly large (40), the central limit theorem tells you that the sampling distribution of sample means is expected to be normal as well, no matter what kind of distribution the underlying individual observations have, as long as the sample size is at least 30.

389. **not able to determine from the information given**

Here, you're not told how large each sample is; you're told only how large each *population* is. The central limit theorem says that sampling distributions of sample means are normal if the sample sizes are sufficiently large ($n \geq 30$) or if the underlying distribution of observations is normal.

In this case, you may be talking about small samples (of less than 30 observations), and three of the four distributions involved are clearly not normal. Therefore, you can't make any general statement about the resulting shape of the sampling distributions of sample means.

390. $n \geq 30$

When data isn't drawn from a normal distribution, the sampling distribution of sample means becomes normal only when sufficiently large samples ($n \geq 30$) are used.

391. **a precise normal distribution**

In this case, you know that the observations themselves are normally distributed, so samples of any size will give rise to a normal sampling distribution of sample means (even samples of size 3).

392. **an approximate normal distribution**

Because the sample sizes are large ($n = 100$, which is much larger than the approximately 30 cases required by the central limit theorem), you know that the sampling distribution of sample means should be approximately normal.

393. **No, because the sample sizes are too small to use the central limit theorem.**

In this case, the original population distribution is unknown, so you can't assume that you have a normal distribution. The central limit theorem can't be invoked because the sample sizes are too small (less than 30).

394. $n = 30$

According to the central limit theorem, if you repeatedly take sufficiently large samples, the distribution of the means from those samples will be approximately normal. For most non-normal populations, you can choose sample sizes of at least 30 from the distribution, which usually leads to a normal sampling distribution of sample means no matter what the underlying shape of the distribution of scores is. In fact, if the underlying distribution of values approximates a normal distribution, it may be possible to achieve a normal sampling distribution of sample means with smaller samples.

For populations with several peaks, wild variation, and/or extreme outliers, you may need larger sample sizes.

395. **The researcher has not violated the condition because the sampling distribution of sample means is approximately normal whenever the sample size is at least 30.**

The central limit theorem says that the sampling distribution of sample means is approximately normal if the sample size is at least 30. The central limit theorem is used to ensure that studies meet the assumptions underlying other tests, which rely on a normal sampling distribution of sampling means. It isn't necessary to draw repeated samples from a distribution to invoke the central limit theorem. It's only necessary to have either an underlying normal distribution of observations or a sample size of $n \geq 30$ in most cases, for the condition to be met.

396. **The sampling distribution of sample means is normal.**

The sampling distribution of sample means is normal whenever the observations come from a normally distributed population of scores, which is true in this case.

397. **C. The central limit theorem can be used because the sample size is large enough and the population distribution is unknown.**

Because the distribution of the population isn't discussed, you can't assume that it's normal. You have to appeal to the central limit theorem. In this case, the condition of the central limit theorem is met: The sample size is well over 30 ($n = 150$), so you can use it to find probabilities about the sample mean.

398. **No, the central limit theorem can't be used to infer a normal sampling distribution of sample means drawn from a non-normal population because there are too few observations in each sample.**

To use the central limit theorem for samples drawn from a population that isn't normally distributed, the sample size must be relatively large ($n \geq 30$). Because $n = 10$ for each sample, the central limit theorem can't be used.

399. **The central limit theorem can be used to infer a normal sampling distribution of sample means.**

Although the observations have a skewed (and hence non-normal) distribution, the central limit theorem does allow you to conclude that the sampling distribution of sample means is normal, because the size of each sample is sufficiently large ($n = 150$ here, which is well more than the 30 or more suggested by the central limit theorem). Note that the number of samples taken isn't relevant here.

400. **Yes, the central limit theorem can be used to infer a normal sampling distribution of sample means.**

This case just barely qualifies to invoke the central limit theorem and infer that the sampling distribution of sample means is normal. The population of observations isn't normal, but the sample size is just large enough to use the central limit theorem. (The larger the sample size, the better the approximation in all situations.)

401. 9.6

Use the formula for finding the mean:

$$\bar{x} = \frac{\sum x_i}{n}$$

where $\sum x_i$ is the sum of the original observations, and n is the sample size.

Substitute the known values into the formula and solve:

$$\bar{x} = \frac{7+6+7+14+9+9+11+11+11+11}{10}$$
$$= \frac{96}{10}$$
$$= 9.6$$

402. D. Larger samples tend to yield more precise estimates of the population mean.

The sample mean is the best unbiased estimate of the population mean, and a larger sample will generally produce a more precise estimate because larger samples change less from sample to sample.

403. E. Choices (B) and (C) (The sample mean is the best estimate of the population mean; larger samples yield more precise estimates of the population mean.)

By definition, the sample mean is the best estimator of the population mean. Larger samples yield more precise estimates of the sample mean because they vary less from one sample to another than do smaller samples.

404. 67.44%

Because the observations are drawn from a normally distributed population, the sample means are also normally distributed.

First, find the z-score equivalent of this sample mean by using the population mean and the standard error of the mean for samples of this size.

$$z = \frac{\bar{x} - \mu_X}{\sigma_X / \sqrt{n}}$$

Here, $\bar{x}$ is the sample mean, μ_X is the population mean, σ_X is the population standard deviation, and n is the sample size.

Now, substitute the known values into the formula and solve:

$$z = \frac{9.6 - 10}{3 / \sqrt{10}}$$
$$= \frac{-0.4}{3 / 3.1623}$$
$$= \frac{-0.4}{0.9487}$$
$$= -0.4216$$

Because the probabilities equate to tail areas in a normal distribution, you start by finding the area under the curve associated with that z-score by using a Z-table (Table

A–1 in the appendix). So, round the z-score to –0.42, and use Table A–1 to find the associated probability of 0.3372.

Now, figure out how much area is in the tails. In the Z-table, areas reflect the proportion of the normal curve in one tail. Because you want to find the probability of a score this far from the mean *in either direction*, you need to double the area you just found to account for the area in the other tail:

$$
\begin{aligned}
\text{Total tail area} &= 2(\text{Single} - \text{tail area}) \\
&= 2(0.3372) \\
&= 0.6744, \text{ or } 67.44\%
\end{aligned}
$$

Finally, report that probability (the proportion of curve area contained by the two tails) as the probability of finding a mean score this extreme or more so.

405. 2.5%

Because the observations are drawn from a normally distributed population, the sample means are also normally distributed.

First, find the z-score equivalent of this sample mean by using the population mean and the standard error of the mean for samples of this size.

$$
z = \frac{\bar{x} - \mu_X}{\sigma_X / \sqrt{n}}
$$

Here, $\bar{x}$ is the sample mean, μ_X is the population mean, σ_X is the population standard deviation, and n is the sample size.

Now, substitute the known values into the formula and solve:

$$
\begin{aligned}
z &= \frac{9.7 - 10}{3 / \sqrt{500}} \\
&= \frac{-0.3}{3 / 22.3607} \\
&= \frac{-0.3}{0.1342} \\
&= -2.2355
\end{aligned}
$$

Because probabilities equate to tail areas in a normal distribution, you start by finding the area under the curve associated with that z-score by using a Z-table (Table A–1 in the appendix). So, round the z-score to –2.24, and use Table A–1 to find the associated probability of 0.0125

Now, figure out how much area is in the tails. In the Z-table, areas reflect the proportion of the normal curve in one tail. Because you want to find the probability of a score this far from the mean *in either direction*, you need to double the area you just found to account for the area in the other tail:

$$
\begin{aligned}
\text{Total tail area} &= 2(\text{Single-tail area}) \\
&= 2(0.0125) \\
&= 0.025, \text{ or } 2.5\%
\end{aligned}
$$

Finally, report that probability (the proportion of curve area contained by the two tails) as the probability of finding a mean score this extreme or more so.

406. 4.88%

Because the sample size is greater than 30, you can use the central limit theorem to solve this problem. Assume a score 10 units below the mean, to simplify calculations.

First, find the z-score equivalent of this sample mean by using the population mean and the standard error of the mean for samples of this size.

$$z = \frac{\bar{x} - \mu_X}{\sigma_X / \sqrt{n}}$$

Here, $\bar{x}$ is the sample mean, μ_X is the population mean, σ_X is the population standard deviation, and n is the sample size.

Now, substitute the known values into the formula and solve:

$$z = \frac{90 - 100}{30 / \sqrt{35}}$$
$$= \frac{-10}{30 / 5.916}$$
$$= \frac{-10}{5.071}$$
$$= -1.972$$

Because probabilities equate to tail areas in a normal distribution, you start by finding the area under the curve associated with that z-score by using a Z-table (Table A-1 in the appendix). So, round the z-score to –1.97, and use Table A-1 to find the associated probability of 0.0244.

Now, figure out how much area is in the tails. In the Z-table, areas reflect the proportion of the normal curve in one tail. Because you want to find the probability of a score this far from the mean *in either direction,* you need to double the area you just found to account for the area in the other tail:

$$\text{Total tail area} = 2(\text{Single-tail area})$$
$$= 2(0.0244)$$
$$= 0.0488, \text{or } 4.88\%$$

Finally, report that probability (the proportion of curve area contained by the two tails) as the probability of finding a mean score this extreme or more so.

407. 4.56%

Because the observations are drawn from a normally distributed population, the sample means are also normally distributed. Assume a score 10 units below the mean, to simplify calculations.

First, find the z-score equivalent of this sample mean by using the population mean and the standard error of the mean for samples of this size.

$$z = \frac{\bar{x} - \mu_X}{\sigma_X / \sqrt{n}}$$

Here, $\bar{x}$ is the sample mean, μ_X is the population mean, σ_X is the population standard deviation, and n is the sample size.

Now, substitute the known values into the formula and solve:

$$z = \frac{90 - 100}{40 / \sqrt{64}}$$
$$= \frac{-10}{40 / 8}$$
$$= \frac{-10}{5}$$
$$= -2.0$$

Because probabilities equate to tail areas in a normal distribution, you start by finding the area under the curve associated with that z-score by using a Z-table (Table A-1 in the appendix). Using Table A-1, you find the associated probability of 0.0228.

Now, figure out how much area is in the tails. In the Z-table, areas reflect the proportion of the normal curve in one tail. Because you want to find the probability of a score this far from the mean *in either direction*, you need to double the area you just found to account for the area in the other tail:

$$\text{Total tail area} = 2(\text{Single-tail area})$$
$$= 2(0.0228)$$
$$= 0.0456, \text{or } 4.56\%$$

Finally, report that probability (the proportion of curve area contained by the two tails) as the probability of finding a mean score this extreme or more so.

408.   less than 0.02%

Because the observations are drawn from a normally distributed population, the sample means are also normally distributed. Assume a score 10 units below the mean, to simplify calculations.

First, find the z-score equivalent of this sample mean by using the population mean and the standard error of the mean for samples of this size.

$$z = \frac{\bar{x} - \mu_X}{\sigma_X / \sqrt{n}}$$

Here, $\bar{x}$ is the sample mean, μ_X is the population mean, σ_X is the population standard deviation, and n is the sample size.

Now, substitute the known values into the formula and solve:

$$z = \frac{40 - 50}{16 / \sqrt{64}}$$
$$= \frac{-10}{16 / 8}$$
$$= \frac{-10}{2}$$
$$= -5$$

Because probabilities equate to tail areas in a normal distribution, you start by finding the area under the curve associated with that z-score by using a Z-table (Table A-1 in the appendix).

Use Table A-1 to find the associated probability; because this value is more extreme than any value included in the table, the probability is less than the smallest table value of 0.0001.

Now, figure out how much area is in the tails. In the Z-table, areas reflect the proportion of the normal curve in one tail. Because you want to find the probability of a score this far from the mean *in either direction,* you need to double the area you just found to account for the area in the other tail:

$$\text{Total tail area} = 2(\text{Single-tail area})$$
$$= 2(0.0001)$$
$$= 0.0002, \text{ or } 0.02\%$$

Finally, report that probability (the proportion of curve area contained by the two tails) as the probability of finding a mean score this extreme or more so.

409. **1.24%**

Because the observations are drawn from a normally distributed population, the sample means are also normally distributed. Assume a score 10 units below the mean, to simplify calculations.

First, find the z-score equivalent of this sample mean by using the population mean and the standard error of the mean for samples of this size.

$$z = \frac{\bar{x} - \mu_X}{\sigma_X / \sqrt{n}}$$

Here, $\bar{x}$ is the sample mean, μ_X is the population mean, σ_X is the population standard deviation, and n is the sample size.

Now, substitute the known values into the formula and solve:

$$z = \frac{40 - 50}{16 / \sqrt{16}}$$
$$= \frac{-10}{16 / 4}$$
$$= \frac{-10}{4}$$
$$= -2.5$$

Because probabilities equate to tail areas in a normal distribution, you start by finding the area under the curve associated with that z-score by using a Z-table (Table A-1 in the appendix). Using Table A-1, you find the associated probability of 0.0062.

Now, figure out how much area is in the tails. In the Z-table, areas reflect the proportion of the normal curve in one tail. Because you want to find the probability of a score this far from the mean *in either direction,* you need to double the area you just found to account for the area in the other tail:

$$\text{Total tail area} = 2(\text{Single-tail area})$$
$$= 2(0.0062)$$
$$= 0.0124, \text{ or } 1.24\%$$

Finally, report that probability (the proportion of curve area contained by the two tails) as the probability of finding a mean score this extreme or more so.

410.   **52.86%**

Because the observations are drawn from a normally distributed population, the sample means are also normally distributed. Assume a score 10 units below the mean, to simplify calculations.

First, find the z-score equivalent of this sample mean by using the population mean and the standard error of the mean for samples of this size.

$$z = \frac{\bar{x} - \mu_X}{\sigma_X / \sqrt{n}}$$

Here, $\bar{x}$ is the sample mean, μ_X is the population mean, σ_X is the population standard deviation, and n is the sample size.

Now, substitute the known values into the formula and solve:

$$z = \frac{-10 - 0}{160 / \sqrt{100}}$$
$$= \frac{-10}{160 / 10}$$
$$= \frac{-10}{16}$$
$$= -0.625, \text{ rounded to } -0.63$$

Because probabilities equate to tail areas in a normal distribution, you start by finding the area under the curve associated with that z-score by using a Z-table (Table A-1 in the appendix). Using Table A-1, you find the associated probability of 0.2643.

Now, figure out how much area is in the tails. In the Z-table, areas reflect the proportion of the normal curve in one tail. Because you want to find the probability of a score this far from the mean *in either direction*, you need to double the area you just found to account for the area in the other tail:

$$\text{Total tail area} = 2(\text{Single-tail area})$$
$$= 2(0.2643)$$
$$= 0.5286, \text{ or } 52.86\%$$

Finally, report that probability (the proportion of curve area contained by the two tails) as the probability of finding a mean score this extreme or more so.

411.   **4.65%**

Before going through the steps for solving this problem, determine whether you can use the central limit theorem. Although the lab technician doesn't know whether the population of white blood cell counts is normally distributed, by taking a sample of 40 independent measurements, the technician sets up a case where the central limit theorem can be used.

First, find the z-score equivalent of this sample mean by using the population mean and the standard error of the mean for samples of this size.

$$z = \frac{\bar{x} - \mu_X}{\sigma_X / \sqrt{n}}$$

Here, $\bar{x}$ is the sample mean, μ_x is the population mean, σ_x is the population standard deviation, and n is the sample size.

Now, substitute the known values into the formula and solve:

$$z = \frac{7{,}616 - 7{,}250}{1{,}375 / \sqrt{40}}$$
$$= \frac{366}{1{,}375 / 6.32456}$$
$$= \frac{366}{217.4064}$$
$$= 1.6835$$

Because probabilities equate to tail areas in a normal distribution, you start by finding the tail area associated with this z-score (rounding to 1.68), using a Z-table (Table A-1 in the appendix).

The table gives tail areas only for negative z-scores, but the normal distribution is symmetrical, so you can look up the left-tail area equivalent to $z = -1.68$, which turns out to be 0.0465.

Now, figure out the total tail area. The problem asks for the probability of getting a sample mean of 7,616 or larger. Because the problem specifically refers to only one tail of the distribution, you should *not* double the probability. There's only one tail specified by the question (scores of 7,616 *or larger*), so you can use 0.0465, or 4.65%, as your answer.

---

412.   49.08%

The distribution of the weights of the cookies is unknown, and you can't assume that it's normal. To answer this question, look to the central limit theorem. The central limit theorem can be applied to this problem because the sample size of 36 is large enough ($n \geq 30$).

First, find the z-score equivalent of this sample mean by using the population mean and the standard error of the mean for samples of this size.

$$z = \frac{\bar{x} - \mu_x}{\sigma_x / \sqrt{n}}$$

Here, $\bar{x}$ is the sample mean, μ_x is the population mean, σ_x is the population standard deviation, and n is the sample size.

Now, substitute the known values into the formula and solve:

$$z = \frac{12.011 - 12}{0.1 / \sqrt{36}}$$
$$= \frac{0.011}{0.1 / 6}$$
$$= \frac{0.011}{0.01667}$$
$$= 0.6599$$

Because probabilities equate to tail areas in a normal distribution, you start by finding the tail area associated with this z-score (rounding to 0.66), using a Z-table (Table A-1 in the appendix).

The table gives tail areas only for negative z-scores, but the normal distribution is symmetrical, so you can look up the left-tail area equivalent to $z = -0.66$, which turns out to be 0.2546.

The question asks for the probability of a value at least this close to the mean, so you want the probability *excluding* the tail area. The total area under the curve is 1, so you double the tail area and subtract from 1:

$$1 - (2)(0.2546)$$
$$= 1 - 0.5092$$
$$= 0.4908, \text{ or } 49.08\%$$

413. **22.06%**

The distribution of the weights of the cookies is unknown, and you can't assume that it's normal. To answer this question, look to the central limit theorem. The central limit theorem can be applied to this problem because the sample size of 49 is large enough ($n \geq 30$).

First, find the z-score equivalent of this sample mean by using the population mean and the standard error of the mean for samples of this size.

$$z = \frac{\bar{x} - \mu_X}{\sigma_X / \sqrt{n}}$$

Here, $\bar{x}$ is the sample mean, μ_X is the population mean, σ_X is the population standard deviation, and n is the sample size.

Now, substitute the known values into the formula and solve:

$$z = \frac{12.004 - 12}{0.1 / \sqrt{49}}$$
$$= \frac{0.004}{0.1 / 7}$$
$$= \frac{0.004}{0.01429}$$
$$= 0.2799$$

Because probabilities equate to tail areas in a normal distribution, you start by finding the tail area associated with this z-score (rounding to 0.28), using a Z-table (Table A-1 in the appendix).

The table gives tail areas only for negative z-scores, but the normal distribution is symmetrical, so you can look up the left-tail area equivalent to $z = -0.28$, which turns out to be 0.3897.

The question asks for the probability of a value at least this close to the mean, so you want the probability *excluding* the tail area.

The total area under the curve is 1, so double the tail area and subtract from 1:

$$1 - (2)(0.3897)$$
$$= 1 - 0.7794$$
$$= 0.2206, \text{ or } 22.06\%$$

414. 76.98%

The distribution of the weights of the cookies is unknown, and you can't assume that it's normal. To answer this question, look to the central limit theorem. The central limit theorem can be applied to this problem because the sample size of 36 is large enough ($n \geq 30$).

First, find the z-score equivalent of this sample mean by using the population mean and the standard error of the mean for samples of this size.

$$z = \frac{\bar{x} - \mu_x}{\sigma_x / \sqrt{n}}$$

Here, $\bar{x}$ is the sample mean, μ_x is the population mean, σ_x is the population standard deviation, and n is the sample size.

Now, substitute the known values into the formula and solve:

$$z = \frac{12.02 - 12}{0.1 / \sqrt{36}}$$
$$= \frac{0.02}{0.1 / 6}$$
$$= \frac{0.02}{0.01667}$$
$$= 1.1998$$

Because probabilities equate to tail areas in a normal distribution, you start by finding the tail area associated with this z-score (rounding to 1.20), using a Z-table (Table A-1 in the appendix).

The table gives tail areas only for negative z-scores, but the normal distribution is symmetrical, so you can look up the left-tail area equivalent to $z = -1.20$, which turns out to be 0.1151.

The question asks for the probability of a value at least this close to the mean, so you want the probability *excluding* the tail area. The total area under the curve is 1, so double the tail area and subtract from 1:

$$1 - (2)(0.1151)$$
$$= 1 - 0.2302$$
$$= 0.7698, \text{ or } 76.98\%$$

415. 83.84%

The distribution of the weights of the cookies is unknown, and you can't assume that it's normal. To answer this question, look to the central limit theorem. The central limit theorem can be applied to this problem because the sample size of 49 is large enough ($n \geq 30$).

First, find the z-score equivalent of this sample mean by using the population mean and the standard error of the mean for samples of this size.

$$z = \frac{\bar{x} - \mu_x}{\sigma_x / \sqrt{n}}$$

Here, $\bar{x}$ is the sample mean, μ_x is the population mean, σ_x is the population standard deviation, and n is the sample size.

Now, substitute the known values into the formula and solve:

$$z = \frac{12.02 - 12}{0.1 / \sqrt{49}}$$

$$= \frac{0.02}{0.1 / 7}$$

$$= \frac{0.02}{0.014286}$$

$$= 1.39997$$

Because probability is associated with tail area when working with the normal distribution, you find the tail area associated with this z-score (rounding to 1.40), using a Z-table (Table A-1 in the appendix).

The table gives tail areas only for negative z-scores, but the normal distribution is symmetrical, so you can look up the left-tail area equivalent to $z = -1.40$, which turns out to be 0.0808.

The question asks for the probability of a value at least this close to the mean, so you want the probability *excluding* the tail area. The total area under the curve is 1, so double the tail area and subtract from 1:

$$1 - (2)(0.0808)$$

$$= 1 - 0.1616$$

$$= 0.8384, \text{ or } 83.84\%$$

416.   0.69

$\hat{p}$ (pronounced "p-hat") is the observed proportion of female zombies in the sample. Because 20 zombies were identified as female and 9 were identified as male, the proportion is

$$\hat{p} = \frac{\text{Number of female zombies in the sample}}{\text{Total number of zombies in the sample}}$$

$$= \frac{20}{20 + 9}$$

$$= \frac{20}{29}$$

$$= 0.689$$

417.   0.50

The population proportion, p, can be an actual proportion observed in a population or a theoretical proportion that should happen under some set of assumptions, such as the assumption that exactly half (or 0.50) of zombies would be female in the absence of bias.

418.   0.0928

$\sigma_{\hat{p}}$ stands for the standard error of a sample proportion. You calculate $\sigma_{\hat{p}}$ with the following formula:

$$\sigma_{\hat{p}} = \sqrt{\frac{p(1-p)}{n}}$$

where p is the population proportion and n is the sample size.

Now, substitute the known values into the formula and solve:

$$\sigma_p = \sqrt{\frac{0.5(1-0.5)}{29}}$$

$$= \sqrt{\frac{0.25}{29}}$$

$$= 0.092848$$

This rounds to 0.0928.

419. **Yes**

The number of female zombies, X, has a binomial distribution with $p = 0.5$ and $n = 29$. You can apply the central limit theorem as long as both np and $n(1-p)$ are greater than 10. (**Note:** You don't use the condition that n is at least 30 like you do for other central limit theorem problems. The binomial is very common and has its own special conditions to check.)

In this case, $np = (29)(0.5) = 14.5$ and $n(1-p) = 29(1-0.5) = 14.5$, so the sample size is large enough to use the central limit theorem.

420. **0.03**

Use the formula for calculating the standard error of a sample proportion

$$\sigma_{\hat{p}} = \sqrt{\frac{p(1-p)}{n}}$$

where p is the population proportion and n is the sample size.

In this case, $p = 0.9$ and $n = 100$, so you get

$$\sigma_p = \sqrt{\frac{0.9(1-0.9)}{100}}$$

$$= \sqrt{\frac{0.09}{100}}$$

$$= 0.03$$

421. **0.03**

$\sigma_{\hat{p}}$ is the standard error of a sample proportion, which you can find by using this formula:

$$\sigma_p = \sqrt{\frac{p(1-p)}{n}}$$

where p is the population proportion and n is the sample size.

In this case, $p = 0.1$ and $n = 100$, so you get

$$\sigma_{\hat{p}} = \sqrt{\frac{0.1(1-0.1)}{100}}$$

$$= \sqrt{\frac{0.09}{100}}$$

$$= 0.03$$

422. 0.05

$\sigma_{\hat{p}}$ is the standard error of a sample proportion, which you can find by using this formula:

$$\sigma_p = \sqrt{\frac{p(1-p)}{n}}$$

where p is the population proportion and n is the sample size.

In this case, $p = 0.5$ and $n = 100$, so you get

$$\sigma_{\hat{p}} = \sqrt{\frac{0.5(1-0.5)}{100}}$$
$$= \sqrt{\frac{0.25}{100}}$$
$$= 0.05$$

423. 0.0607

$\sigma_{\hat{p}}$ is the standard error of the sample proportion, which you can find by using this formula:

$$\sigma_p = \sqrt{\frac{p(1-p)}{n}}$$

where p is the population proportion and n is the sample size.

In this case, $p = 0.67$ and $n = 60$, so you get

$$\sigma_p = \sqrt{\frac{0.67(1-0.67)}{n}}$$
$$= \sqrt{\frac{0.2211}{60}}$$
$$= 0.0607$$

424. 100

To use the central limit theorem for proportions, both np and $n(1-p)$ must be 10 or greater, where n is the sample size and p is the population proportion. (*Note:* You don't use the condition that n is at least 30 like you do for other central limit theorem problems. The binomial is very common and has its own special conditions to check.)

In this case, $p < (1-p)$, so you need to ensure that np is at least 10, which you can write as $np \geq 10$ and rearrange as

$$n \geq \frac{10}{p}$$

(Because $p > 0$, you don't need to reverse the inequality here.) Now substitute 0.1 for p to get

$$n \geq \frac{10}{0.1}$$
$$n \geq 100$$

The sample size must be at least 100.

425. 20

To use the central limit theorem for proportions, both np and $n(1-p)$ must be 10 or greater, where n is the sample size and p is the population proportion. (*Note:* You don't use the condition that n is at least 30 like you do for other central limit theorem problems. The binomial is very common and has its own special conditions to check.)

Because p and $(1-p)$ are the same (0.5), solving for either np or $n(1-p)$ will work.

So, you can rearrange the statement $np \geq 10$ as

$$n \geq \frac{10}{p}$$

Now substitute 0.5 for p to get

$$n \geq \frac{10}{0.5}$$
$$n \geq 20$$

The sample size must be at least 20.

426. −1.58

Use the formula for a z-score for proportions:

$$z = \frac{\hat{p} - p}{\sigma_{\hat{p}}}$$

where $\hat{p}$ is the sample proportion, p is the population proportion, and $\sigma_{\hat{p}}$ is the standard error of the sample proportion.

Now, substitute the known values into the formula and solve:

$$z = \frac{0.25 - 0.5}{0.1581}$$
$$= \frac{-0.25}{0.1581}$$
$$= -1.5813$$

427. −2.5

Use the formula for a z-score for proportions:

$$z = \frac{\hat{p} - p}{\sigma_{\hat{p}}}$$

where $\hat{p}$ is the sample proportion, p is the population proportion, and $\sigma_{\hat{p}}$ is the standard error of the sample proportion.

Now, substitute the known values into the formula and solve:

$$z = \frac{0.25 - 0.5}{0.1}$$
$$= \frac{-0.25}{0.1}$$
$$= -2.5$$

428. 0

You can find the answer to this problem quickly, without doing the math. If you notice that the observed proportion (0.25) is the same as the population proportion (0.25), you know that the z-score is 0. You can see why by looking at the equation for a z-score:

$$z = \frac{\hat{p} - p}{\sigma_{\hat{p}}}$$

Here, $\hat{p}$ is the sample proportion, p is the population proportion, and $\sigma_{\hat{p}}$ is the standard error of the sample proportion. If the numerator is 0 ($0.25 - 0.25 = 0$), then z is also 0.

429. cannot use the central limit theorem here

For the central limit theorem to work here, you need to verify that np and $n(1-p)$ are both 10 or greater. In this case, $np = (10)(0.5) = 5$, so you can't use the central limit theorem.

Note: You can compute the exact probability by using the binomial probability formula or table.

430. 0.62%

For the central limit theorem to work here, you need to verify that np and $n(1-p)$ are both 10 or greater. In this case, $np = (25)(0.5) = 12.5$ and $n(1-p) = 25(1-0.5) = 12.5$, so you can use the central limit theorem.

First, use the formula for finding the standard error:

$$\sigma_p = \sqrt{\frac{p(1-p)}{n}}$$

where p is the population proportion and n is the sample size.

$$\sigma_p = \sqrt{\frac{0.5(1-0.5)}{25}}$$
$$= \sqrt{\frac{0.25}{25}}$$
$$= \sqrt{0.01}$$
$$= 0.1$$

Then convert the information you have to a z-score, using the following formula:

$$z = \frac{\hat{p} - p}{\sigma_{\hat{p}}}$$

where $\hat{p}$ is the sample proportion, p is the population proportion, and $\sigma_{\hat{p}}$ is the standard error of the sample proportion.

In this case, $\hat{p} = 0.25$, $p = 0.5$, and $\sigma_p = 0.1$, so you get

$$z = \frac{0.25 - 0.5}{0.1}$$
$$= \frac{-0.25}{0.1}$$
$$= -2.5$$

Because probabilities equate to tail areas in a normal distribution, Table A-1 in the appendix shows the probability in the tail below any z-value you look up. From the table, you can see that the area under the curve below a z-score of -2.5 is 0.0062, or 0.62%, which is the approximate probability needed by the central limit theorem.

431.   50%

Before diving into the calculations with this problem, think it through a little. You can see that the sample probability and the population probability are both the same (0.25). So, this question is really just asking you how likely it is that the sample probability is less than the population probability. Because the sample probability is an unbiased estimator of the population probability, it's equally likely to be above or below p, so the probability that it's below p is 50%. You should also confirm that the central limit theorem applies here by computing $np = (40)(0.25) = 10$ and $p(1-p) = 40(1-0.25) = 30$. Both are at least 10, so the central limit theorem is applicable.

432.   15.87%

Because the coin is fair, the probability of heads on each toss is $p = 0.5$. The 36 tosses form a sample of $n = 36$.

Check to see whether you can use the normal approximation to the binomial by making sure that both np and $n(1-p)$ equal at least 10. In this case, $n = 36$ and $p = 0.5$, so $np = (36)(0.5) = 18$ and $n(1-p) = (36)(1-0.5) = (36)(0.5) = 18$.

Now convert 21 heads to proportions by dividing by the sample size:

$$\hat{p} = \frac{21}{36} = 0.5833$$

Next, find the standard error, using this formula:

$$\sigma_{\hat{p}} = \sqrt{\frac{p(1-p)}{n}}$$

where p is the population proportion and n is the sample size.

Substitute the known values into the formula and solve:

$$\sigma_p = \sqrt{\frac{0.5(1-0.5)}{36}}$$
$$= \sqrt{\frac{0.25}{36}}$$
$$= 0.0833$$

Then convert the sample proportion to a z-value with this formula:

$$z = \frac{\hat{p} - p}{\sigma_{\hat{p}}}$$

where $\hat{p}$ is the sample proportion, p is the population proportion, and $\sigma_{\hat{p}}$ is the standard deviation.

$$z = \frac{\hat{p} - p}{\sigma_{\hat{p}}}$$

$$= \frac{0.5833 - 0.5}{0.0833}$$

$$= \frac{0.0833}{0.0833}$$

$$= 1$$

Because tail area is the same as probability when you're dealing with the normal distribution, use Table A-1 in the appendix to find the area to the left of $z = +1.0$ is 0.8413; to find the area to the right, subtract this from 1 (because total probability is always 1):

$$1 - 0.8413 = 0.1587, \text{ or } 15.87\%$$

433. >99.99%

Because the coin is fair, the probability of heads on each toss is $p = 0.5$. The 50 tosses form a sample of $n = 50$.

Check to see whether you can use the normal approximation to the binomial by making sure that both np and $n(1 - p)$ equal at least 10. In this case, $n = 50$ and $p = 0.5$, so $np = (50)(0.5) = 25$ and $n(1 - p) = (50)(1 - 0.5) = (50)(0.5) = 25$.

Now convert ten heads to proportions by dividing by the sample size:

$$\hat{p} = \frac{10}{50} = 0.2$$

Next, find the standard error, using this formula:

$$\sigma_{\hat{p}} = \sqrt{\frac{p(1 - p)}{n}}$$

where p is the population proportion and n is the sample size.

Substitute the known values into the formula and solve:

$$\sigma_{\hat{p}} = \sqrt{\frac{0.5(1 - 0.5)}{50}}$$

$$= \sqrt{\frac{0.25(0.25)}{50}}$$

$$= 0.07071$$

Then convert each sample proportion to a z-value with this formula:

$$z = \frac{\hat{p} - p}{\sigma_{\hat{p}}}$$

where $\hat{p}$ is the sample proportion, p is the population proportion, and $\sigma_{\hat{p}}$ is the standard deviation.

$$z = \frac{\hat{p} - p}{\sigma_{\hat{p}}}$$

$$= \frac{0.2 - 0.5}{0.07071}$$

$$= \frac{-0.3}{0.07071}$$

$$= -4.2427$$

To find the probability of these results, find the probability of the z-score using Table A-1. The smallest value for a z-score in Table A-1 is -3.69; the probability of a z-score lower than this value is 0.0001. The probability of a z-score of -4.2427 is less than this, because -4.2427 is farther from 0 than -3.69. So, you can say that the probability of getting at least ten heads in 50 tosses of a fair coin is greater than the following:

$$1 - 0.0001 = 0.9999 \text{ or } 99.99\%$$

434. 2.28%

Start by finding the observed proportion of interest. The question asks about getting more than 60 heads, so you're interested in sample probabilities greater than $60/100 = 0.6$, or $\hat{p} = 0.6$.

Note that the value of the population proportion (p) used in the following formulas is 0.5, the chance of getting heads on one flip of a fair coin.

Next, find the standard error, using this formula:

$$\sigma_{\hat{p}} = \sqrt{\frac{p(1-p)}{n}}$$

where p is the population proportion and n is the sample size.

Substitute the known values into the formula and solve:

$$\sigma_{\hat{p}} = \sqrt{\frac{0.5(1-0.5)}{100}}$$
$$= \sqrt{\frac{0.25}{100}}$$
$$= 0.05$$

Then convert the sample proportion to a z-value with this formula:

$$z = \frac{\hat{p} - p}{\sigma_{\hat{p}}}$$

where $\hat{p}$ is the sample proportion, p is the population proportion, and $\sigma_{\hat{p}}$ is the standard deviation.

$$z = \frac{0.6 - 0.5}{0.05}$$
$$= 2.0$$

So, you're interested in the probability of getting a z-score greater than 2.0. Table A-1 in the appendix gives the probability of getting a z-score less than 2.0 as 0.9722.

You can subtract this from 1 to get the probability of a score greater than 2.0: $1 - 0.9722 = 0.0228$.

435. 5.59%

First, identify the information you have in symbolic notation, converting from percentages to proportions as necessary:

$p = 0.01$ (the population proportion)

$n = 1,000$ (the sample size)

$\hat{p} = 0.015$ (the sample proportion)

Find out whether you can use the central limit theorem by making sure that both np and $n(1-p)$ are at least 10. Because np is $(1,000)(0.01) = 10$ and $1,000(1-0.01) = 1,000(0.99) = 990$, you can proceed.

To get the appropriate probability, you need to find the standard error and then convert the sample proportion to a z-value. The formula for standard error is

$$\sigma_{\hat{p}} = \sqrt{\frac{p(1-p)}{n}}$$

where p is the population proportion and n is the sample size.

Substitute the known values in the formula and solve:

$$\sigma_{\hat{p}} = \sqrt{\frac{0.01(1-0.01)}{1,000}}$$

$$= \sqrt{\frac{0.0099}{1,000}}$$

$$= 0.003146$$

Then convert the sample proportion to a z-value, using this formula:

$$z = \frac{\hat{p} - p}{\sigma_{\hat{p}}}$$

where $\hat{p}$ is the sample proportion, p is the population proportion, and $\sigma_{\hat{p}}$ is the standard deviation.

$$z = \frac{0.015 - 0.01}{0.003146}$$

$$= \frac{0.005}{0.003146}$$

$$= 1.58931$$

You can find the probability of observing a z-value of 1.589 or less by using Table A-1 in the appendix (rounding z to 1.59): 0.9411.

To answer the question, you need the probability of observing a z-value this high or greater, which is found by subtracting from 1 (the total area under the curve):

$$1 - 0.9441 = 0.0559 = 5.59\%$$

436. **19.02%**

First, find out whether you can use the normal approximation to the binomial by making sure that both np and $n(1-p)$ are at least 10. Here, $np = 36(0.3) = 10.8$, and $n(1-p) = 36(1-0.3) = 36(0.7) = 25.2$, so you can proceed.

To get the appropriate probability, you need to find the standard error, using this formula:

$$\sigma_{\hat{p}} = \sqrt{\frac{p(1-p)}{n}}$$

where p is the population proportion and n is the sample size.

Substitute the known values into the formula and solve:

$$\sigma_{\hat{p}} = \sqrt{\frac{0.3(1-0.3)}{36}}$$

$$= \sqrt{\frac{0.21}{36}}$$

$$= 0.07638$$

Then convert the observed proportions to z-scores with this formula:

$$z = \frac{\hat{p} - p}{\sigma_{\hat{p}}}$$

where $\hat{p}$ is the sample proportion, p is the population proportion, and $\sigma_{\hat{p}}$ is the standard deviation.

$$z_1 = \frac{\hat{p}_1 - p}{\sigma_{\hat{p}}}$$

$$= \frac{0.2 - 0.3}{0.07638}$$

$$= \frac{-0.1}{0.07638}$$

$$= -1.3092$$

$$z_2 = \frac{\hat{p}_2 - p}{\sigma_{\hat{p}}}$$

$$= \frac{0.4 - 0.3}{0.07638}$$

$$= \frac{0.1}{0.07638}$$

$$= 1.3092$$

Because probabilities equate to tail areas in a normal distribution, you can find the area in the left tail by using the Z-table (Table A-1 in the appendix) and looking up the area to the left of $Z = -1.3092$, which is 0.0951.

Doubling that gives a total tail area of 0.1902, or 19.02%.

437. **100%**

A proportion must be in the range of 0 to 1, so there's no probability of observing a proportion greater than or equal to 2. Therefore, the probability of an observed proportion of less than 2 is 100%.

438. **90.49%**

First, determine whether you can use the central limit theorem by making sure that both np and $n(1-p)$ are at least 10. Because $np = 36(0.3) = 10.8$ and $n(1-p) = 36(1-0.3) = 36(0.7) = 25.2$, you can proceed.

Next, find the standard error, using this formula:

$$\sigma_{\hat{p}} = \sqrt{\frac{p(1-p)}{n}}$$

where p is the population proportion and n is the sample size.

Substitute the known values into the formula and solve:

$$\sigma_{\hat{p}} = \sqrt{\frac{0.3(1-0.3)}{36}}$$
$$= \sqrt{\frac{0.21}{36}}$$
$$= 0.07638$$

Then convert the observed proportions to z-scores with this formula:

$$z = \frac{\hat{p} - p}{\sigma_{\hat{p}}}$$

where $\hat{p}$ is the sample proportion, p is the population proportion, and $\sigma_{\hat{p}}$ is the standard deviation.

$$z = \frac{\hat{p}_1 - p}{\sigma_{\hat{p}}}$$
$$= \frac{0.4 - 0.3}{0.07638}$$
$$= \frac{0.1}{0.07638}$$
$$= 1.3092$$

Because probabilities equate to tail areas in a normal distribution, you can find the approximate area in the left tail by using the Z-table (Table A-1 in the appendix) and looking up the area to the left of $z = 1.3092$ (rounded to 1.31), which is 0.9049, or 90.49%.

439. **5%**

First, determine whether you can use the central limit theorem by making sure that both np and $n(1-p)$ are at least 10. Because $np = 81(0.3) = 24.3$ and $n(1-p) = 81(1-0.3) = 81(0.7) = 56.7$ you can proceed.

Next, find the standard error, using this formula:

$$\sigma_{\hat{p}} = \sqrt{\frac{p(1-p)}{n}}$$

where p is the population proportion and n is the sample size.

Substitute the known values into the formula and solve:

$$\sigma_{\hat{p}} = \sqrt{\frac{0.3(1-0.3)}{36}}$$
$$= \sqrt{\frac{0.21}{36}}$$
$$= 0.07638$$

Then convert the observed proportions to z-scores with this formula:

$$z = \frac{\hat{p} - p}{\sigma_{\hat{p}}}$$

where $\hat{p}$ is the sample proportion, p is the population proportion, and $\sigma_{\hat{p}}$ is the standard deviation.

$$z_1 = \frac{\hat{p}_1 - p}{\sigma_{\hat{p}}}$$

$$= \frac{0.2 - 0.3}{0.05092}$$

$$= \frac{-0.1}{0.05092}$$

$$= -1.96386$$

$$z_2 = \frac{\hat{p}_2 - p}{\sigma_{\hat{p}}}$$

$$= \frac{0.4 - 0.3}{0.05092}$$

$$= \frac{0.1}{0.05092}$$

$$= 1.96386$$

Because probabilities equate to tail areas in a normal distribution, you can find the area in the left tail by rounding the smaller z-value to two decimal places and using the Z-table (Table A-1 in the appendix) and looking up the area to the left of $z = -1.96$, which is 0.025.

Doubling that gives you a total tail area of 0.05, or 5%.

440.   99.12%

First, determine whether you can use the normal approximation to the binomial by making sure that both np and $n(1-p)$ are at least 10. Because $np = 144(0.3) = 43.2$ and $n(1-p) = 144(1-0.3) = 144(0.7) = 100.8$, you can proceed.

Next, find the standard error, using this formula:

$$\sigma_{\hat{p}} = \sqrt{\frac{p(1-p)}{n}}$$

where p is the population proportion and n is the sample size.

Substitute the known values into the formula and solve:

$$\sigma_{\hat{p}} = \sqrt{\frac{0.3(1-0.3)}{144}}$$

$$= \sqrt{\frac{0.21}{144}}$$

$$= 0.03819$$

Then convert the observed proportions to z-scores with this formula:

$$z = \frac{\hat{p} - p}{\sigma_{\hat{p}}}$$

where $\hat{p}$ is the sample proportion, p is the population proportion, and $\sigma_{\hat{p}}$ is the standard deviation.

$$z_1 = \frac{\hat{p}_1 - p}{\sigma_p}$$

$$= \frac{0.2 - 0.3}{0.03819}$$

$$= \frac{-0.1}{0.03819}$$

$$= -2.6185$$

$$z_2 = \frac{\hat{p}_2 - p}{\sigma_{\hat{p}}}$$

$$= \frac{0.4 - 0.3}{0.03819}$$

$$= \frac{0.1}{0.03819}$$

$$= 2.6185$$

Because probabilities equate to tail areas in a normal distribution, you can find the area in the left tail by rounding the smaller z-value to two decimal places and using the Z-table (Table A-1 in the appendix) and looking up the area to the left of $z = -2.62$ which is 0.0044.

Doubling that gives a total tail area of 0.0088. This is the proportion of probability in the range. To find the proportion outside the range, subtract from 1:

$$1 - 0.0088 = 0.9912, \text{ or } 99.12\%$$

441. It tells you how precise you can expect the results to be, across many random samples of the same size.

The basic idea in surveys is that sample results vary. *Margin of error* measures how much you expect your sample results could change if you took many different samples of the same size from this population. Here, the survey result of 60% is based on one sample, and the 4% margin of error means that, with a certain level of confidence, this value of 60% can change by as much as 4% on either side if different samples of the same size are taken.

Note: You assume that all samples were chosen at random in this case, or the margin of error means nothing. Margin of error assumes that the samples were randomly selected and only measures how much the results can change from sample to sample.

442. the confidence level

Any statistical result involving a margin of error is basically calculating a confidence interval, which is the sample statistic plus or minus the margin of error. The claim requires a sample result, the margin of error, and the level of confidence. Here, the sample is 1,000, the sample statistic is 0.93, but the confidence level is missing.

443. The election is too close to call.

You can use the poll to conclude that 54% of the voters in this sample would vote for Garcia, and when you project the results to the population, you add a margin of error of $\pm$ 5%. That means that the proportion voting for Garcia is estimated to be between $54\% - 5\% = 49\%$ and $54\% + 5\% = 59\%$ in the population with 95% confidence.

You can also use the poll to conclude that 46% of the voters in this sample would vote for Smith, and when you project the results to the population, you add a margin of error of ± 5%. That means that the proportion voting for Smith is estimated to be between 46% – 5% = 41% and 46% + 5% = 51% in the population with 95% confidence (over many samples).

Garcia's confidence interval is 49% to 59%, and Smith's confidence interval is 41% to 51%. Because the confidence intervals overlap, the election is too close to call.

444.   ± 2.94

The formula to find the margin of error when estimating a population mean is

$$MOE = \pm z^* \left(\frac{\sigma}{\sqrt{n}} \right)$$

where z^* is the value from Table A-4 for a given confidence level (95% in this case, or 1.96), σ is the population standard deviation (15), and n is the sample size (100).

Now, substitute these values into the formula and solve:

$$MOE = \pm 1.96 \left(\frac{15}{\sqrt{100}} \right)$$
$$= \pm 1.96(1.5)$$
$$= \pm 2.94$$

The margin of error for a 95% confidence interval for the mean is ± 2.94.

445.   ± 0.438

The formula for margin of error when estimating a population mean is

$$MOE = \pm z^* \left(\frac{\sigma}{\sqrt{n}} \right)$$

where z^* is the value from Table A-4 for a given confidence level (95% in this case, or 1.96), σ is the population standard deviation (5), and n is the sample size (500).

Now, substitute the values into the formula and solve:

$$MOE = \pm 1.96 \left(\frac{5}{\sqrt{500}} \right)$$
$$= \pm 1.96(0.2236)$$
$$= \pm 0.438$$

The margin of error for a 95% confidence interval for the population mean is ± 0.438.

446.   ± $3,099.03

The formula for margin of error when estimating a population mean is

$$MOE = \pm z^* \left(\frac{\sigma}{\sqrt{n}} \right)$$

where z^* is the value from Table A-4 for a given confidence level (95% in this case, or 1.96), σ is the population standard deviation ($10,000), and n is the sample size (40).

Now, substitute the values into the formula and solve:

$$MOE = \pm 1.96 \left(\frac{\$10,000}{\sqrt{40}} \right)$$

$$= \pm 1.96(\$1,581.13883)$$

$$= \pm \$3,099.03$$

The margin of error for a 95% confidence interval for the population mean is ± \$3,099.03.

447. ± **\$3.72**

The formula for margin of error when estimating a population mean is

$$MOE = \pm z^* \left(\frac{\sigma}{\sqrt{n}} \right)$$

where z^* is the value from Table A-4 for a given confidence level (99% in this case, or 2.58), σ is the standard deviation (\$25), and n is the sample size (300).

Now, substitute the values into the formula and solve:

$$MOE = \pm 2.58 \left(\frac{\$25}{\sqrt{300}} \right)$$

$$= \pm 2.58(\$1.4434)$$

$$= \pm \$3.72$$

The margin of error for a 99% confidence interval for the population mean is ± \$3.72.

448. ± **\$2.83**

The formula for margin of error when estimating a population mean is

$$MOE = \pm z^* \left(\frac{\sigma}{\sqrt{n}} \right)$$

where z^* is the value from Table A-4 for a given confidence level (95% in this case, or 1.96), σ is the standard deviation (\$25), and n is the sample size (300).

Now, substitute the values into the formula and solve:

$$MOE = \pm 1.96 \left(\frac{\$25}{\sqrt{300}} \right)$$

$$= \pm 1.96(\$1.4434)$$

$$= \pm \$2.83$$

The margin of error for a 95% confidence interval for the population mean is ± \$2.83.

449. **\$83.15**

To find the lower limit for the 80% confidence interval, you first have to find the margin of error. The formula for the margin of error when estimating a population mean is

$$MOE = \pm z^* \left(\frac{\sigma}{\sqrt{n}} \right)$$

where z^* is the value from Table A-4 for a given confidence level (80% in this case, or 1.28), σ is the standard deviation ($25), and n is the sample size (300).

Now, substitute the values into the formula and solve:

$$\text{MOE} = \pm 1.28 \left(\frac{\$25}{\sqrt{300}} \right)$$
$$= \pm 1.28(\$1.4434)$$
$$= \pm \$1.85$$

Next, subtract the MOE from the sample mean to find the lower limit: $85.00 - \$1.85 = \$83.15$.

450. 1.96

First off, if you look at Table A-4, you see that the number you need for z^* for a 95% confidence interval is 1.96. However, when you look up 1.96 on Table A-1 in the appendix, you get a probability of 0.975. Why?

In a nutshell, Table A-1 shows only the probability below a certain z-value, and you want the probability between two z-values, $-z$ and z. If 95% of the values must lie between $-z$ and z, you expand this idea to notice that a combined 5% of the values lie above z and below $-z$. So, 2.5% of the values lie above z, and 2.5% of the values lie below $-z$. To get the total area below this z-value, take the 95% between $-z$ and z plus the 2.5% below $-z$, and you get 97.5%. That's the z-value with 97.5% area below it. It's also the number with 95% lying between two z-values, $-z$ and z.

To avoid all these extra steps and headaches, Table A-4 has already done this conversion for you. So, when you look up 1.96 on Table A-4, you automatically find 95% (not 97.5%).

451. 2.58

Table A-4 shows the answer: A 99% confidence level has a z^*-value of 2.58.

452. 1.28

Table A-4 shows the answer: An 80% confidence level has a z^*-value of 1.28.

453. ± 0.83 years

The formula for margin of error when estimating a population mean is

$$\text{MOE} = \pm z^* \left(\frac{\sigma}{\sqrt{n}} \right)$$

where z^* is the value from Table A-4 for a given confidence level (95% in this case, or 1.96), σ is the standard deviation (3 years), and n is the sample size (50).

Now, substitute the numbers into the formula and solve:

$$\text{MOE} = \pm 1.96\left(\frac{3\,\text{years}}{\sqrt{50}}\right)$$
$$= \pm 1.96(0.4243\,\text{years})$$
$$= \pm 0.83\,\text{years}$$

With 50 women in the sample, the sociologist has a margin of error of ± 0.83 years.

---

454. **± 0.59 years**

The formula for margin of error when estimating a population mean is

$$\text{MOE} = \pm z^*\left(\frac{\sigma}{\sqrt{n}}\right)$$

where z^* is the value from Table A-4 for a given confidence level (95% in this case, or 1.96), σ is the standard deviation (3 years), and n is the sample size (100).

Now, substitute the numbers into the formula and solve:

$$\text{MOE} = \pm 1.96\left(\frac{3\,\text{years}}{\sqrt{100}}\right)$$
$$\pm 1.96(0.3\,\text{years})$$
$$= \pm 0.59\,\text{years}$$

With this sample size, the margin of error is ± 0.59 years.

---

455. **9**

To solve this problem, you have to do a little algebra, using the general formula for margin of error when estimating a population mean:

$$\text{MOE} = \pm z^*\left(\frac{\sigma}{\sqrt{n}}\right)$$

Here, z^* is the value from Table A-4 for a given confidence level (95% in this case, or 1.96), σ is the population standard deviation (3 years), and n is the sample size.

To find the sample size, n, you have to rearrange the formula for margin of error: Multiply both sides by the square root of n, divide both sides by the margin of error, and square both sides (note that a sample size can only be positive). Here are the steps:

$$\text{MOE} = \pm z^*\left(\frac{\sigma}{\sqrt{n}}\right)$$
$$\text{MOE}\sqrt{n} = \pm z^*(\sigma)$$
$$(\sqrt{n}) = \pm \frac{z^*(\sigma)}{\text{MOE}}$$
$$n = \left(\frac{z^*(\sigma)}{\text{MOE}}\right)^2$$

Now to solve the problem at hand, substitute the values into the formula:

$$n = \left(\frac{(1.96)(3)}{2} \right)^2$$
$$= 8.643$$

You can't have a fraction of a participant and you need to make sure the margin of error is less than or equal to two years, so round up to nine participants.

Note: Even if the result was a number that you'd normally round down (like 8.123), you still always round up to the next greatest integer when solving for sample size.

456. ± 0.00182 mm

To calculate the margin of error (MOE) for estimating a population mean, use this formula:

$$\text{MOE} = \pm z^* \left(\frac{\sigma}{\sqrt{n}} \right)$$

Here, z^* is the value from Table A-4 for a given confidence level (99% in this case, or 2.58), σ is the standard deviation (0.01 millimeters), and n is the sample size (200).

Now, substitute all the known values into the formula and solve:

$$\text{MOE} = \pm 2.58 \left(\frac{0.01 \text{ mm}}{\sqrt{200}} \right)$$
$$= \pm 2.58(0.000707 \text{ mm})$$
$$= \pm 0.00182 \text{ mm}$$

457. ± 0.00912 mm

To calculate the margin of error (MOE) for estimating a population mean, use this formula:

$$\text{MOE} = \pm z^* \left(\frac{\sigma}{\sqrt{n}} \right)$$

Here, z^* is the value from Table A-4 for a given confidence level (99% in this case, or 2.58), σ is the population standard deviation (0.05 millimeters), and n is the sample size (200).

Now, substitute the known values into the formula and solve:

$$\text{MOE} = \pm 2.58 \left(\frac{0.05 \text{ mm}}{\sqrt{200}} \right)$$
$$= \pm 2.58(0.003536 \text{ mm})$$
$$= \pm 0.00912 \text{ mm}$$

458. ± 0.0182 mm

To calculate the margin of error (MOE) for estimating a population mean, use this formula:

$$\text{MOE} = \pm z^* \left(\frac{\sigma}{\sqrt{n}} \right)$$

Here, z^* is the value from Table A-4 for a given confidence level (99% in this case, or 2.58), σ is the population standard deviation (0.10 millimeters), and n is the sample size (200).

Substitute the known values into the formula and solve:

$$\text{MOE} = \pm 2.58 \left(\frac{0.1 \text{ mm}}{\sqrt{200}} \right)$$
$$= \pm 2.58 (0.00707 \text{ mm})$$
$$= \pm 0.0182 \text{ mm}$$

459. **Quadruple the original sample size.**

The basic formula for margin of error when estimating a population mean is

$$\text{MOE} = \pm z^* \left(\frac{\sigma}{\sqrt{n}} \right)$$

where z^* is the value from Table A-4 for a given confidence level (99% in this case, or 2.58), σ is the population standard deviation, and n is the sample size (200). You don't need to know the units of measurement or the value of the standard deviation to answer this question.

Note that the sample size, n, appears in the denominator here. If you want to cut the margin of error in half, the sample size will change. Halving the MOE is equivalent to dividing the MOE by 2; to maintain the integrity of the equation, you have to divide the other half of the equation by 2 also. Because the denominator of the right-hand side of the equation is the square root of n, dividing by the square root of 4 is equivalent to dividing by 2, because 2 is the square root of 4.

$$\text{MOE} = \pm z^* \left(\frac{\sigma}{\sqrt{n}} \right)$$
$$\frac{\text{MOE}}{2} = \pm z^* \left(\frac{\sigma}{2\sqrt{n}} \right)$$
$$\frac{\text{MOE}}{2} = \pm z^* \left(\frac{\sigma}{\sqrt{4n}} \right)$$

Therefore, with all else held constant, increasing n fourfold will cut the MOE in half.

Note that, because of the distributive law, the following is true:

$$\sqrt{4n} = (\sqrt{4})(\sqrt{n}) = 2\sqrt{n}$$

460. **converted a margin of error at 99% confidence to a margin of error at 80% confidence**

Note that you don't need to know what part or what dimension of the part is involved; you can solve this problem through your knowledge of the formula for margin of error. Switching from a 99% confidence level to an 80% confidence level will change the margin of error by a little more than half.

Consider the formula for calculating margin of error:

$$\text{MOE} = z^* \left(\frac{\sigma}{\sqrt{n}} \right)$$

where z^* is the value from Table A-1 corresponding to the margin of error, σ is the standard deviation, and n is the sample size. To preserve the equation but reduce the MOE by half, you must divide both sides by 2 (or multiply both by ½).

$$\text{MOE} = z^* \left(\frac{\sigma}{\sqrt{n}} \right)$$

$$\left(\frac{1}{2} \right)\text{MOE} = \left(\frac{1}{2} \right) z^* \left(\frac{\sigma}{\sqrt{n}} \right)$$

You can't change the sample size or standard deviation; the only thing you can change is z^*. You need two values, one of which is about twice the other. The value of z^* for a 99% confidence interval is 2.58, which is about twice that for an 80% confidence interval (1.28). The best conclusion is that the factory owner changed the width of the confidence interval.

461.   800

The sample size, n, appears in the denominator of the formula for margin of error.

$$\text{MOE} = \pm z^* \left(\frac{\sigma}{\sqrt{n}} \right)$$

The researcher wants to cut the MOE to ⅓ of its current value, which is equivalent to dividing by 3. To retain the integrity of the equation, you must divide both sides by 3. Note that the denominator of the right-hand side of the equation is the square root of n; dividing by the square root of 9 is equivalent to dividing by 3, because 3 is the square root of 9. So, to divide the MOE by 3, holding everything else constant, you must increase the sample size to 9 times its current value.

$$\text{MOE} = \pm z^* \left(\frac{\sigma}{\sqrt{n}} \right)$$

$$\frac{\text{MOE}}{3} = \pm z^* \left(\frac{\sigma}{3\sqrt{n}} \right)$$

$$\frac{\text{MOE}}{3} = \pm z^* \left(\frac{\sigma}{\sqrt{9n}} \right)$$

Note that, because of the distributive law, the following is true:

$$\sqrt{9n} = (\sqrt{9})(\sqrt{n}) = 3\sqrt{n}$$

Because $n = 100$ and $9n = (9)(100) = 900$, the market researcher needs 800 more participants to get the desired margin of error (a total of 900 participants is required).

462. ± 0.978%

The formula to calculate the margin of error (MOE) for a population proportion is

$$\text{MOE} = \pm z^* \sqrt{\frac{\hat{p}(1-\hat{p})}{n}}$$

where z^* is the value from Table A-4 for a given confidence level (95% in this case, or 1.96), $\hat{p}$ is the sample proportion (0.53), and n is the sample size (10,000).

You convert 53% to the proportion 0.53 by dividing the percentage by 100: $53/100 = 0.53$.

Now, substitute the known values in the formula and solve:

$$\text{MOE} = \pm 1.96 \sqrt{\frac{0.53(1-0.53)}{10,000}}$$

$$= \pm 1.96 \sqrt{\frac{0.2491}{10,000}}$$

$$= \pm 1.96 \sqrt{0.00002491}$$

$$= \pm 1.96(0.00499)$$

$$= \pm 0.00978$$

Convert this proportion to a percentage by multiplying by 100%: $(0.00978)(100\%) = 0.978\%$

The margin of error is therefore ± 0.978%.

You estimate that 53% ± 0.978% of all Europeans are unhappy with the euro, based on these survey results with 95% confidence.

463. ± 7.69%

The formula for the margin of error (MOE) for a proportion is

$$\text{MOE} = \pm z^* \sqrt{\frac{\hat{p}(1-\hat{p})}{n}}$$

where z^* is the value from Table A-4 for a given confidence level (99% in this case, or 2.58), $\hat{p}$ is the sample proportion (0.48), and n is the sample size (281).

To compute the sample proportion, divide the number who responded that they intend to vote by the total number polled: $135/281 = 0.48$.

Now, substitute the known values into the formula and solve:

$$\text{MOE} = \pm 2.58 \sqrt{\frac{0.48(1-0.48)}{281}}$$

$$= \pm 2.58 \sqrt{\frac{0.2496}{281}}$$

$$= \pm 2.58(0.0298)$$

$$= \pm 0.0769$$

Convert the proportion to a percentage by multiplying by 100%: $(0.0769)(100\%) = 7.69\%$.

Using a 99% confidence level, the county election office has achieved a margin of error of ± 7.69%.

464. **90% confidence**

The confidence level tells you how many standard errors to add and subtract to get the margin of error you want, which is quantified by the z-value. To find the confidence level, you first need to solve for z by rearranging the margin of error formula for a population proportion:

$$\text{MOE} = \pm z^* \sqrt{\frac{\hat{p}(1-\hat{p})}{n}}$$

Here, z^* is the value from Table A-4 for a given confidence level, $\hat{p}$ is the sample proportion (0.46), and n is the sample size (922).

Rearrange this formula as follows:

$$\text{MOE} = \pm z^* \sqrt{\frac{\hat{p}(1-\hat{p})}{n}}$$

$$\frac{\text{MOE}}{\sqrt{\frac{\hat{p}(1-\hat{p})}{n}}} = \pm z^*$$

Note that you can drop the $\pm$ symbol because the standard normal distribution is symmetrical. Also, by convention, this formula is usually written with z^* on the left side (because the two sides are equivalent, it doesn't matter which way you write the equation):

$$z^* = \frac{\text{MOE}}{\sqrt{\frac{\hat{p}(1-\hat{p})}{n}}}$$

Then substitute the known values into the formula and solve. (First, convert the margin of error to a proportion by dividing by 100%: $2.7\% / 100\% = 0.027$.)

$$z^* = \frac{0.027}{\sqrt{\frac{0.46(1-0.46)}{922}}}$$

$$= \frac{0.027}{\sqrt{\frac{0.2484}{922}}}$$

$$= \frac{0.027}{0.01641}$$

$$= 1.64534$$

Rounding to three decimal places, you get 1.645. Now you can find the confidence level by looking at Table A-4 and seeing that 1.645 corresponds to the 90% confidence level z^*-value. So, you can safely say that the confidence level is 90%.

465. **B. If the same study were repeated many times, about 95% of the time, the confidence interval would contain the average money spent for all the customers.**

The average money spent for all the customers is an unknown value, called a *population parameter*. The average money spent for the 100 customers in the sample is a known value, $45, which is called a *statistic*.

The store is using a sample statistic to estimate a population parameter. Because samples vary from sample to sample, they know the sample mean may not correspond exactly to the population mean, so they use confidence intervals to state a plausible range of values for the population mean. If the same experiment were repeated many times (drawing a sample of the same size from the same population and calculating the sample average), the population mean would be expected to be contained in 95% of the confidence intervals created.

466. **E. Choices (A) and (C) (The store studied a sample of sales records rather than the entire population of sales records; because sample results vary, the sample mean is not expected to correspond exactly to the population mean, so a range of likely values is required.)**

The store studied a sample of records to estimate a population parameter, and because sample results vary (called *sampling error*), the sample mean isn't expected to correspond exactly to the population mean. If another sample of the same size were drawn from the population, the sample mean would be expected to be somewhat different, so a range of likely values for the population mean (that is, a confidence interval) is required.

467. **D. Choices (B) and (C) (The larger sample will produce a more precise estimate of the population mean; the 95% confidence interval calculated from the larger sample will be narrower.)**

A larger sample drawn from the same population will tend to produce a narrower confidence interval and a more precise estimate of the population mean. The amount of bias isn't measured by the confidence interval.

468. **E. None of the above**

The sample mean isn't expected to be exactly the same as the population mean nor is another sample of the same size drawn from the same population expected to have exactly the same mean. That number by itself isn't a "good" number to use to estimate the population mean; you need a margin of error to go with it. The sample mean changes if another sample were taken.

469. **Sample D**

For samples of the same size, greater variability in the data will tend to produce a wider confidence interval. In this case, Sample A has no variability, Sample B has limited variability, Sample C has some variability, and Sample D has the greatest variability due to one extremely high value (20). You can conduct calculations on these data sets to confirm this notion.

470. **one with a confidence level of 99%**

The confidence level is the amount of confidence you have that a confidence interval will contain the actual population parameter (in this case, the population mean) if you repeated the process over and over.

If the confidence level is higher, the interval needs to be wider to include a wider range of likely values for the population parameter. Therefore, the confidence interval with the highest confidence level will be the widest.

471. **The width of the confidence interval will increase.**

Increasing the confidence level will increase the width of the confidence interval because to be more confident in your process of trying to estimate the mean of the population, you must include a wider range of possible values for it (assuming that all the other parts involved in the confidence interval stay the same).

472. **A higher confidence level means that the margin of error is increased, requiring a wider confidence interval.**

With the factors remaining equal, increasing the confidence level means that you're offering a wider range of possible values for the population parameter. This increases the width of the confidence interval.

473. **± 5%**

The margin of error is the number that is added or subtracted from the sample statistic to produce the confidence interval. In this case, the sample statistic is 65%. To start with 65% and end up with a confidence interval of 60% to 70%, you must add and subtract 5%:

$$65\% + 5\% = 70\%$$
$$65\% - 5\% = 60\%$$

So, the margin of error is ± 5%.

474. **A. $n = 100$**

All else held constant, the smaller the sample size is, the larger the margin of error will be. You don't have as much precision when you have a smaller data set. Less precision means a larger margin of error, which means that the confidence interval will be wider. Sets of small samples have greater variability than sets of large samples.

475. **E. $n = 5,000$**

All else held constant, the larger the sample size is, the more precise your results and the smaller the margin of error will be. A smaller margin of error means that the confidence interval is narrower. Sets of large samples don't have as much variability as sets of small samples.

476. **E. Choices (B) and (D) (The sample of 500 will have a narrower 95% confidence interval; the sample of 500 will produce a more precise estimate of the population mean.)**

All else being equal, a larger sample will produce a narrower 95% confidence interval and a more precise estimate of the population mean because sets of large samples don't have as much variability as sets of small samples.

477. **For a higher confidence level, the confidence interval will be wider.**

When calculating confidence intervals for different confidence levels from the same sample, a higher confidence level will produce a wider confidence interval.

478. **C. The confidence interval related to Population A is expected to be narrower.**

All else being equal, a sample from a less-variable population can be expected to produce a narrower confidence interval. That's because the samples don't change as much when they come from a population whose values are more similar (less variable).

479. **E. Choices (B) and (C) (A confidence level of 80% will produce a narrower confidence interval than a confidence level of 90%; a sample of 300 students will produce a narrower confidence interval than a sample of 150 students.)**

A larger sample and a lower confidence level will both produce a narrower confidence interval when dealing with random samples drawn from the same population.

480. **Sample B**

For the same sample size and confidence level, greater variability in the sample will produce a wider confidence interval. Sample B has the greatest variability and will have the widest confidence interval for a given confidence level.

481. **D. all workers ages 22 to 30 who live in North America (Canada, the U.S., and Mexico), adjusted in U.S. dollars**

The variability of the income of this population is the greatest because it includes workers from countries with substantially different mean incomes. This causes larger margin of error when calculating a confidence interval.

482. **E. adults ages 55 to 65**

When sample size and confidence level are held constant, the sample from the least variable population will be expected to product the least variable sample and, hence, the smallest confidence interval. The population with the least variability is expected to be that of adults who have attained their full adult height (adults ages 55 to 65) but have not yet become subject to decreases in height due to osteoporosis.

483. **B. one with confidence level 95%, $n = 200$, and $\sigma = 12.5$**

The margin of error increases as the confidence level and the standard deviation increase. The margin of error also increases with a smaller sample.

484. **D. Choices (A) and (B) (Increase the sample size from 200 to 1,000 subjects; decrease the confidence level from 95% to 90%.)**

A larger sample size and a lower confidence level will both decrease the margin of error of a confidence interval.

485. **E. The 95% confidence interval for the mean height of all the boys is between 5 feet 5 inches and 6 feet 1 inch.**

The confidence interval is constructed as the statistic (point estimate) plus or minus the margin of error. In this case, the 95% confidence interval is the point estimate of

5 feet 9 inches plus or minus the margin of error of 4 inches, which is 5 feet 5 inches to 6 feet 1 inch.

There can be a wide distribution of individual heights of the boys. A confidence interval and a confidence level don't say anything about the height of an individual boy or the total range of heights overall.

486. **The 95% confidence interval for the mean summer income of all college students is $4,100 to $4,900.**

The confidence interval is constructed as the sample statistic (point estimate) plus or minus the margin of error. In this case, the 95% confidence interval is the point estimate of $4,500 plus or minus the margin of error of $400, giving a confidence interval of $4,100 to $4,900.

487. **C. The margin is used to calculate a range of likely values for a population parameter, based on a sample.**

The margin of error isn't due to an error in the sample or survey. It factors in the fact that sample results vary from sample to sample and gives you a measure of how much you expect them to vary with a certain level of confidence.

488. **B. The margin of error measures the amount by which your sample results could change, with 99% confidence.**

The confidence interval is constructed as the point estimate plus or minus the margin of error. In this case, the 99% confidence interval is the point estimate of $450 plus or minus the margin of error of $50, giving a confidence interval of $400 to $500.

489. **The election is too close to predict. The actual support for either candidate could be above 50%.**

The margin of error is used to construct the confidence interval, which is a range of likely values for the population parameter (here, the parameter is the percentage of all voters who would vote for a candidate). To calculate a confidence interval, you take the result from the sample and add and subtract the margin of error.

In this case, the 98% confidence interval for the proportion of all voters for Candidate Smith is 48% (from the sample) plus or minus 3% (the margin of error), which is a range of 45% to 51%. For Candidate Jones, the 98% confidence interval is 52% (from the sample) plus or minus 3% (the margin of error), which is a range of 49% to 55%. Both confidence intervals contain possible values above 50%, so either candidate could win; therefore, the results are too close to call.

490. **A. The survey has a built-in bias.**

This survey is biased because it wasn't conducted with a random sample of Americans but rather with a sample of fans who attended this particular football game. It's plausible that these fans don't represent the taste preferences of all Americans.

Taking a higher sample of fans won't make the survey less biased, and having a narrower confidence interval won't solve the problem, because bias isn't measured by the margin of error (it measures only how the results of a random sample would change from sample to sample).

491. **E. Choices (B) and (D) (The survey is biased because it was based only on first-year employees, who may feel differently about their jobs than other employees. The sample size is only 30; the margin of error must be higher than 3% based on the size of the sample and the confidence level.)**

First, the sample has bias because even though it was a random sample, it was based only on first-year employees, who may feel differently about their jobs than employees who have been on the job longer.

Second, with a sample of only 30 employees, the margin of error can't be as small as 3%. The formula for margin of error for a population proportion is

$$\text{MOE} = z^* \sqrt{\frac{\hat{p}(1-\hat{p})}{n}}$$

Where z^* is the value from Table A-1 for the selected confidence level, $\hat{p}$ is the sample proportion, and n is the sample size.

The sample size is 30, $\hat{p}$ is 0.8, and z^* is 1.96 for this problem.

$$\begin{aligned}
\text{MOE} &= 1.96 \sqrt{\frac{0.8(1-0.8)}{30}} \\
&= 1.96 \sqrt{\frac{(0.8)(0.2)}{30}} \\
&= 1.96(0.0730) \\
&= 0.1431
\end{aligned}$$

Convert the proportion to a percent by multiplying by 100%:

$$0.1431(100\%) = 14.31\%$$

This value is much larger than the reported MOE of 3%.

492. **B. The survey can be used only by the company as part of its analysis of the Internet spending habits of all visitors to its website during the last three months.**

The random sample can be used only in analysis for that specific website for that specific time period. It can't be extended to a longer period of time or to a larger group of customers. But it can be used to draw conclusions to the visitors because that's what the random sample is based on. The statistics for the mean, margin of error, and confidence level are realistic with a sample size of 1,000.

493. **A. The results are invalid because the survey was done at a movie theater.**

The survey is asking only those teenagers who are already at a theater. This means that each response would be at least 1, but certainly it's possible that someone visited a theater 0 times in the past 12 months. A valid random sample would include all teenagers.

494. **E. Choices (B), (C), and (D) (The sample isn't based on a representative sample of the magazine's readers. Because the magazine is based on Colorado, more readers**

are likely to buy the magazine who are from Colorado; therefore, they'd be more likely to vote it as the best place to live. The sample results are likely biased because the respondents had to make the effort to mail back the survey.)

The survey results are almost certain to be biased. A mail-in survey isn't a valid way to draw a random sample, because those who choose to take part in such a survey are probably not representative of the magazine's entire readership. It's also biased because the magazine is based on Colorado and, hence, is more likely to be purchased by people living in Colorado, who are more likely to choose Colorado as their favorite place to live.

A larger sized sample can't reduce bias that already exists.

495. **A. With 95% confidence, the average points scored by all intramural basketball players is between 7.3 and 8.7 points.**

Use the formula for finding the confidence interval for a population when the standard deviation is known:

$$\bar{x} \pm \text{MOE} = \bar{x} \pm z^* \left(\frac{\sigma}{\sqrt{n}} \right)$$

where $\bar{x}$ is the sample mean, σ is the population standard deviation, n is the sample size, and z^* represents the appropriate z^*-value from the standard normal distribution for your desired confidence level. The data has to come from a normal distribution, or n has to be large enough (a standard rule of thumb is at least 30 or so), for the central limit theorem to apply. You can find z^*-values from Table A-1 or Table A-4 in the appendix. The z^*-value is 1.96 for a two-tailed confidence interval with a confidence level of 95%.

Next, substitute the values into the formula:

$$\text{MOE} = 1.96 \left(\frac{2.5}{\sqrt{50}} \right) \approx 0.693$$

The 95% confidence interval is 8 ± 0.7 (rounded to the nearest tenth), or 7.3 to 8.7 points scored.

496. **The 99% confidence interval for the average SAT math score for all students at the high school is between 624.2 and 678.8.**

Use the formula for finding the confidence interval for a population when the standard deviation is known:

$$\bar{x} \pm \text{MOE} = \bar{x} \pm z^* \left(\frac{\sigma}{\sqrt{n}} \right)$$

where $\bar{x}$ is the sample mean, σ is the population standard deviation, n is the sample size, and z^* represents the appropriate z^*-value from the standard normal distribution for your desired confidence level. The data has to come from a normal distribution, or n has to be large enough (a standard rule of thumb is at least 30 or so), for the central limit theorem to apply. You can find z^*-values from Table A-1 or Table A-4 in the appendix. The z^*-value for a two-tailed confidence interval with a confidence level of 99% is 2.58.

Next, substitute the values into the formula:

$$\text{MOE} = 2.58 \left(\frac{100}{\sqrt{100}} \right) = 25.8$$

The confidence interval is 650 ± 25.8 (rounded to the nearest tenth), or 624.2 to 678.8.

497. **The 99% confidence interval for the average weight of all apples from the ten trees is between 6.5 and 7.5 ounces.**

Use the formula for finding the confidence interval for a population when the standard deviation is known:

$$\bar{x} \pm \text{MOE} = \bar{x} \pm z^* \left(\frac{\sigma}{\sqrt{n}} \right)$$

where $\bar{x}$ is the sample mean, σ is the population standard deviation, n is the sample size, and z^* represents the appropriate z^*-value from the standard normal distribution for your desired confidence level. The data has to come from a normal distribution, or n has to be large enough (a standard rule of thumb is at least 30 or so), for the central limit theorem to apply. You can find z^*-values from Table A-1 or Table A-4 in the appendix. The z^*-value for a two-tailed confidence interval with a confidence level of 99% is 2.58.

Next, substitute the values into the formula:

$$\text{MOE} = 2.58 \left(\frac{1.5}{\sqrt{50}} \right) \approx 0.5473$$

The confidence interval is 7 ± 0.5 (rounded to the nearest tenth), or 6.5 to 7.5 ounces.

498. **The 90% confidence interval for the average time students at the university spend doing homework each day is between 2.88 and 3.12 hours.**

Use the formula for finding the confidence interval for a population when the standard deviation is known:

$$\bar{x} \pm \text{MOE} = \bar{x} \pm z^* \left(\frac{\sigma}{\sqrt{n}} \right)$$

where $\bar{x}$ is the sample mean, σ is the population standard deviation, n is the sample size, and z^* represents the appropriate z^*-value from the standard normal distribution for your desired confidence level. The data has to come from a normal distribution, or n has to be large enough (a standard rule of thumb is at least 30 or so), for the central limit theorem to apply. You can find z^*-values from Table A-1 or Table A-4 in the appendix. The z^*-value for a two-tailed confidence interval with a confidence level of 90% is 1.645.

Next, substitute the values into the formula:

$$\text{MOE} = 1.645 \left(\frac{1}{\sqrt{200}} \right) \approx 0.1163$$

The confidence interval is 3 ± 0.12 (rounded to the nearest hundredth), or 2.88 to 3.12 hours of homework.

499. **The 95% confidence interval for the average spending of people ages 18 to 22 on a typical outing with a friend is between \$30.42 and \$34.58.**

Use the formula for finding the confidence interval for a population when the standard deviation is known:

$$\bar{x} \pm \mathrm{MOE} = \bar{x} \pm z^* \left(\frac{\sigma}{\sqrt{n}} \right)$$

where $\bar{x}$ is the sample mean, σ is the population standard deviation, n is the sample size, and z^* represents the appropriate z^*-value from the standard normal distribution for your desired confidence level. The data has to come from a normal distribution, or n has to be large enough (a standard rule of thumb is at least 30 or so), for the central limit theorem to apply. You can find z^*-values from Table A-1 or Table A-4 in the appendix. The z^*-value for a two-tailed confidence interval with a confidence level of 95% is 1.96.

Next, substitute the values into the formula:

$$\mathrm{MOE} = 1.96 \left(\frac{15.00}{\sqrt{200}} \right) \approx 2.0789$$

The confidence interval is 32.50 ± 2.08 (rounded to the nearest hundredth), or \$30.42 to \$34.58 spent.

500. **The 90% confidence interval for the average time people above age 17 spend doing vigorous exercise is 28 to 32 minutes per day.**

Use the formula for finding the confidence interval for a population when the standard deviation is known:

$$\bar{x} \pm \mathrm{MOE} = \bar{x} \pm z^* \left(\frac{\sigma}{\sqrt{n}} \right)$$

where $\bar{x}$ is the sample mean, σ is the population standard deviation, n is the sample size, and z^* represents the appropriate z^*-value from the standard normal distribution for your desired confidence level. The data has to come from a normal distribution, or n has to be large enough (a standard rule of thumb is at least 30 or so), for the central limit theorem to apply. You can find z^*-values from Table A-1 or Table A-4 in the appendix. The z^*-value for a two-tailed confidence interval with a confidence level of 90% is 1.645.

Next, substitute the values into the formula:

$$\mathrm{MOE} = 1.645 \left(\frac{15}{\sqrt{150}} \right) \approx 2.0147$$

The confidence interval is 30 ± 2.0 (rounded to the nearest tenth), or 28 to 32 minutes.

501. **The 95% confidence interval for the average income of all first-year college graduates is between \$34,891 and \$37,109.**

Use the formula for finding the confidence interval for a population when the standard deviation is known:

$$\bar{x} \pm \text{MOE} = \bar{x} \pm z^* \left(\frac{\sigma}{\sqrt{n}} \right)$$

where $\bar{x}$ is the sample mean, σ is the population standard deviation, n is the sample size, and z^* represents the appropriate z^*-value from the standard normal distribution for your desired confidence level. The data has to come from a normal distribution, or n has to be large enough (a standard rule of thumb is at least 30 or so), for the central limit theorem to apply. You can find z^*-values from Table A-1 or Table A-4 in the appendix. The z^*-value for a two-tailed confidence interval with a confidence level of 95% is 1.96.

Next, substitute the values into the formula:

$$\text{MOE} = 1.96 \left(\frac{8,000}{\sqrt{200}} \right) \approx 1,108.7434$$

The confidence interval is $36,000 \pm 1,109$, or \$34,891 to \$37,109.

502. **The 95% confidence interval for the average amount of time of all bus trips along this route is between 44:40 and 45:20 minutes.**

Use the formula for finding the confidence interval for a population when the standard deviation is known:

$$\bar{x} \pm \text{MOE} = \bar{x} \pm z^* \left(\frac{\sigma}{\sqrt{n}} \right)$$

where $\bar{x}$ is the sample mean, σ is the population standard deviation, n is the sample size, and z^* represents the appropriate z^*-value from the standard normal distribution for your desired confidence level. The data has to come from a normal distribution, or n has to be large enough (a standard rule of thumb is at least 30 or so), for the central limit theorem to apply. You can find z^*-values from Table A-1 or Table A-4 in the appendix. The z^*-value for a two-tailed confidence interval with a confidence level of 95% is 1.96.

Next, substitute the values into the formula:

$$\text{MOE} = 1.96 \left(\frac{3}{\sqrt{300}} \right) \approx 0.3395$$

The confidence interval is 45 ± 0.34 minutes (rounded to the nearest hundredth), or 44.66 minutes to 45.34 minutes. Convert the fractional minutes to the easier-to-use unit of seconds: 44 minutes and 40 seconds to 45 minutes and 20 seconds, or 44:40 to 45:20.

503. **The 98% confidence interval for the average amount of time among all requests for an airline itinerary to be displayed online is between 4.36 and 4.64 seconds.**

Use the formula for finding the confidence interval for a population when the standard deviation is known:

$$\bar{x} \pm \text{MOE} = \bar{x} \pm z^* \left(\frac{\sigma}{\sqrt{n}} \right)$$

where $\bar{x}$ is the sample mean, σ is the population standard deviation, n is the sample size, and z^* represents the appropriate z^*-value from the standard normal distribution for your desired confidence level. The data has to come from a normal distribution, or n has to be large enough (a standard rule of thumb is at least 30 or so), for the central limit theorem to apply. You can find z^*-values from Table A-1 or Table A-4 in the appendix. The z^*-value for a two-tailed confidence interval with a confidence level of 98% is 2.33.

Next, substitute the values into the formula:

$$\text{MOE} = 2.33 \left(\frac{2}{\sqrt{1,100}} \right) \approx 0.1405$$

The confidence interval is 4.5 ± 0.14 (rounded to the nearest hundredth), or 4.36 to 4.64 seconds.

504. **The 95% confidence interval for the average amount of time to assemble an MP3 player is between 11.95 and 12.55 minutes.**

Use the formula for finding the confidence interval for a population when the standard deviation is known:

$$\bar{x} \pm \text{MOE} = \bar{x} \pm z^* \left(\frac{\sigma}{\sqrt{n}} \right)$$

where $\bar{x}$ is the sample mean, σ is the population standard deviation, n is the sample size, and z^* represents the appropriate z^*-value from the standard normal distribution for your desired confidence level. The data has to come from a normal distribution, or the n has to be large enough (a standard rule of thumb is at least 30 or so) for the central limit theorem to apply. You can find z^*-values from Table A-1 or Table A-4 in the appendix. The z^*-value for a two-tailed confidence interval with a confidence level of 95% is 1.96.

Next, substitute the values into the formula:

$$\text{MOE} = 1.96 \left(\frac{2.15}{\sqrt{200}} \right) \approx 0.2980$$

The confidence interval is 12.25 ± 0.30 (rounded to the nearest hundredth), or 11.95 to 12.55 minutes.

505. **The 90% confidence interval for the average of all such distances to the university from a student's hometown is between 121.2 and 128.8 miles.**

Use the formula for finding the confidence interval for a population when the standard deviation is known:

$$\bar{x} \pm \text{MOE} = \bar{x} \pm z^* \left(\frac{\sigma}{\sqrt{n}} \right)$$

where $\bar{x}$ is the sample mean, σ is the population standard deviation, n is the sample size, and z^* represents the appropriate z^*-value from the standard normal distribution for your desired confidence level. The data has to come from a normal distribution, or n has to be large enough (a standard rule of thumb is at least 30 or so), for the central limit theorem to apply. You can find z^*-values from Table A-1 or Table A-4 in the appendix. The z^*-value for a two-tailed confidence interval with a confidence level of 90% is 1.645.

Next, substitute the values into the formula:

$$\text{MOE} = 1.645 \left(\frac{40}{\sqrt{300}} \right) \approx 3.7990$$

The confidence interval is 125 ± 3.8 (rounded to the nearest tenth), or 121.2 to 128.8 miles.

506. **The 95% confidence interval for the average number of errors among data specialists is between 2.53 and 2.87 errors in 10,000 entries.**

Use the formula for finding the confidence interval for a population when the standard deviation is known:

$$\bar{x} \pm \text{MOE} = \bar{x} \pm z^* \left(\frac{\sigma}{\sqrt{n}} \right)$$

where $\bar{x}$ is the sample mean, σ is the population standard deviation, n is the sample size, and z^* represents the appropriate z^*-value from the standard normal distribution for your desired confidence level. The data has to come from a normal distribution, or n has to be large enough (a standard rule of thumb is at least 30 or so), for the central limit theorem to apply. You can find z^*-values from Table A-1 or Table A-4 in the appendix. The z^*-value for a two-tailed confidence interval with a confidence level of 95% is 1.96.

Next, substitute the values into the formula:

$$\text{MOE} = 1.96 \left(\frac{0.75}{\sqrt{75}} \right) \approx 0.1697$$

The confidence interval is 2.7 ± 0.17 (rounded to the nearest hundredth), or 2.53 to 2.87 errors.

507. **The 99% confidence level for the average length of all major league 38-inch bats is between 38.009 and 38.011 inches.**

Use the formula for finding the confidence interval for a population when the standard deviation is known:

$$\bar{x} \pm \text{MOE} = \bar{x} \pm z^* \left(\frac{\sigma}{\sqrt{n}} \right)$$

where $\bar{x}$ is the sample mean, σ is the population standard deviation, n is the sample size, and z^* represents the appropriate z^*-value from the standard normal distribution for your desired confidence level. The data has to come from a normal distribution, or n has to be large enough (a standard rule of thumb is at least 30 or so), for the central limit theorem to apply. You can find z^*-values from Table A-1 or Table A-4 in the appendix. The z^*-value for a two-tailed confidence interval with a confidence level of 99% is 2.58.

Next, substitute the values into the formula:

$$\text{MOE} = 2.58\left(\frac{0.01}{\sqrt{500}}\right) \approx 0.00115$$

The confidence interval is 38.01 ± 0.001 (rounded to the nearest thousandth), or 38.009 to 38.011 inches.

508. **The 99% confidence interval for the average length of all special valve engine parts is between 3.2536 and 3.2564 centimeters.**

Use the formula for finding the confidence interval for a population when the standard deviation is known:

$$\bar{x} \pm \text{MOE} = \bar{x} \pm z^*\left(\frac{\sigma}{\sqrt{n}}\right)$$

where $\bar{x}$ is the sample mean, σ is the population standard deviation, n is the sample size, and z^* represents the appropriate z^*-value from the standard normal distribution for your desired confidence level. The data has to come from a normal distribution, or n has to be large enough (a standard rule of thumb is at least 30 or so), for the central limit theorem to apply. You can find z^*-values from Table A-1 or Table A-4 in the appendix. The z^*-value for a two-tailed confidence interval with a confidence level of 99% is 2.58.

Next, substitute the values into the formula:

$$\text{MOE} = 2.58\left(\frac{0.025}{\sqrt{2,000}}\right) \approx 0.001442$$

The confidence interval is 3.2550 ± 0.0014 (rounded to the nearest ten-thousandth), or 3.2536 to 3.2564 centimeters.

509. **The 95% confidence interval for the average cost of medium-quality hardwood during the 12-month period had an average cost of between \$0.743 and \$0.817 per board foot.**

Use the formula for finding the confidence interval for a population when the standard deviation is known:

$$\bar{x} \pm \text{MOE} = \bar{x} \pm z^*\left(\frac{\sigma}{\sqrt{n}}\right)$$

where $\bar{x}$ is the sample mean, σ is the population standard deviation, n is the sample size, and z^* represents the appropriate z^*-value from the standard normal distribution for your desired confidence level. The data has to come from a normal distribution, or n has to be large enough (a standard rule of thumb is at least 30 or so), for the central limit theorem to apply. You can find z^*-values from Table A-1 in the appendix or Table A-4. The z^*-value for a two-tailed confidence interval with a confidence level of 95% is 1.96.

Next, substitute the values into the formula:

$$MOE = 1.96\left(\frac{0.12}{\sqrt{40}}\right) \approx 0.0371$$

The confidence interval is 0.78 ± 0.037 (rounded to the nearest thousandth), or $0.743 to $0.817.

---

510. **The 95% confidence interval for the average first-year college student score for the math test is between 81.9% and 86.1%.**

Use the formula for the confidence interval when the population's standard deviation is unknown and n is small (less than 30):

$$\bar{x} \pm MOE = \bar{x} \pm t^*_{n-1}\left(\frac{s}{\sqrt{n}}\right)$$

where $\bar{x}$ is the sample mean, t^*_{n-1} is the critical t^*-value from the t-distribution with $n - 1$ degrees of freedom (where n is the sample size) and the confidence level desired, and s is the sample standard deviation (if the population's standard deviation is known, substitute it for s).

Use the t-table (Table A-2 in the appendix) to find the t^*-value for 95% with degrees of freedom for a sample size of 25 ($n-1=25$ $-1=24$). Find 95% in the CI row at the bottom of the table, and move up the column to intersect with the df/p row labeled 24: $t^* = 2.06390$.

Next, substitute the values into the formula:

$$MOE = 2.06390\left(\frac{5}{\sqrt{25}}\right) = 2.0639$$

The confidence interval is 84 ± 2.1 (rounded to nearest tenth), or 81.9 and 86.1.

---

511. **The 90% confidence interval for the average household size is between 3.13 and 3.67 people.**

Use the formula for the confidence interval when the population's standard deviation is unknown and n is small (less than 30):

$$\bar{x} \pm MOE = \bar{x} \pm t^*_{n-1}\left(\frac{s}{\sqrt{n}}\right)$$

where $\bar{x}$ is the sample mean, t^*_{n-1} is the critical t^*-value from the t-distribution with $n - 1$ degrees of freedom (where n is the sample size) and the confidence level desired, and s is the sample standard deviation (if the population's standard deviation is known, substitute it for s).

Use the t-table (Table A-2 in the appendix) to find the t^*-value for 90% with degrees of freedom for a sample size of 25 ($n-1=25\ -1=24$). Find 90% in the CI row at the bottom of the table, and move up the column to intersect with the df/p row labeled 24: $t^* = 1.710882$.

Next, substitute the values into the formula:

$$\text{MOE} = 1.710882\left(\frac{0.8}{\sqrt{25}}\right) \approx 0.2737$$

The confidence interval is 3.4 ± 0.27 (rounded to the nearest hundredth), or 3.13 to 3.67 people.

512. **The 95% confidence interval for teenagers' average number of social network friends is 66 to 104.**

Use the formula for the confidence interval when the population's standard deviation is unknown.

$$\bar{x} \pm \text{MOE} = \bar{x} \pm t^*_{n-1}\left(\frac{s}{\sqrt{n}}\right)$$

where $\bar{x}$ is the sample mean, t^*_{n-1} is the critical t^*-value from the t-distribution with $n-1$ degrees of freedom (where n is the sample size) and the confidence level desired, and s is the sample standard deviation (if the population's standard deviation is known, substitute it for s).

Use the t-table (Table A-2 in the appendix) to find the t^*-value for 95% with degrees of freedom for a sample size of 30 ($n-1=30-1=29$). Find 95% in the CI row at the bottom of the table, and move up the column to intersect with the df/p row labeled 29: $t^* = 2.04523$.

Next, substitute the values into the formula:

$$\text{MOE} = 2.04523\left(\frac{50}{\sqrt{30}}\right) \approx 18.6703$$

The confidence interval is 85 ± 19 (rounded to the nearest whole number), or 66 to 104 friends.

513. **The 95% confidence interval for the average length of the longest trip first-year college students took last year was between 273 and 527 miles.**

Use the formula for the confidence interval when the population's standard deviation is unknown and n is small (less than 30):

$$\bar{x} \pm \text{MOE} = \bar{x} \pm t^*_{n-1}\left(\frac{s}{\sqrt{n}}\right)$$

where $\bar{x}$ is the sample mean, t^*_{n-1} is the critical t^*-value from the t-distribution with $n-1$ degrees of freedom (where n is the sample size) and the confidence level desired, and s is the sample standard deviation (if the population's standard deviation is known, substitute it for s).

Use the t-table (Table A-2 in the appendix) to find the t^*-value for 95% with a degree of freedom for a sample size of 24 ($n-1 = 24-1 = 23$). Find 95% in the *CI* row at the bottom of the table, and move up the column to intersect with the *df/p* row labeled 23: $t^* = 2.06866$.

Next, substitute the values into the formula:

$$\text{MOE} = 2.06866 \left(\frac{300}{\sqrt{24}} \right) \approx 126.6790$$

The confidence interval is 400 ± 127 (rounded to the nearest whole number), or 273 to 527 miles.

514. **The 95% confidence interval for the average amount of purchases by mall shoppers that day is between \$54.75 and \$102.25.**

Use the formula for the confidence interval when the population's standard deviation is unknown and n is small (less than 30):

$$\bar{x} \pm \text{MOE} = \bar{x} \pm t^*_{n-1} \left(\frac{s}{\sqrt{n}} \right)$$

where $\bar{x}$ is the sample mean, t^*_{n-1} is the critical t^*-value from the t-distribution with $n-1$ degrees of freedom (where n is the sample size) and the confidence level desired, and s is the sample standard deviation (if the population's standard deviation is known, substitute it for s).

Use the t-table (Table A-2 in the appendix) to find the t^*-value for 95% with degrees of freedom for a sample size of 20 ($n-1 = 20-1 = 19$). Find 95% in the *CI* row at the bottom of the table, and move up the column to intersect with the *df/p* row labeled 19: $t^* = 2.09302$.

Next, substitute the values into the formula:

$$\text{MOE} = 2.09302 \left(\frac{50.75}{\sqrt{20}} \right) = 23.7517$$

The confidence interval is 78.50 ± 23.75 (rounded to the nearest hundredth), or \$54.75 to \$102.25.

515. **The 90% confidence interval for the average amount of time visitors spent in the museum that day is between 2.6 and 3.4 hours.**

Use the formula for the confidence interval when the population's standard deviation is unknown and n is small (less than 30):

$$\bar{x} \pm \text{MOE} = \bar{x} \pm t^*_{n-1} \left(\frac{s}{\sqrt{n}} \right)$$

where $\bar{x}$ is the sample mean, t^*_{n-1} is the critical t^*-value from the t-distribution with $n-1$ degrees of freedom (where n is the sample size) and the confidence level desired, and s is the sample standard deviation (if the population's standard deviation is known, substitute it for s).

Use the t-table (Table A-2 in the appendix) to find the t^*-value for 90% with degrees of freedom for a sample size of 20 ($n-1=20-1=19$). Find 90% in the *CI* row at the bottom of the table, and move up the column to intersect with the *df/p* row labeled 19: $t^* = 1.729133$.

Next, substitute the values into the formula:

$$\text{MOE} = 1.729133\left(\frac{1}{\sqrt{20}}\right) \approx 0.3866$$

The confidence interval is 3 ± 0.4 (rounded to the nearest tenth), or 2.6 to 3.4 hours (or between 2 hours 36 minutes and 3 hours and 24 minutes).

516. The 90% confidence interval for the average weight of all 1-pound loaves of bread is between 17.65 and 18.35 ounces.

Use the formula for the confidence interval when the population's standard deviation is unknown and n is small (less than 30):

$$\bar{x} \pm \text{MOE} = \bar{x} \pm t_{n-1}^*\left(\frac{s}{\sqrt{n}}\right)$$

where $\bar{x}$ is the sample mean, t^*_{n-1} is the critical t^*-value from the t-distribution with $n-1$ degrees of freedom (where n is the sample size) and the confidence level desired, and s is the sample standard deviation (if the population's standard deviation is known, substitute it for s).

Use the t-table (Table A-2 in the appendix) to find the t^*-value for 90% with degrees of freedom for a sample size of 50 ($n-1=50-1=49$). Find 90% in the *CI* row at the bottom of the table, and move up the column to the z-value (because when $df > 30$, the t- and z-values are almost the same), or 1.644854. For ease of calculation, you can use 1.645 (which is also the t-table value rounded to three decimal places).

Next, substitute the values into the formula:

$$\text{MOE} = 1.644854\left(\frac{1.5}{\sqrt{50}}\right) \approx 0.3489$$

The confidence interval is 18 ± 0.35 (rounded to the nearest hundredth), or 17.65 to 18.35 ounces.

517. The 80% confidence limit for the average number of visits by a person to the store each month is between 1.93 and 3.67 visits.

Use the formula for the confidence interval when the population's standard deviation is unknown and n is small (less than 30):

$$\bar{x} \pm \text{MOE} = \bar{x} \pm t_{n-1}^*\left(\frac{s}{\sqrt{n}}\right)$$

where $\bar{x}$ is the sample mean, t^*_{n-1} is the critical t^*-value from the t-distribution with $n-1$ degrees of freedom (where n is the sample size) and the confidence level desired, and s is the sample standard deviation (if the population's standard deviation is known, substitute it for s).

Use the t-table (Table A-2 in the appendix) to find the t^*-value for 80% with degrees of freedom for a sample size of 10 ($n-1=10-1=9$). Find 80% in the CI row at the bottom of the table, and move up the column to intersect with the df/p row labeled 9: $t^* = 1.383029$.

Next, substitute the values into the formula:

$$\text{MOE} = 1.383029\left(\frac{2}{\sqrt{10}}\right) \approx 0.8747$$

The confidence interval is 2.8 ± 0.87 (rounded to the nearest hundredth), or 1.93 to 3.67 visits.

---

518. **The 90% confidence interval for the average number of movies per month watched by first-year university students is 3.8 to 6.2 movies.**

Use the formula for the confidence interval when the population's standard deviation is unknown and n is small (less than 30):

$$\bar{x} \pm \text{MOE} = \bar{x} \pm t^*_{n-1}\left(\frac{\sigma}{\sqrt{n}}\right)$$

where $\bar{x}$ is the sample mean, t^*_{n-1} is the critical t^*-value from the t-distribution with $n-1$ degrees of freedom (where n is the sample size) and the confidence level desired, and s is the sample standard deviation (if the population's standard deviation is known, substitute it for s).

Use the t-table (Table A-2 in the appendix) to find the t^*-value for 90% with degrees of freedom for a sample size of 18 ($n-1=18-1=17$). Find 90% in the CI row at the bottom of the table, and move up the column to intersect with the df/p row labeled 17: $t^* = 1.739607$.

Next, substitute the values into the formula:

$$\text{MOE} = 1.739607\left(\frac{3}{\sqrt{18}}\right) \approx 1.2301$$

The confidence interval is 5 ± 1.2 (rounded to the nearest tenth), or 3.8 to 6.2 movies.

---

519. **The 99% confidence interval for the average spending by a park visitor that day is $28.64 to $35.36.**

Use the formula for the confidence interval when the population's standard deviation is unknown and n is small (less than 30):

$$\bar{x} \pm \text{MOE} = \bar{x} \pm t^*_{n-1}\left(\frac{\sigma}{\sqrt{n}}\right)$$

where $\bar{x}$ is the sample mean, t^*_{n-1} is the critical t^*-value from the t-distribution with $n-1$ degrees of freedom (where n is the sample size) and the confidence level desired, and s is the sample standard deviation (if the population's standard deviation is known, substitute it for s).

Use the t-table (Table A-2 in the appendix) to find the t^*-value for 99% with degrees of freedom for a sample size of 25 ($n-1=25-1=24$). Find 99% in the CI row at the bottom of the table, and move up the column to intersect with the df/p row labeled 24: $t^* = 2.79694$.

Next, substitute the values into the formula:

$$\text{MOE} = 2.79694\left(\frac{6.00}{\sqrt{25}}\right) \approx 3.3563$$

The confidence interval is $32.00 ± $3.36 (rounded to the nearest hundredth), or $28.64 to $35.36.

520. **D. Choices (A) and (B) (117.2; 117.6)**

Fractional sample sizes are always rounded up, even if the fractional is less than 0.5, so 117.2 and 117.6 are both rounded up to 118, while 118.1 is rounded up to 119. In other words, if you need a fraction of an individual to be included in the sample, no matter how small the fraction is, you need the whole individual to meet the conditions for the margin of error.

521. **122, 122, 132, 132**

Fractional results in sample size calculations are always rounded up, even if the fractional part is less than 0.5. Therefore, 121.1 and 121.5 both round up to 122, and 131.2 and 131.6 both round up to 132. In other words, if you need a fraction of an individual to be included in the sample, no matter how small the fraction is, you need the whole individual to meet the conditions for the margin of error.

522. **E. Choices (A), (B), and (C) (A larger sample often means greater costs; it may be difficult to recruit a larger sample [for example, if you're studying people with a rare disease]; at some point, increasing the sample size may not significantly improve precision [for instance, increasing the sample size from 3,000 to 3,500])**

Although it's important to have an adequate sample size when conducting a study, costs and difficulties with recruiting subjects can both limit the size of samples a researcher can work with. In addition, increasing the size of a large sample produces proportionately less of an increase in precision than increasing the size of a small sample so that the relatively small gains in precision may not be worth the added effort and cost.

523. **35 records**

The formula to calculate the sample size needed for a confidence interval for a sample mean is

$$n \geq \left(\frac{z^* \sigma}{\text{MOE}}\right)^2$$

where n is the sample size required, z^* is the value from Table A-1 for the chosen confidence level, σ is the population standard deviation, and MOE is the margin of error. For this example, z^* is 1.96, σ is 1.6, and the MOE is 1.5.

$$n \geq \left(\frac{z^* \sigma}{\text{MOE}} \right)^2$$

$$\geq \left(\frac{(1.96)(4.5)}{1.5} \right)^2$$

$$\geq 34.5744$$

Samples sizes are always rounded up to the nearest integer, so the sample size must be at least 35.

524. **50 records**

The formula to calculate sample size to estimate a mean, when the standard deviation is known, is

$$n \geq \left(\frac{z^* \sigma}{\text{MOE}} \right)^2$$

where n is the sample size required, z^* is the value from Table A-1 for the chosen confidence level, σ is the population standard deviation, and MOE is the margin of error. For this example, z^* is 1.96, σ is 1.6, and the MOE is 0.5.

$$n \geq \left(\frac{z^* \sigma}{\text{MOE}} \right)^2$$

$$\geq \left(\frac{1.96(1.8)}{0.5} \right)^2$$

$$\geq 49.787136$$

They need a sample of at least 50 to achieve the desired precision.

525. **The sample size must be at least 35 to produce a margin of error of ± 1 inch.**

The formula to find the required sample size based on a desired margin of error is

$$n \geq \left(\frac{z^* \sigma}{\text{MOE}} \right)^2$$

Here, MOE is the margin of error, z^* is the z^*-value corresponding to your desired confidence level, and σ is the population standard deviation. If σ is unknown, you can do a small pilot study to find the standard deviation of the sample (including making a conservative adjustment to the sample standard deviation to be safe).

Substitute the known values into the formula:

$$n \geq \left(\frac{1.96(3)}{1} \right)^2 \approx 34.57$$

Always round up the answer to the nearest whole number to be sure the sample size is large enough to give the margin of error needed. So, $n \geq 35$. That means that you need at least 35 boys in your sample to get a margin of error of no more than 1 inch for average height.

Note that a sample size of 35 will give you the margin of error you want; a higher sample size will give you an even lower margin of error.

526. **The sample size (number of students) must be at least 107 to produce a margin of error of ± $5.**

The formula to find the required sample size based on a desired margin of error is

$$n \geq \left(\frac{z^* \sigma}{\text{MOE}} \right)^2$$

Here, MOE is the margin of error, z^* is the z^*-value corresponding to your desired confidence level, and σ is the population standard deviation. If σ is unknown, you can do a small pilot study to find the standard deviation of the sample (including making a conservative adjustment to the sample standard deviation to be safe).

Substitute the known values into the formula:

$$n \geq \left(\frac{2.58(20)}{5} \right)^2 = 106.5$$

Always round up the answer to the nearest whole number to be sure the sample size is large enough to give the margin of error needed. So, $n \geq 107$. That means that you need at least 107 students in your sample to get a margin of error of ± $5 when finding the mean weekly earnings.

Note that a sample size of 107 will give you the margin of error you want; a higher sample size will give you an even lower margin of error.

527. **The sample size (number of students) must be at least 82 to produce a margin of error no more than ± $10.**

The formula to find the required sample size based on a desired margin of error is

$$n \geq \left(\frac{z^* \sigma}{\text{MOE}} \right)^2$$

Here, MOE is the margin of error, z^* is the z^*-value corresponding to your desired confidence level, and σ is the population standard deviation. If σ is unknown, you can do a small pilot study to find the standard deviation of the sample (including making a conservative adjustment to the sample standard deviation to be safe).

Substitute the known values into the formula:

$$n \geq \left(\frac{1.645(55)}{10} \right)^2 \approx 81.86$$

Always round up the answer to the nearest whole number to be sure the sample size is large enough to give the margin of error needed. So, $n \geq 82$. That means that you need at least 82 students in your sample to have a margin of error of no more than ± $10 for the mean weekly earnings.

Note that a sample size of 82 will give you the margin of error you want; a higher sample size will give you an even lower margin of error.

528. **The sample size (number of students) must be at least 89 to produce a margin of error no more than ± 0.25 hours.**

The formula to find the required sample size based on a desired margin of error is

$$n \geq \left(\frac{z^* \sigma}{\text{MOE}} \right)^2$$

Here, MOE is the margin of error, z^* is the z^*-value corresponding to your desired confidence level, and σ is the population standard deviation. If σ is unknown, you can do a small pilot study to find the standard deviation of the sample (including making a conservative adjustment to the sample standard deviation to be safe).

Substitute the known values into the formula:

$$n \geq \left(\frac{1.96(1.2)}{0.25} \right)^2 \approx 88.5105$$

Always round up the answer to the nearest whole number to be sure the sample size is large enough to give the margin of error needed. So, $n \geq 89$. That means that you need at least 89 students in your sample to have a margin of error of ± 0.25 hours for the average amount of sleep per night.

Note that a sample size of 89 will give you the margin of error you want; a higher sample size will give you an even lower margin of error.

529. **The sample size (number of games) must be at least 32 to produce a margin of error of no more than ± 800 people.**

The formula to find the required sample size based on a desired margin of error is

$$n \geq \left(\frac{z^* \sigma}{\text{MOE}} \right)^2$$

Here, MOE is the margin of error, z^* is the z^*-value corresponding to your desired confidence level, and σ is the population standard deviation. If σ is unknown, you can do a small pilot study to find the standard deviation of the sample (including making a conservative adjustment to the sample standard deviation to be safe).

Substitute the known values into the formula:

$$n \geq \left(\frac{1.96(2,300)}{800} \right)^2 \approx 31.7532$$

Always round up the answer to the nearest whole number to be sure the sample size is large enough to give the margin of error needed. So, $n \geq 32$. That means that you need at least 32 games in your sample to have a margin of error of ± 800 for average game attendance.

Note that a sample size of 32 will give you the margin of error you want; a higher sample size will give you an even lower margin of error.

530. **D. Choices (B) and (C) (You are using sample data to estimate a parameter; if you drew a different sample of the same size, you would expect the results to be slightly different.)**

Your goal is to estimate the proportion of all students in the university who are thinking of changing their major. However, you're working with a sample rather than the entire university population. Thus, you don't expect your parameter estimate to be exactly the same as the true parameter, and you realize that if you drew another sample of the same size, your parameter estimate would probably be slightly different. The confidence interval expresses this uncertainty.

531. **E. Choices (C) and (D) (an estimate of the proportion of all students at the university who are thinking of changing their major; the proportion of students in the sample of 100 who are thinking of changing their major)**

The value 0.38 represents the proportion of students in the sample who are thinking of changing their major. In other words, in your sample of 100, 38 students indicated they were thinking about changing their major, so $38/100 = 0.38$.

It is also an estimate of the proportion of all students in the university who are thinking of changing their major.

532. **Yes, because $n\hat{p}$ and $n(1-\hat{p})$ are both greater than 10.**

To use the normal approximation to the binomial to calculate a confidence interval, both $n\hat{p}$ and $n(1-\hat{p})$ must be greater than 10, where n is the sample size and $\hat{p}$ is the sample proportion.

In this case, $n = 100$ and $\hat{p} = 0.38$. So, plugging in the numbers, you get $n\hat{p} = (100)0.38 = 38$ and $n(1-\hat{p}) = (100)(1-0.38) = 62$.

Both 38 and 62 are greater than 10, so you can use the normal approximation.

533. **0.0485**

The formula for the standard error of a proportion is

$$SE = \sqrt{\frac{\hat{p}(1-\hat{p})}{n}}$$

where n is the sample size and $\hat{p}$ is the sample proportion.

In this example, $n = 100$ and $\hat{p} = 0.38$. So, simply plug in the numbers and solve for the standard error:

$$SE = \sqrt{\frac{0.38(1-0.38)}{100}}$$
$$= \sqrt{\frac{0.2356}{100}}$$
$$\approx 0.0485$$

534. **E. 99%**

Without doing any calculations, you know that the 99% confidence interval will be widest, because it includes the largest set of samples taken from the population. In other words, it includes the largest range of plausible guesses for the population parameter.

535. between 30% and 46%

To determine a confidence interval (CI) for one population proportion, use this formula:

$$CI = \hat{p} \pm z^* \sqrt{\frac{\hat{p}(1-\hat{p})}{n}}$$

Here, $\hat{p}$ is the sample proportion, n is the sample size, and z^* is the appropriate value from the standard normal distribution for your desired confidence level (see Table A-4 in the appendix for various confidence levels).

First, you have to find the sample proportion, $\hat{p}$, by dividing the number of "successes" (38 in this case) by the sample size (100):

$$\hat{p} = \frac{\text{Number of successes}}{n}$$

$$= \frac{38}{100} = 0.38$$

Then confirm whether you can use the normal approximation to the binomial. To use the normal approximation to the binomial to calculate a confidence interval, both $n\hat{p}$ and s $n(1-\hat{p})$ must be greater than 10, where n is the sample size and $\hat{p}$ is the sample proportion.

In this case, $n = 100$ and $\hat{p} = 0.38$. So, plugging in the numbers, you get $n\hat{p} = (100)0.38 = 38$ and $n(1-\hat{p}) = (100)(1-0.38) = 62$.

Both 38 and 62 are greater than 10, so you can use the normal approximation to the binomial.

Next, substitute the known values into the formula for the confidence interval and solve:

$$CI = 0.38 \pm 1.645 \sqrt{\frac{0.38(1-0.38)}{100}}$$

$$= 0.38 \pm 1.645 \sqrt{\frac{0.2356}{100}}$$

$$= 0.38 \pm 1.645 \sqrt{0.002356}$$

$$= 0.38 \pm 1.645(0.04854)$$

$$= 0.38 \pm 0.0798$$

Add and subtract to/from the sample proportion to find the range:

$$0.38 - 0.0798 = 0.3002$$
$$0.38 + 0.0798 = 0.4598$$

Then convert to percentages by multiplying by 100%:

$$0.3040(100\%) = 30.02\%$$
$$0.4598(100\%) = 45.98\%$$

Round to the nearest whole percentage point, so the 90% confidence interval for the proportion of all students thinking of changing their major is 30% to 46%.

536. between 28% and 48%

To determine a confidence interval (CI) for one population proportion, use this formula:

$$CI = \hat{p} \pm z^* \sqrt{\frac{\hat{p}(1-\hat{p})}{n}}$$

Here, $\hat{p}$ is the sample proportion, n is the sample size, and z^* is the appropriate value from the standard normal distribution for your desired confidence level (see Table A-4 in the appendix for various confidence levels).

First, you have to find the sample proportion, $\hat{p}$, by dividing the number of "successes" (38 in this case) by the sample size (100):

$$\hat{p} = \frac{\text{Number of successes}}{n}$$
$$= \frac{38}{100} = 0.38$$

Then confirm whether you can use the normal approximation to the binomial. To use the normal approximation to the binomial to calculate a confidence interval, both $n\hat{p}$ must and $n(1-\hat{p})$ be greater than 10, where n is the sample size and $\hat{p}$ is the sample proportion.

In this case, $n = 100$ and $\hat{p} = 0.38$. So, plugging in the numbers, you get $n\hat{p} = (100)0.38 = 38$ and $n(1-\hat{p}) = (100)(1-0.38) = 62$.

Both 38 and 62 are greater than 10, so you can use the normal approximation to the binomial.

Next, substitute the known values into the formula for the confidence interval and solve:

$$CI = 0.38 \pm 1.96 \sqrt{\frac{0.38(1-0.38)}{100}}$$
$$= 0.38 \pm 1.96 \sqrt{\frac{0.38(0.62)}{100}}$$
$$= 0.38 \pm 1.96 \sqrt{\frac{0.2356}{100}}$$
$$= 0.38 \pm 1.96 \sqrt{0.002356}$$
$$= 0.38 \pm 1.96(0.04854)$$
$$= 0.38 \pm 0.0951$$

Add and subtract to/from the sample proportion to find the range:

$$0.38 + 0.0951 = 0.4751$$
$$0.38 - 0.0951 = 0.2849$$

Then convert to percentages by multiplying by 100%:

$$0.4751(100\%) = 47.51\%$$
$$0.2849(100\%) = 28.49\%$$

Round to the nearest percentage point, so the 95% confidence interval for the proportion of all students thinking of changing their major is 28% to 48%.

537. 0.75, 0.03

Find the sample proportion, $\hat{p}$ by dividing the number of "successes" (75 in this case) by the sample size (200):

$$\hat{p} = \frac{\text{Number of successes}}{n}$$

$$= \frac{38}{100} = 0.38$$

The sample proportion represents the proportion of customers in the sample who are satisfied with their online purchases.

Then use the following formula to find the standard error (SE):

$$SE = \sqrt{\frac{\hat{p}(1-\hat{p})}{n}}$$

where $\hat{p}$ is the sample proportion and n is the sample size:

$$SE = \sqrt{\frac{0.75(1-0.75)}{200}}$$

$$= \sqrt{\frac{0.75(0.25)}{200}}$$

$$= \sqrt{\frac{0.1875}{200}}$$

$$\approx 0.03$$

So, the standard error for the sample proportion in this example is 0.03.

538. 0.06

Use the formula for finding the margin of error (MOE):

$$MOE = z^* \sqrt{\frac{\hat{p}(1-\hat{p})}{n}}$$

Here, $\hat{p}$ is the sample proportion, n is the sample size, and z^* is the appropriate value from the standard normal distribution for your desired confidence level (see Table A-4 in the appendix for various confidence levels). For a 95% confidence level, the z^*-value is 1.96. Note that the margin of error is the z^*-value times the standard error.

Now, plug in the known values and solve:

$$MOE = 1.96 \sqrt{\frac{0.75(0.25)}{200}}$$

$$= 1.96 \sqrt{\frac{0.1875}{200}}$$

$$= 1.96(0.0306)$$

$$\approx 0.06$$

With 95% confidence, the margin of error is ± 0.06 for estimating the proportion of all customers who purchased products online in the past 12 months.

539. 0.67 to 0.83

To determine a confidence interval (CI) for one population proportion, use this formula:

$$CI = \hat{p} \pm z^* \sqrt{\frac{\hat{p}(1-\hat{p})}{n}}$$

Here, $\hat{p}$ is the sample proportion, n is the sample size, and z^* is the appropriate value from the standard normal distribution for your desired confidence level (see Table A-4 for various confidence levels). For a confidence level of 99%, $z^* = 2.58$.

First, you have to find the sample proportion, $\hat{p}$, by dividing the number of "successes" (150 in this case) by the sample size (200):

$$\hat{p} = \frac{\text{Number having the characteristic}}{n}$$

$$= \frac{150}{200} = 0.75$$

Then confirm whether you can use the normal approximation to the binomial. To use the normal approximation to the binomial to calculate a confidence interval, both $n(1-\hat{p})$ and must be greater than 10, where n is the sample size and $\hat{p}$ is the sample proportion.

In this case, $n = 200$ and $\hat{p} = 0.75$. So, plugging in the numbers, you get $n\hat{p} = (200)(0.75) = 150$ and $n(1-\hat{p}) = (200)(1-0.75) = 50$.

Both 150 and 50 are greater than 10, so you can use the normal approximation to the binomial.

Next, substitute the known values into the formula for the confidence interval and solve:

$$CI = 0.75 \pm 2.58 \sqrt{\frac{0.75(1-0.75)}{200}}$$

$$= 0.75 \pm 2.58 \sqrt{\frac{0.75(0.25)}{200}}$$

$$= 0.75 \pm 2.58 \sqrt{\frac{0.1875}{200}}$$

$$= 0.75 \pm 2.58 \sqrt{0.0009375}$$

$$= 0.75 \pm 2.58(0.0306)$$

$$\approx 0.75 \pm 0.0789$$

Add and subtract to/from the sample proportion to find the range:

$$0.75 - 0.0789 = 0.6711$$
$$0.75 + 0.0789 = 0.8289$$

Round to two decimal places, so the 99% confidence interval for the proportion of all customers who purchased products online in the past 12 months is 0.67 to 0.83.

540. **Yes, because $n\hat{p}$ and $n(1-\hat{p})$ are both greater than 10.**

To use the normal approximation to the binomial, both $n\hat{p}$ and $n(1-\hat{p})$ must be greater than 10.

For this example, n=80 and $\hat{p}=0.15$ So, plugging in the numbers, you find that $n\hat{p}=80(0.15)=12$ and $n(1-\hat{p})=80(0.85)=68$.

Both 12 and 68 are greater than 10, so you can use the normal approximation for this data.

541. **0.0989 to 0.2011**

To use the normal approximation to the binomial, both $n\hat{p}$ and $n(1-\hat{p})$ must be greater than 10.

For this example, $n=80$ and $\hat{p}=0.15$. So, plugging in the numbers, you find that $n\hat{p}=80(0.15)=12$ and $n(1-\hat{p})=80(0.85)=68$.

Both 12 and 68 are greater than 10, so you can use the normal approximation.

To calculate a confidence interval (CI) for a proportion, use this formula:

$$CI = \hat{p} \pm z^* \sqrt{\frac{\hat{p}(1-\hat{p})}{n}}$$

where $\sqrt{\frac{\hat{p}(1-\hat{p})}{n}}$ is the standard error, and z^* is the appropriate value from the standard normal distribution for your desired confidence level (see Table A-4 in the appendix for various confidence levels). For a confidence level of 80%, $z^*=1.28$.

Calculate the standard error:

$$SE = \sqrt{\frac{0.15(1-0.15)}{80}}$$
$$= \sqrt{\frac{0.1275}{80}}$$
$$= 0.03992$$

Plug in the known values into the formula for the confidence interval and solve:

$$CI = 0.15 \pm 1.28(0.03992)$$
$$= 0.15 \pm 0.0510976$$
$$= 0.15 \pm 0.0511$$

To find the limits of the confidence interval, add and subtract 0.0511 from the sample proportion:

$$0.15 - 0.0511 = 0.0989$$
$$0.15 + 0.0511 = 0.2011$$

The 80% confidence interval for the population proportion is 0.0989 to 0.2011.

542. **0.0843 to 0.2175**

To use the normal approximation to the binomial, both $n\hat{p}$ and $n(1-\hat{p})$ must be greater than 10.

For this example, $n = 80$ and $\hat{p} = 0.15$. So, plugging in the numbers, you find that $n\hat{p} = 80(0.15) = 12$ and $n(1 - \hat{p}) = 80(0.85) = 68$.

Both 12 and 68 are greater than 10, so you can use the normal approximation.

To calculate a confidence interval (CI) for a proportion, use this formula:

$$CI = \hat{p} \pm z^* \sqrt{\frac{\hat{p}(1 - \hat{p})}{n}}$$

where $\sqrt{\frac{\hat{p}(1 - \hat{p})}{n}}$ is the standard error, and z^* is the appropriate value from the standard normal distribution for your desired confidence level (see Table A-4 in the appendix for various confidence levels). For a confidence level of 90%, $z^* = 1.645$.

Calculate the standard error:

$$SE = \sqrt{\frac{0.15(1 - 0.15)}{80}}$$
$$= \sqrt{\frac{0.1275}{80}}$$
$$= 0.03992$$

Plug in the known values into the formula for the confidence interval and solve:

$$CI = 0.15 \pm 1.645(0.03992)$$
$$\approx 0.15 \pm 0.0657$$

To find the limits of the confidence interval, add and subtract 0.0657 from the sample proportion:

$$0.15 - 0.0657 = 0.0843$$
$$0.15 + 0.0657 = 0.2175$$

The 90% confidence interval for the population proportion is 0.0843 to 0.2175.

543. 0.0718 to 0.2282

To use the normal approximation to the binomial, both $n\hat{p}$ and $n(1 - \hat{p})$ must be greater than 10.

For this example, $n = 80$ and $\hat{p} = 0.15$. So, plugging in the numbers, you find that $n\hat{p} = 80(0.15) = 12$ and $n(1 - \hat{p}) = 80(0.85) = 68$.

Both 12 and 68 are greater than 10, so you can use the normal approximation.

To calculate a confidence interval (CI) for a proportion, use this formula:

$$CI = \hat{p} \pm z^* \sqrt{\frac{\hat{p}(1 - \hat{p})}{n}}$$

where $\sqrt{\frac{\hat{p}(1 - \hat{p})}{n}}$ is the standard error, and z^* is the appropriate value from the standard normal distribution for your desired confidence level (see Table A-4 in the appendix for various confidence levels). For a confidence level of 95%, $z^* = 1.96$.

Calculate the standard error:

$$SE = \sqrt{\frac{0.15(1-0.15)}{80}}$$

$$= \sqrt{\frac{0.1275}{80}}$$

$$= 0.03992$$

Plug in the known values into the formula for the confidence interval and solve:

$$CI = 0.15 \pm 1.96(0.03992)$$

$$\approx 0.15 \pm 0.0782$$

To find the limits of the confidence interval, add and subtract 0.0782 from the sample proportion:

$$0.15 - 0.0782 = 0.0718$$
$$0.15 + 0.0782 = 0.2282$$

The 95% confidence interval for the population proportion is 0.0718 to 0.2282.

544. **0.0570 to 0.2430**

To use the normal approximation to the binomial, both $n\hat{p}$ and $n(1-\hat{p})$ must be greater than 10.

For this example, $n = 80$ and $\hat{p} = 0.15$. So, plugging in the numbers, you find that $n\hat{p} = 80(0.15) = 12$ and $n(1-\hat{p}) = 80(0.85) = 68$.

Both 12 and 68 are greater than 10, so you can use the normal approximation.

To calculate a confidence interval (CI) for a proportion, use this formula:

$$CI = \hat{p} \pm z^* \sqrt{\frac{\hat{p}(1-\hat{p})}{n}}$$

where $\sqrt{\frac{\hat{p}(1-\hat{p})}{n}}$ is the standard error, and z* is the appropriate value from the standard normal distribution for your desired confidence level (see Table A-4 in the appendix for various confidence levels). For a confidence level of 98%, $z^* = 2.33$.

Calculate the standard error:

$$SE = \sqrt{\frac{0.15(1-0.15)}{80}}$$

$$= \sqrt{\frac{0.1275}{80}}$$

$$= 0.03992$$

Plug in the known values into the formula for the confidence interval and solve:

$$CI = 0.15 \pm 2.33(0.03992)$$

$$\approx 0.15 \pm 0.0930$$

To find the limits of the confidence interval, add and subtract 0.0930 from the sample proportion:

$$0.15 - 0.0930 = 0.0570$$
$$0.15 + 0.0930 = 0.2430$$

The 98% confidence interval for the population proportion is 0.0570 to 0.2430.

545. 0.0470 to 0.2530

To use the normal approximation to the binomial, both $n\hat{p}$ and $n(1-\hat{p})$ must be greater than 10.

For this example, $n = 80$ and $\hat{p} = 0.15$. So, plugging in the numbers, you find that $n\hat{p} = 80(0.15) = 12$ and $n(1-\hat{p}) = 80(0.85) = 68$.

Both 12 and 68 are greater than 10, so you can use the normal approximation.

To calculate a confidence interval (CI) for a proportion, use this formula:

$$CI = \hat{p} \pm z^* \sqrt{\frac{\hat{p}(1-\hat{p})}{n}}$$

where $\sqrt{\frac{\hat{p}(1-\hat{p})}{n}}$ is the standard error, and z^* is the appropriate value from the standard normal distribution for your desired confidence level (see Table A-4 in the appendix for various confidence levels). For a confidence level of 99%, $z^* = 2.58$.

Calculate the standard error:

$$SE = \sqrt{\frac{0.15(1-0.15)}{80}}$$
$$= \sqrt{\frac{0.1275}{80}}$$
$$= 0.03992$$

Plug in the known values into the formula for the confidence interval and solve:

$$CI = 0.15 \pm 2.58(0.03992)$$
$$\approx 0.15 \pm 0.1030$$

To find the limits of the confidence interval, add and subtract 0.1030 from the sample proportion:

$$0.15 - 0.1030 = 0.0470$$
$$0.15 + 0.1030 = 0.2530$$

The 99% confidence interval for the population proportion is 0.0470 to 0.2530.

546. A. 0.15 to 0.35

For the same data, a 98% confidence interval will be wider than a 95% confidence interval.

547. **D. Choice (A) or (B) (80% or 90%)**

For the same data, a 99% confidence interval will be wider than a 95% confidence interval, while an 80% and 90% confidence interval will be narrower. The confidence interval 0.22 to 0.28 is narrower than the 95% confidence interval of 0.20 to 0.30, so it must represent a lower level of confidence.

548. **0.55**

If you had only one number to use to estimate the population proportion, you'd use the sample proportion. To find the sample proportion, $\hat{p}$, divide the number of "successes" (88 in this case) by the sample size (160):

$$\hat{p} = \frac{\text{Number of successes}}{n}$$
$$= \frac{88}{160} = 0.55$$

However, note that a confidence interval is actually the best estimate of a population proportion because you know that the sample proportion changes with each new sample. So, using the sample proportion of 0.55, plus or minus a margin of error, gives you the best possible estimate.

549. **0.0393**

The formula for the standard error of a sample proportion is

$$SE = \sqrt{\frac{\hat{p}(1-\hat{p})}{n}}$$

where n is the sample size, and $\hat{p}$ is the sample proportion. In this example, $n = 160$ and $\hat{p} = 88/160$ (number of success divided by sample size) $= 0.55$.

Now, plug in the known values and solve for the standard error:

$$SE = \sqrt{\frac{0.55(1-0.55)}{160}}$$
$$= \sqrt{\frac{0.2475}{160}}$$
$$\approx 0.0393$$

550. **0.0770**

To find the margin of error (MOE), use the formula

$$MOE = z^* \sqrt{\frac{\hat{p}(1-\hat{p})}{n}}$$

where $\hat{p}$ is the sample proportion, n is the sample size, and z^* is the appropriate value from the standard normal distribution for your desired confidence level (see Table A-4 in the appendix for various confidence levels). For a 95% confidence level, the z^*-value is 1.96. Note that the margin of error is the z^*-value times the standard error.

First, you have to find the sample proportion, $\hat{p}$, by dividing the number of "successes" (in this case, 88) by the sample size (160):

$$\hat{p} = \frac{\text{Number of successes}}{n}$$

$$= \frac{88}{160} = 0.55$$

Then, plug in the known values to the formula for the margin of error:

$$\text{MOE} = 1.96\sqrt{\frac{0.55(1-0.55)}{160}}$$

$$= 1.96\sqrt{\frac{0.55(0.45)}{160}}$$

$$= 1.96\sqrt{\frac{0.2475}{160}}$$

$$= 1.96(0.0393)$$

$$\approx 0.0770$$

With a 95% confidence level, 0.0770 is the margin of error for estimating the proportion of all adults in the city who favor the new tax.

551. 0.1014

To find the margin of error (MOE), use the formula

$$\text{MOE} = z^*\sqrt{\frac{\hat{p}(1-\hat{p})}{n}}$$

where $\hat{p}$ is the sample proportion, n is the sample size, and z^* is the appropriate value from the standard normal distribution for your desired confidence level (see Table A-4 in the appendix for various confidence levels). For a 99% confidence level, the z^*-value is 2.58. Note that the margin of error is the z^*-value times the standard error.

First, you have to find the sample proportion, $\hat{p}$, by dividing the number of "successes" (in this case, 88) by the sample size (160):

$$\hat{p} = \frac{\text{Number of successes}}{n}$$

$$= \frac{88}{160} = 0.55$$

Then, plug in the known values to the formula for the margin of error:

$$\text{MOE} = 2.58\sqrt{\frac{0.55(1-0.55)}{160}}$$

$$= 2.58\sqrt{\frac{0.55(0.45)}{160}}$$

$$= 2.58\sqrt{\frac{0.2475}{160}}$$

$$= 2.58(0.0393)$$

$$\approx 0.1014$$

With a 99% confidence level, 0.1014 is the margin of error for estimating the proportion of all adults in the city who favor the new tax.

552. 0.0503

To find the margin of error (MOE), use the formula

$$MOE = z^* \sqrt{\frac{\hat{p}(1-\hat{p})}{n}}$$

where $\hat{p}$ is the sample proportion, n is the sample size, and z^* is the appropriate value from the standard normal distribution for your desired confidence level (see Table A-4 for various confidence levels). For an 80% confidence level, the z^*-value is 1.28. Note that the margin of error is the z^*-value times the standard error.

First, you have to find the sample proportion, $\hat{p}$, by dividing the number of "successes" (in this case, 88) by the sample size (160):

$$\hat{p} = \frac{\text{Number of successes}}{n}$$
$$= \frac{88}{160} = 0.55$$

Then, plug in the known values to the formula for the margin of error:

$$MOE = 1.28 \sqrt{\frac{0.55(1-0.55)}{160}}$$
$$= 1.28 \sqrt{\frac{0.55(0.45)}{160}}$$
$$= 1.28 \sqrt{\frac{0.2475}{160}}$$
$$= 1.28(0.0393)$$
$$\approx 0.0503$$

With an 80% confidence level, the margin of error for estimating the proportion of all adults in the city who favor the new tax is 0.0503.

553. 0.30

If you could use only one number to estimate the difference between two population proportions, you'd use the difference in the two sample proportions. So, calling the population of males Population 1 and the population of females Population 2, for this specific data set from these samples taken from the populations, you get

$$\hat{p}_1 - \hat{p}_2 = 0.55 - 0.25$$
$$= 0.30$$

Note, however, that a confidence interval is the best estimate for the difference in two population proportions because you know the sample proportions change as soon as the sample changes, and a confidence interval provides a range of likely values rather than just one number for the population parameter. So, using 0.30 plus or minus a margin of error gives you the best possible estimate.

554. 0.0660

To calculate the standard error for the estimated difference in two population proportions, use the formula

$$SE = \sqrt{\frac{\hat{p}_1(1-\hat{p}_1)}{n_1} + \frac{\hat{p}_2(1-\hat{p}_2)}{n_2}}$$

where $\hat{p}_1$ and n_1 are the sample proportion and sample size of the sample from Population 1, and $\hat{p}_2$ and n_2 are the sample proportion and sample size of the sample from Population 2.

Treating the sample of males from Population 1 as Sample 1 and the sample of females from Population 2 as Sample 2, plug in the numbers and solve:

$$SE = \sqrt{\frac{0.55(1-0.55)}{100} + \frac{0.25(1-0.25)}{100}}$$
$$= \sqrt{\frac{0.55(0.45)}{100} + \frac{0.25(0.75)}{100}}$$
$$= \sqrt{0.002475 + 0.001875}$$
$$= \sqrt{0.00435}$$
$$\approx 0.0660$$

So, the standard error for the estimate of the difference in proportions in the male and female populations is 0.0660.

555. between 17% and 43%

To find a confidence interval when estimating the difference of two population proportions, use the formula

$$CI = (\hat{p}_1 - \hat{p}_2) \pm z^* \sqrt{\frac{\hat{p}_1(1-\hat{p}_1)}{n_1} + \frac{\hat{p}_2(1-\hat{p}_2)}{n_2}}$$

where $\hat{p}_1$ and n_1 are the sample proportion and sample size of the sample taken from Population 1, $\hat{p}_2$ and n_2 are the sample proportion and sample size of the sample taken from Population 2, and z^* is the appropriate value from the standard normal distribution for your desired confidence level (see Table A-4 in the appendix for various confidence levels).

To solve, follow these steps:

1. Use the confidence level to find the appropriate z^*-value by referring to Table A-4. The z^*-value for a confidence level of 95% is 1.96.

2. To make the calculations somewhat easier, label the group of males as "Sample 1" and the group of females as "Sample 2."

3. For each proportion, divide the number having the attribute by the sample size:

$$\hat{p} = \frac{\text{Number having the characteristic}}{n}$$

Sample 1: $\frac{55}{100} = 0.55$

Sample 2: $\frac{25}{100} = 0.25$

4. Substitute the values into the formula for the confidence interval and solve:

$$CI = (0.55 - 0.25) \pm 1.96 \sqrt{\frac{0.55(1-0.55)}{100} + \frac{0.25(1-0.25)}{100}}$$

$$= 0.30 \pm 1.96 \sqrt{\frac{0.55(0.45)}{100} + \frac{0.25(0.75)}{100}}$$

$$= 0.30 \pm 1.96 \sqrt{0.002475 + 0.001875}$$

$$= 0.30 \pm 1.96 \sqrt{0.00435}$$

$$= 0.30 \pm 1.96(0.0660), \text{ rounded}$$

$$\approx 0.30 \pm 0.12936$$

5. Subtract and add the margin of error:

$$0.30 - 0.12936 = 0.17064$$
$$0.30 + 0.12936 = 0.42936$$

6. Convert to percentages by multiplying by 100%:

$$0.17064(100\%) = 17.064\%$$
$$0.42936(100\%) = 42.936\%$$

Round to the nearest whole percentage point so the 95% confidence interval is 17% to 43%.

This is a 95% confidence interval for the difference in the percentage of all males and females favoring Johnson among all likely voters. Because you subtracted the sample proportion of females from the sample proportion of males to get these results, you can conclude that the males are the ones with a higher likelihood to vote for candidate Johnson.

556. between 19% and 41%

To find a confidence interval when estimating the difference of two population proportions, use the formula

$$CI = (\hat{p}_1 - \hat{p}_2) \pm z^* \sqrt{\frac{\hat{p}_1(1-\hat{p}_1)}{n_1} + \frac{\hat{p}_2(1-\hat{p}_2)}{n_2}}$$

where $\hat{p}_1$ and n_1 are the sample proportion and sample size of the sample taken from Population 1, $\hat{p}_2$ and n_2 are the sample proportion and sample size of the sample taken from Population 2, and z^* is the appropriate value from the standard normal distribution for your desired confidence level (see Table A-4 in the appendix for various confidence levels).

To solve, follow these steps:

1. Use the confidence level to find the appropriate z^*-value by referring to Table A-4. The z^*-value for a confidence level of 90% is 1.645.

2. To make the calculations somewhat easier, label the sample of males taken from Population 1 as "Sample 1" and the sample of females taken from Population 2 as "Sample 2."

3. For each proportion, divide the number having the attribute by the sample size:

$$\hat{p} = \frac{\text{Number having the characteristic}}{n}$$

Sample 1: $\frac{55}{100} = 0.55$

Sample 2: $\frac{25}{100} = 0.25$

4. Substitute the values into the formula for the confidence interval and solve:

$$CI = (0.55 - 0.25) \pm 1.96\sqrt{\frac{0.55(1-0.55)}{100} + \frac{0.25(1-0.25)}{100}}$$

$$= 0.30 \pm 1.96\sqrt{\frac{0.55(0.45)}{100} + \frac{0.25(0.75)}{100}}$$

$$= 0.30 \pm 1.96\sqrt{0.002475 + 0.001875}$$

$$= 0.30 \pm 1.96\sqrt{0.00435}$$

$$= 0.30 \pm 1.96(0.0660), \text{ rounded}$$

$$\approx 0.30 \pm 0.12936$$

5. Subtract and add the margin of error:

$$0.30 - 0.10857 = 0.19143$$
$$0.30 + 0.10857 = 0.40857$$

6. Convert to percentages by multiplying by 100%:

$$0.19143(100\%) = 19.143\%$$
$$0.40857(100\%) = 40.857\%$$

Round to the nearest whole percentage point so the 90% confidence interval is 19% to 41%.

This is a 90% confidence interval for the difference in the percentage of all males and females favoring Johnson among all likely voters. Because you subtracted the sample proportion of females from the sample proportion of males to get these results, you can conclude that the males are the ones with a higher likelihood to vote for candidate Johnson.

557. between 13% and 47%

To find a confidence interval when estimating the difference of two population proportions, use the formula

$$CI = (\hat{p}_1 - \hat{p}_2) \pm z^*\sqrt{\frac{\hat{p}_1(1-\hat{p}_1)}{n_1} + \frac{\hat{p}_2(1-\hat{p}_2)}{n_2}}$$

where $\hat{p}_1$ and n_1 are the sample proportion and sample size of the sample taken from Population 1, $\hat{P}_2$ and n_2 are the sample proportion and sample size of the sample taken from Population 2, and z* is the appropriate value from the standard normal distribution for your desired confidence level (see Table A-4 in the appendix for various confidence levels).

To solve, follow these steps:

1. Use the confidence level to find the appropriate z*-value by referring to Table A-4. The z*-value for a confidence level of 99% is 2.58.

2. To make the calculations somewhat easier, label the sample of males taken from Population 1 as "Sample 1" and the sample of females taken from Population 2 as "Sample 2."

3. For each proportion, divide the number having the attribute by the sample size:

$$\hat{p} = \frac{\text{Number having the characteristic}}{n}$$

Sample 1: $\frac{55}{100} = 0.55$

Sample 2: $\frac{25}{100} = 0.25$

4. Substitute the values into the formula and solve:

$$CI = (0.55 - 0.25) \pm 1.645\sqrt{\frac{0.55(1-0.55)}{100} + \frac{0.25(1-0.25)}{100}}$$

$$= 0.30 \pm 1.645\sqrt{\frac{0.55(0.45)}{100} + \frac{0.25(0.75)}{100}}$$

$$= 0.30 \pm 1.645\sqrt{0.002475 + 0.001875}$$

$$= 0.30 \pm 1.645\sqrt{0.00435}$$

$$= 0.30 \pm 1.645(0.0660), \text{ rounded}$$

$$\approx 0.30 \pm 0.10857$$

5. Subtract and add the margin of error:

$$0.30 - 0.17028 = 0.12972$$
$$0.30 + 0.17028 = 0.47028$$

6. Convert to percentages by multiplying by 100%:

$$0.12972(100\%) = 12.972\%$$
$$0.47028(100\%) = 47.028\%$$

Round to the nearest whole percentage point so the 99% confidence interval is 13% to 47%.

This is a 99% confidence interval for the difference in the percentage of all males and females favoring Johnson among all likely voters. Because you subtracted the sample proportion of females from the sample proportion of males to get these results, you can conclude that the males are the ones with a higher likelihood to vote for candidate Johnson.

558.   0.3333

If you could choose only one number to estimate the difference in two population proportions, you'd use the difference between two sample proportions (one from the large cites and one from the small cities).

For cities with more than 1 million in population, 220 out of 300 adults wanted increased funding, so the proportion wanting increased funding is 220/300 = 0.7333.

For cities with fewer than 100,000 in population, 120 of 300 adults wanted increased funding, so the proportion wanting increased funding is 120/300 = 0.4.

The difference in the sample proportions is 0.7333 – 0.4 = 0.3333 for this sample of data. Because of the order of subtraction (large cities minus small cities), the value of 0.3333 means the proportion in favor from large cities is larger than the proportion in favor for small cities for these samples.

Note, however, that a confidence interval is the best estimate for the difference in population proportions because you know that the sample proportions change as soon as the sample changes, and a confidence interval provides a range of likely values rather than just one number for the population parameter. So, adding and subtracting a margin of error to the value of 0.3333 gives you the best possible estimate.

559. **0.2706 to 0.3960**

To find a confidence interval for the difference of two population proportions, use the formula

$$CI = (\hat{p}_1 - \hat{p}_2) \pm z^* \sqrt{\frac{\hat{p}_1(1-\hat{p}_1)}{n_1} + \frac{\hat{p}_2(1-\hat{p}_2)}{n_2}}$$

where $\hat{p}_1$ and n_1 are the sample proportion and sample size of the sample taken from Population 1, $\hat{p}_2$ and n_2 are the sample proportion and sample size of the sample taken from Population 2, and z^* is the appropriate value from the standard normal distribution for your desired confidence level (see Table A-4 in the appendix for various confidence levels).

To solve, follow these steps:

1. Use the confidence level to find the appropriate z^*-value by referring to Table A-4. The z^*-value for a confidence level of 90% is 1.645.

2. To make the calculations somewhat easier, label the sample of adults taken from large cities as "Sample 1" and the sample of adults taken from small cities as "Sample 2."

3. For each proportion, divide the number having the attribute by the sample size:

$$\hat{p} = \frac{\text{Number having the characteristic}}{n}$$

Sample 1: $\frac{220}{300} \approx 0.7333$

Sample 2: $\frac{120}{300} = 0.4$

4. Substitute the values into the formula and solve:

$$CI = (\hat{p}_1 - \hat{p}_2) \pm z^* \sqrt{\frac{\hat{p}_1(1-\hat{p}_1)}{n_1} + \frac{\hat{p}_2(1-\hat{p}_2)}{n_2}}$$

$$= (0.7333 - 0.4) \pm 1.645 \sqrt{\frac{0.7333(1-0.7333)}{300} + \frac{0.4(1-0.4)}{300}}$$

$$= 0.3333 \pm 1.645 \sqrt{\frac{0.19557}{300} + \frac{0.24}{300}}$$

$$= 0.3333 \pm 1.645 \sqrt{0.0006519 + 0.0008}$$

$$= 0.3333 \pm 1.645(0.038104)$$

$$= 0.3333 \pm 0.06268$$

This rounds to 0.0627.

5. Subtract and add the margin of error:

$$0.3333 - 0.0627 = 0.2706$$
$$0.3333 + 0.0627 = 0.3960$$

This is a 90% confidence interval for the difference in the proportion of all adults who are in favor of public transportation, comparing large cities and small cities.

Note that because you took the sample proportion from large cities (as Population 1) minus the sample proportion from small cities (as Population 2) and your confidence interval contains all positive values, you can conclude that the large cities are the ones more in favor of public transportation.

560. **0.2485 to 0.3821**

To find a confidence interval for the difference of two population proportions, use the formula

$$CI = (\hat{p}_1 - \hat{p}_2) \pm z^* \sqrt{\frac{\hat{p}_1(1-\hat{p}_1)}{n_1} + \frac{\hat{p}_2(1-\hat{p}_2)}{n_2}}$$

Where $\hat{p}_1$ and n_1 are the sample proportion and sample size of the sample taken from Population 1, $\hat{P}_2$ and n_2 are the sample proportion and sample size of the sample taken from Population 2, and z^* is the appropriate value from the standard normal distribution for your desired confidence level (see Table A-4 in the appendix for various confidence levels).

To solve, follow these steps:

1. Use the confidence level to find the appropriate z^*-value by referring to Table A-4. The z^*-value for a confidence level of 80% is 1.28.

2. To make the calculations somewhat easier, label the sample of adults taken from large cities as "Sample 1" and the sample of adults taken from small cities as "Sample 2."

3. For each proportion, divide the number having the attribute by the sample size:

$$\hat{p} = \frac{\text{Number having the characteristic}}{n}$$

Sample 1: $\frac{220}{300} \approx 0.7333$

Sample 2: $\frac{120}{300} = 0.4$

4. Substitute the known values into the formula for the confidence interval and solve.

(**Remember:** The standard error, $\sqrt{\frac{\hat{p}_1(1-\hat{p}_1)}{n_1} + \frac{\hat{p}_2(1-\hat{p}_2)}{n_2}}$ is 0.0381.)

$$CI = (0.7333 - 0.4) \pm 1.28(0.0381)$$
$$\approx 0.3333 \pm 0.0488$$

5. Subtract and add the margin of error:

$$0.3333 - 0.0488 = 0.2845$$
$$0.3333 + 0.0488 = 0.3821$$

This is an 80% confidence interval for the difference in the proportion of all adults who are in favor of public transportation, comparing large cities and small cities.

Note that because you took the sample proportion from large cities (as Population 1) minus the sample proportion from small cities (as Population 2) and your confidence interval contains all positive values, you can conclude that the large cities are the ones more in favor of public transportation.

561. 3.35 to 4.65 inches

To find the confidence interval for the difference of two population means, where the population standard deviations are known, use the following formula:

$$CI = (\bar{x}_1 - \bar{x}_2) \pm z^* \sqrt{\frac{\sigma_1^2}{n_1} + \frac{\sigma_2^2}{n_2}}$$

Here, $\bar{x}_1$ and n_1 are the mean and the size of the sample taken from Population 1, whose population standard deviation, σ_1 is given (known); $\bar{x}_2$ and n_2 are the mean and the size of the sample taken from Population 2, whose population standard deviation, σ_2 is given (known).

Follow these steps to solve:

1. Use the confidence level to find the appropriate z^*-value by referring to Table A-4 in the appendix for various confidence levels. The z^*-value for a 95% confidence level is 1.96.

2. Substitute the known values into the equation and solve, making sure to follow the order of operations:

$$CI = (71 - 67) \pm 1.96 \sqrt{\frac{2^2}{70} + \frac{1.8^2}{60}}$$
$$= 4 \pm 1.96 \sqrt{\frac{4}{70} + \frac{3.24}{60}}$$
$$= 4 \pm 1.96 \sqrt{0.05714 + 0.054}$$
$$= 4 \pm 1.96 \sqrt{0.11114}$$
$$= 4 \pm 1.96(0.3334)$$
$$= 4 \pm 0.653464$$

3. Find the *lower end* of the confidence interval by subtracting the margin of error from the difference of the means:

$$4 - 0.653464 = 3.346536$$

4. Find the *upper end* of the confidence interval by adding the margin of error to the difference of the means:

$$4 + 0.653464 = 4.653464$$

5. Round to the nearest hundredth, so the 95% confidence interval is 3.35 to 4.65 inches.

This is a 95% confidence interval for the difference in heights of all boys and girls in these populations.

Note that because you took the sample mean of the sample of boys taken from Population 1 minus the sample mean of the sample of girls taken from Population 2, and the confidence interval contains all positive values, you can conclude that the boys are the ones with the higher average height.

562. **3.57 to 4.43 inches**

To find the confidence interval for the difference of two population means, where the population standard deviations are known, use the following formula:

$$CI = (\bar{x}_1 - \bar{x}_2) \pm z^* \sqrt{\frac{\sigma_1^2}{n_1} + \frac{\sigma_2^2}{n_2}}$$

Here, x_1 and n_1 are the mean and the size of the sample taken from Population 1, whose populations standard deviation, σ_1 is given (known); $\dot{x}_2$ and n_2 are the mean and the size of the sample taken from Population 2, whose population standard deviation, σ_2 is given (known).

Follow these steps to solve:

1. Use the confidence level to find the appropriate z^*-value by referring to Table A-4 in the appendix for various confidence levels. The z^*-value for an 80% confidence level is 1.28.

2. Substitute the known values into the equation and solve, making sure to follow the order of operations:

$$CI = (71-67) \pm 1.28 \sqrt{\frac{2^2}{70} + \frac{1.8^2}{60}}$$
$$= 4 \pm 1.28 \sqrt{\frac{4}{70} + \frac{3.24}{60}}$$
$$= 4 \pm 1.28 \sqrt{0.05714 + 0.054}$$
$$= 4 \pm 1.28 \sqrt{0.11114}$$
$$= 4 \pm 1.28(0.3334)$$
$$= 4 \pm 0.426752$$

3. Find the *lower end* of the confidence interval by subtracting the margin of error from the difference of the means:

$$4 - 0.426752 = 3.573248$$

4. Find the *upper end* of the confidence interval by adding the margin of error to the difference of the means:

$$4 + 0.426752 = 4.426752$$

5. Round to the nearest hundredth, so the 80% confidence interval is 3.57 to 4.43 inches.

This is a 80% confidence interval for the difference in heights of all boys and girls in these populations.

Note that because you took the sample mean of the sample of boys taken from Population 1 minus the sample mean of the sample of girls taken from Population 2, and the confidence interval contains all positive values, you can conclude that the boys are the ones with the higher average height.

563. **3.14 to 4.86 inches**

To find the confidence interval for the difference of two population means, where the population standard deviations are known, use the following formula:

$$CI = (\bar{x}_1 - \bar{x}_2) \pm z^* \sqrt{\frac{\sigma_1^2}{n_1} + \frac{\sigma_2^2}{n_2}}$$

Here, $\bar{x}_1$ and n_1 are the mean and the size of the sample taken from Population 1, whose population standard deviation, σ_1 is given (known); $\bar{x}_2$ and n_2 are the mean and the size of the sample taken from Population 2, whose population standard deviation, σ_2 is given (known).

Follow these steps to solve:

1. Use the confidence level to find the appropriate z^*-value by referring to Table A-4 in the appendix for various confidence levels. The z^*-value for a 99% confidence level is 2.58.

2. Substitute the known values into the equation and solve, making sure to follow the order of operations:

$$CI = (71 - 67) \pm 2.58 \sqrt{\frac{2^2}{70} + \frac{1.8^2}{60}}$$

$$= 4 \pm 2.58 \sqrt{\frac{4}{70} + \frac{3.24}{60}}$$

$$= 4 \pm 2.58 \sqrt{0.05714 + 0.054}$$

$$= 4 \pm 2.58 \sqrt{0.11114}$$

$$= 4 \pm 2.58(0.3334)$$

$$= 4 \pm 0.860172$$

3. Find the *lower end* of the confidence interval by subtracting the margin of error from the difference of the means:

$$4 - 0.860172 = 3.139828$$

4. Find the *upper end* of the confidence interval by adding the margin of error to the difference of the means:

$$4 + 0.860172 = 4.860172$$

5. Round to the nearest hundredth, so the 99% confidence interval is 3.14 to 4.86 inches.

This is a 99% confidence interval for the difference in heights of all boys and girls in these populations.

Note that because you took the sample mean of the sample of boys taken from Population 1 minus the sample mean of the sample of girls taken from Population 2, and the confidence interval contains all positive values, you can conclude that the boys are the ones with the higher average height.

564. **3.22 to 4.78 inches**

To find the confidence interval for the difference of two population means, where the population standard deviations are known, use the following formula:

$$CI = (\bar{x}_1 - \bar{x}_2) \pm z^* \sqrt{\frac{\sigma_1^2}{n_1} + \frac{\sigma_2^2}{n_2}}$$

Here, $\bar{x}_1$ and n_1 are the mean and the size of the sample taken from Population 1, whose population standard deviation, σ_1 is given (known); $\bar{x}_2$ and n_2 are the mean and the size of the sample taken from Population 2, whose population standard deviation, σ_2 is given (known).

Follow these steps to solve:

1. Use the confidence level to find the appropriate z^*-value by referring to Table A-4 in the appendix for various confidence levels. The z^*-value for a 98% confidence level is 2.33.

2. Substitute the known values into the equation and solve, making sure to follow the order of operations:

$$CI = (71 - 67) \pm 2.33 \sqrt{\frac{2^2}{70} + \frac{1.8^2}{60}}$$
$$= 4 \pm 2.33 \sqrt{\frac{4}{70} + \frac{3.24}{60}}$$
$$= 4 \pm 2.33 \sqrt{0.05714 + 0.054}$$
$$= 4 \pm 2.33 \sqrt{0.11114}$$
$$= 4 \pm 2.33(0.3334)$$
$$= 4 \pm 0.776822$$

3. Find the *lower end* of the confidence interval by subtracting the margin of error from the difference of the means:

$$4 - 0.776822 = 3.223178$$

4. Find the *upper end* of the confidence interval by adding the margin of error to the difference of the means:

$$4 + 0.776822 = 4.776822$$

5. Round to the nearest hundredth, so the 98% confidence interval is 3.22 to 4.78 inches.

This is a 98% confidence interval for the difference in heights of all boys and girls in these populations.

Note that because you took the sample mean of the sample of boys taken from Population 1 minus the sample mean of the sample of girls taken from Population 2, and the confidence interval contains all positive values, you can conclude that the boys are the ones with the higher average height.

565. **If you switch the order of the populations (treating the population of girls as Population 1 and population of boys as Population 2), the mean difference would be negative, but the margin of error would be the same.**

This result is clear from the formula for a confidence interval for the difference in two means:

$$CI = (x_1 - x_2) \pm z^* \sqrt{\frac{\sigma_1^2}{n_1} + \frac{\sigma_2^2}{n_2}}$$

You use the means only to calculate the estimated difference in means, not the margin of error. So, switching girls and boys switches the order in which the means are subtracted, changing the difference from positive (boys – girls) to negative (girls – boys). A negative difference means the results from Group 1 are smaller than the results from Group 2. (For example, if you take $67 - 71$ you get a negative number, meaning 2 is smaller than 4.)

However, switching Populations 1 and 2 doesn't change the margin of error, mainly because the values are squared and summed in this formula rather than subtracted. So, in the end, when you switch the population names, the difference in means changes sign, but the margin of error stays the same. The confidence intervals don't change in their width, but their possible values have different signs. Bottom line: It's important to know which population is designated Population 1 and which population is designated Population 2.

---

566.   1.2

The margin of error (MOE) is the quantity added or subtracted from the sample mean when calculating a confidence interval. For a confidence interval of the difference in two population means, when the population standard deviations are known, the formula for the MOE is

$$\text{MOE} = z^* \sqrt{\frac{\sigma_1^2}{n_1} + \frac{\sigma_2^2}{n_2}}$$

where n_1 is the sample size of the sample taken from Population 1, whose population standard deviation is σ_1 and n_2 is the sample size of the sample taken from Population 2, whose population standard deviation is σ_2.

To solve, follow these steps:

1. Use the confidence level to find the appropriate z^*-value by referring to Table A-4 in the appendix for various confidence levels. The z^*-value for a 90% confidence level is 1.645.

2. Substitute the known values into the equation and solve, making sure to follow the order of operations:

$$\text{MOE} = 1.645\sqrt{\frac{7^2}{120} + \frac{4^2}{130}}$$

$$= 1.645\sqrt{\frac{49}{120} + \frac{16}{130}}$$

$$= 1.645\sqrt{0.4083 + 0.1231}$$

$$= 1.645\sqrt{0.5314}$$

$$= 1.645(0.7290)$$

$$= 1.199205$$

Rounded to one decimal place, the margin of error is 1.2 hours.

This is the margin of error for the estimated difference in average time spent on homework for college physics majors versus college English majors for a 90% confidence level.

567. 0.9

The margin of error (MOE) is the quantity added or subtracted from the sample mean when calculating a confidence interval. For a confidence interval of the difference in two population means, when the population standard deviations are known, the formula for the MOE is

$$\text{MOE} = z^* \sqrt{\frac{\sigma_1^2}{n_1} + \frac{\sigma_2^2}{n_2}}$$

where n_1 is the sample size of the sample taken from Population 1, whose population standard deviation is σ_1 and n_2 is the sample size of the sample taken from Population 2, whose population standard deviation is σ_2.

To solve, follow these steps:

1. Use the confidence level to find the appropriate z*-value by referring to Table A-4 in the appendix for various confidence levels. The z*-value for an 80% confidence level is 1.28.

2. Substitute the known values into the equation and solve, making sure to follow the order of operations:

$$
\begin{aligned}
\text{MOE} &= 1.28 \sqrt{\frac{7^2}{120} + \frac{4^2}{130}} \\
&= 1.28 \sqrt{\frac{49}{120} + \frac{16}{130}} \\
&= 1.28 \sqrt{0.4083 + 0.1231} \\
&= 1.28 \sqrt{0.5314} \\
&= 1.28(0.7290) \\
&= 0.93312
\end{aligned}
$$

Rounded to one decimal place, the margin of error is 0.9 hours.

This is the margin of error for the estimated difference in average time spent on homework for college physics majors versus college English majors with an 80% confidence level.

568. 5.6 to 8.4 hours

To find the confidence interval for the difference of two population means, where the population standard deviations are known, use the following formula:

$$\text{CI} = (\bar{x}_1 - \bar{x}_2) \pm z^* \sqrt{\frac{\sigma_1^2}{n_1} + \frac{\sigma_2^2}{n_2}}$$

Here, $\bar{x}_1$ and n_1 are the mean and the size of the sample taken from Population 1, whose population standard deviation, σ_1 is given (known); x_2 and n_2 are the mean and size of the sample taken from Population 2, whose population standard deviation, σ_2 is given (known).

To solve, follow these steps:

1. Use the confidence level to find the appropriate z^*-value by referring to Table A-4 in the appendix for various confidence levels. The z^*-value for a 95% confidence level is 1.96.

2. Substitute the known values into the equation and solve, making sure to follow the order of operations:

$$Cl = (25 - 18) \pm 1.96 \sqrt{\frac{7^2}{120} + \frac{4^2}{130}}$$

$$= 7 \pm 1.96 \sqrt{\frac{49}{120} + \frac{16}{130}}$$

$$= 7 \pm 1.96 \sqrt{0.4083 + 0.1231}$$

$$= 7 \pm 1.96 \sqrt{0.5314}$$

$$= 7 \pm 1.96(0.7290)$$

$$= 7 \pm 1.42884$$

3. Find the *lower end* of the confidence interval by subtracting the margin of error from the difference of the two sample means:

$$7 - 1.42884 = 5.57116$$

4. Find the *upper end* of the confidence interval by adding the margin of error to the difference of the two means:

$$7 + 1.42884 = 8.42884$$

5. Round to the nearest tenth to get 5.6 to 8.4 hours.

So, a 95% confidence interval for the difference in average study time is 5.6 to 8.4 hours. Because you treated the population of physics majors as Population 1, and all the values in the confidence interval are positive, you can conclude that the physics majors are the ones with the higher average homework time.

569. 5.1 to 8.9 hours

To find the confidence interval for the difference of two population means, where the population standard deviations are known, use the following formula:

$$Cl = (x_1 - x_2) \pm z^* \sqrt{\frac{\sigma_1^2}{n_1} + \frac{\sigma_2^2}{n_2}}$$

Here, $\bar{x}_1$ and n_1 are the mean and the size of the sample taken from Population 1, whose population standard deviation, σ_1 is given (known); $\bar{x}_2$ and n_2 are the mean and size of the sample taken from Population 2, whose population standard deviation, σ_2 is given (known).

To solve, follow these steps:

1. Use the confidence level to find the appropriate z^*-value by referring to Table A-4 for various confidence levels. The z^*-value for a 99% confidence level is 2.58.

2. Substitute the known values into the equation and solve, making sure to follow the order of operations:

$$CI = (25 - 18) \pm 2.58 \sqrt{\frac{7^2}{120} + \frac{4^2}{130}}$$

$$= 7 \pm 2.58 \sqrt{\frac{49}{120} + \frac{16}{130}}$$

$$= 7 \pm 2.58 \sqrt{0.4083 + 0.1231}$$

$$= 7 \pm 2.58 \sqrt{0.5314}$$

$$= 7 \pm 2.58(0.7290)$$

$$= 7 \pm 1.88082$$

3. Find the *lower end* of the confidence interval by subtracting the MOE from the difference in sample means:

$$7 - 1.88082 = 5.11918$$

4. Find the *upper end* of the confidence interval by adding the margin of error to the difference in sample means:

$$7 + 1.88082 = 8.88082$$

5. Round to the nearest tenth to get 5.1 to 8.9 hours.

So, a 99% confidence interval for the difference in average homework time is 5.1 to 8.9 hours. Because you treated the population of physics majors as Population 1, and all the values in the confidence interval are positive, you can conclude that the physics majors are the ones with the higher average homework time.

570. **D. Choices (A) and (B) (You would use t^* from a t-distribution rather than z^* from the standard normal distribution; you would use the sample standard deviations rather than the population standard deviations.)**

If you're estimating the difference in two population means and don't know the population standard deviations, you use t^* rather than z^* and the sample standard deviations when calculating a confidence interval.

571. **You would use a t^*-value rather than a z^*-value.**

If you're estimating the difference in two population means and one or both of your sample sizes are less than 30, you use a t^*-value from a t-distribution rather than a z^*-value from a standard normal distribution when calculating a confidence interval.

572. **0.6**

The margin of error (MOE) is the quantity added or subtracted from the sample mean when calculating a confidence interval. For a confidence interval of the difference in two population means, when the population standard deviations are known, the formula for the MOE is

$$MOE = z^* \sqrt{\frac{\sigma_1^2}{n_1} + \frac{\sigma_2^2}{n_2}}$$

where n_1 is the size of the sample taken from Population 1, whose population standard deviation is σ_1 and n_2 is the size of the sample taken from Population 2, whose population standard deviation is σ_2.

To solve, follow these steps:

1. Use the confidence level to find the appropriate z^*-value by referring to Table A-4 in the appendix for various confidence levels. The z^*-value for an 80% confidence level is 1.28.

2. Substitute the known values into the equation and solve, making sure to follow the order of operations:

$$\begin{aligned}
\text{MOE} &= 1.28\sqrt{\frac{6^2}{200} + \frac{4^2}{220}} \\
&= 1.28\sqrt{\frac{36}{200} + \frac{16}{220}} \\
&= 1.28\sqrt{0.18 + 0.0727} \\
&= 1.28\sqrt{0.2527} \\
&= 1.28(0.5027) \\
&= 0.643456
\end{aligned}$$

Rounded to one decimal place, the margin of error is 0.6 years.

For an 80% confidence level, the margin of error for the estimate of the difference in average age at first marriage for men and women is ± 0.6 years.

573.   0.8

The margin of error (MOE) is the quantity added or subtracted from the sample mean when calculating a confidence interval. For a confidence interval of the difference in two population means, when the population standard deviations are known, the formula for the MOE is

$$\text{MOE} = z^*\sqrt{\frac{\sigma_1^2}{n_1} + \frac{\sigma_2^2}{n_2}}$$

where n_1 is the size of the sample taken from Population 1, whose population standard deviation is σ_1, n_2 is the size of the sample taken from Population 2, whose population standard deviation is σ_2 and z^* is found by using Table A-4.

To solve, follow these steps:

1. Use the confidence level to find the appropriate z^*-value by referring to Table A-4 for various confidence levels. The z^*-value for a 90% confidence level is 1.645.

2. Substitute the known values into the equation and solve, making sure to follow the order of operations:

$$\begin{aligned}
\text{MOE} &= 1.645\sqrt{\frac{6^2}{200} + \frac{4^2}{220}} \\
&= 1.645\sqrt{\frac{36}{200} + \frac{16}{220}} \\
&= 1.645\sqrt{0.18 + 0.0727} \\
&= 1.645\sqrt{0.2527} \\
&= 1.645(0.5027) \\
&= 0.8269415
\end{aligned}$$

Round to one decimal place to get 0.8 years.

So, the margin of error (MOE) for the difference in average ages between men and women at the time of their first marriage at a confidence level of 90% is ± 0.8 years.

574. **2.0 to 4.0 years**

To find the confidence interval for the difference of two population means, where the population standard deviations are known, use the following formula:

$$CI = (\bar{x}_1 - \bar{x}_2) \pm z^* \sqrt{\frac{\sigma_1^2}{n_1} + \frac{\sigma_2^2}{n_2}}$$

Here, $\bar{x}_1$ and n_1 are the mean and the size of the sample taken from Population 1, whose population standard deviation, σ_1 is given (known); x_2 and n_2 are the mean and size of the sample taken from Population 2, whose population standard deviation, σ_2 is given (known).

To solve, follow these steps:

1. Use the confidence level to find the appropriate z^*-value by referring to Table A-4 in the appendix for various confidence levels. The z^*-value for a 95% confidence level is 1.96.

2. Substitute the known values into the equation and solve, making sure to follow the order of operations:

$$CI = (29 - 26) \pm 1.96\sqrt{\frac{6^2}{200} + \frac{4^2}{220}}$$

$$= 3 \pm 1.96\sqrt{\frac{36}{200} + \frac{16}{220}}$$

$$= 3 \pm 1.96\sqrt{0.18 + 0.0727}$$

$$= 3 \pm 1.96\sqrt{0.2527}$$

$$= 3 \pm 1.96(0.5027)$$

$$= 3 \pm 0.9853$$

3. Find the *lower end* of the confidence interval by subtracting the margin of error from the difference of the two means:

$$3 - 0.9853 = 2.0147$$

4. Find the *upper end* of the confidence interval by adding the margin of error to the difference of the two means:

$$3 + 0.9853 = 3.9853$$

5. Round to the nearest tenth to get 2.0 to 4.0 years.

For a 95% confidence level, the confidence interval for the difference in average age at first marriage between men and women is 2.0 to 4.0 years. Because you treated men as Population 1, and all the values in the confidence interval are positive, you can conclude that the males are the ones with the higher average age at first marriage.

575. **38**

Because both sample sizes are less than 30, you'll use a t^*-value rather than a z^*-value to calculate the confidence interval. The formula to calculate the degrees of freedom (*df*) for a difference in means is

$$df = n_1 + n_2 - 2$$

where n_1 is the first sample size and n_2 is the second sample size. In this example, n_1 (12th graders) is 20 and n_2 is also 20, so you get $20 + 20 - 2 = 38$.

576. 12

Because both sample sizes are less than 30, you'll use a t^*-value to calculate the margin of error. The formula for the margin of error (MOE) is

$$\text{MOE} = t^*_{n_1+n_2-2}\left(\sqrt{\frac{(n_1-1)s_1^2+(n_2-1)s_2^2}{n_1+n_2-2}}\right)\left(\sqrt{\frac{1}{n_1}+\frac{1}{n_2}}\right)$$

Here, t^* is the critical value from the t-table (Table A-2 in the appendix) with $n_1 + n_2 - 2$ degrees of freedom, n_1 and n_2 are the two sample sizes respectively, and s_1 and s_2 are the two sample standard deviations.

First, calculate the degrees of freedom, using the formula $df = n_1 + n_2 - 2 = 20 + 20 - 2 = 38$.

This value is larger than any df value in Table A-2, so use the z row. For confidence intervals, use the CI row at the bottom of the table. So, for a 99% level of confidence, the t^*-value (as estimated by a z-value) is 2.57583. Because you're rounding to whole numbers, two decimal places are sufficient for your calculations, so this rounds to 2.58.

Now, plug the values into the formula and solve:

$$\text{MOE} = 2.58\left(\sqrt{\frac{(20-1)18^2+(20-1)12^2}{20+20-2}}\right)\left(\sqrt{\frac{1}{20}+\frac{1}{20}}\right)$$

$$= 2.58\left(\sqrt{\frac{(19)(324)+(19)(144)}{38}}\right)\sqrt{0.05+0.05}$$

$$= 2.58\left(\sqrt{\frac{6.156+2,736}{38}}\right)(0.3162)$$

$$= 2.58\left(\sqrt{\frac{8,892}{38}}\right)(0.3162)$$

$$= 2.58(\sqrt{234})(0.3162)$$

$$= 2.58(15.297)(0.3162)$$

$$= 12.4792$$

Rounded to the nearest whole number of pounds, the margin of error is 12 for a 99% confidence level.

577. 6

Because both sample sizes are less than 30, you'll use a t^*-value to calculate the margin of error. The formula for the margin of error (MOE) is

$$\text{MOE} = t^*_{n_1+n_2-2}\left(\sqrt{\frac{(n_1-1)s_1^2+(n_2-1)s_2^2}{n_1+n_2-2}}\right)\left(\sqrt{\frac{1}{n_1}+\frac{1}{n_2}}\right)$$

Here, t^* is the critical value from the t-table (Table A-2 in the appendix) with $n_1 + n_2 - 2$ degrees of freedom, n_1 and n_2 are the two sample sizes respectively, and s_1 and s_2 are the two sample standard deviations.

First, calculate the degrees of freedom, using the formula $df = n_1 + n_2 - 2 = 20 + 20 - 2 = 38$.

This value is larger than any df value in Table A-2, so use the z row. For confidence intervals, use the CI row at the bottom of the table. For an 80% level of confidence, the t^*-value (as estimated by a z-value) is 1.281552. Because you're rounding to whole numbers, two decimal places are sufficient for your calculations, so this rounds to 1.28.

Now, plug the values into the formula and solve:

$$\text{MOE} = 1.28 \left(\sqrt{\frac{(20-1)18^2 + (20-1)12^2}{20 + 20 - 2}} \right) \left(\sqrt{\frac{1}{20} + \frac{1}{20}} \right)$$

$$= 1.28 \left(\sqrt{\frac{(19)(324) + (19)(144)}{38}} \right) (\sqrt{0.05 + 0.05})$$

$$= 1.28 \left(\sqrt{\frac{6.156 + 2{,}736}{38}} \right) (0.3162)$$

$$= 1.28 \left(\sqrt{\frac{8{,}892}{38}} \right) (0.3162)$$

$$= 1.28 (\sqrt{234})(0.3162)$$

$$= 1.28 (15.297)(0.3162)$$

$$= 6.1912$$

Rounded to the nearest whole number of pounds, the margin of error is 6 for an 80% confidence level.

---

578. 8

Because both sample sizes are less than 30, you'll use a t^*-value to calculate the margin of error. The formula for the margin of error (MOE) is

$$\text{MOE} = t^*_{n_1 + n_2 - 2} \left(\sqrt{\frac{(n_1 - 1)s_1^2 + (n_2 - 1)s_2^2}{n_1 + n_2 - 2}} \right) \left(\sqrt{\frac{1}{n_1} + \frac{1}{n_2}} \right)$$

Here, t^* is the critical value from the t-table (Table A-2 in the appendix) with $n_1 + n_2 - 2$ degrees of freedom, n_1 and n_2 are the two sample sizes respectively, and s_1 and s_2 are the two sample standard deviations.

First, calculate the degrees of freedom, using the formula $df = n_1 + n_2 - 2 = 20 + 20 - 2 = 38$.

This value is larger than any df value in Table A-2, so use the z row. For confidence intervals, use the CI row at the bottom of the table. For a 90% level of confidence, the t^*-value (as estimated by a z-value) is 1.644854. This rounds to 1.645.

Now, plug the values into the formula and solve:

$$\text{MOE} = 1.645 \left(\sqrt{\frac{(20-1)18^2 + (20-1)12^2}{20+20-2}} \right) \left(\sqrt{\frac{1}{20} + \frac{1}{20}} \right)$$

$$= 1.645 \left(\sqrt{\frac{(19)(324) + (19)(144)}{38}} \right) \left(\sqrt{0.05 + 0.05} \right)$$

$$= 1.645 \left(\sqrt{\frac{6.156 + 2{,}736}{38}} \right) (0.3162)$$

$$= 1.645 \left(\sqrt{\frac{8{,}892}{38}} \right) (0.3162)$$

$$= 1.645 \left(\sqrt{234} \right) (0.3162)$$

$$= 1.645 (15.297)(0.3162)$$

$$= 7.9567$$

Rounded to the nearest whole number of pounds, the margin of error is 8 for a 90% confidence level.

579.   21 to 39 pounds

Use the formula for creating a confidence interval for the difference of two population means when the population standard deviation isn't known and/or the sample sizes are small (less than 30) and you can't be sure whether your data came from a normal distribution.

$$\text{CI} = \left(\bar{x}_1 - \bar{x}_2 \right) \pm t^*_{n_1+n_2-2} \left(\sqrt{\frac{(n_1-1)s_1^2 + (n_2-1)s_2^2}{n_1+n_2-2}} \right) \left(\sqrt{\frac{1}{n_1} + \frac{1}{n_2}} \right)$$

Here, t^* is the critical value from the t-table (Table A-2 in the appendix) with $n_1 + n_2 - 2$ degrees of freedom, n_1 and n_2 are the two sample sizes respectively, $\bar{x}_1$ and $\bar{x}_2$ are the two sample means, and s_1 and s_2 are the two sample standard deviations.

Follow these steps to solve:

1. Determine the t^*-value in the t-table by finding the number in the df row that intersects with the given confidence level (or CI).

   In this question, you have a 95% confidence level and $df = n_1 + n_2 - 2 = 20 + 20 - 2 = 38$. Because the degrees of freedom is more than 30, use the number in row z, so $t^* \approx 1.95996$. Two decimal places are sufficient because you're rounding to whole numbers, so use 1.96.

2. Substitute all the values into the formula and solve:

$$\text{CI} = (170 - 140) \pm 1.96 \left(\sqrt{\frac{(20-1)18^2 + (20-1)12^2}{20+20-2}} \right) \left(\sqrt{\frac{1}{20} + \frac{1}{20}} \right)$$

$$= 30 \pm 1.96 \left(\sqrt{\frac{(19)(324) + (19)(144)}{38}} \right) \left(\sqrt{0.05 + 0.05} \right)$$

$$= 30 \pm 1.96 \left(\sqrt{\frac{6.156 + 2{,}736}{38}} \right) (0.3162)$$

$$= 30 \pm 1.96 \left(\sqrt{\frac{8{,}892}{38}} \right) (0.3162)$$

$$= 30 \pm 1.96 \left(\sqrt{234} \right) (0.3162)$$

$$= 30 \pm 1.96 (15.297)(0.3162)$$

$$= 30 \pm 9.4803$$

3. Subtract and add the margin of error:

$$30 - 9.4803 = 20.5197$$
$$30 + 9.4803 = 39.4803$$

4. Round to the nearest whole number to get 21 to 39 pounds.

So, a 95% confidence interval for the true difference in the mean weights of all the 12th-grade and 9th-grade boys at this school is 21 to 39 pounds. Because you treated the population of 12th graders as Population 1 and all the values in the confidence interval are positive, you can conclude that the 12th graders are the ones with the higher average weight.

580. 19 to 41 pounds

Use the formula for creating a confidence interval for the difference of two population means when the population standard deviation isn't known and/or the sample sizes are small (less than 30) and you can't be sure whether your data came from a normal distribution.

$$CI = \left(\bar{x}_1 - \bar{x}_2\right) \pm t^*_{n_1+n_2-2} \left(\sqrt{\frac{(n_1-1)s_1^2 + (n_2-1)s_2^2}{n_1+n_2-2}}\right)\left(\sqrt{\frac{1}{n_1} + \frac{1}{n_2}}\right)$$

Here, t^* is the critical value from the t-table (Table A-2 in the appendix) with $n_1 + n_2 - 2$ degrees of freedom, n_1 and n_2 are the two sample sizes respectively, $\bar{x}_1$ and $\bar{x}_2$ are the two sample means, and s_1 and s_2 are the two sample standard deviations.

Follow these steps to solve:

1. Determine the t^*-value in the t-table by finding the number in the df row that intersects with the given confidence level (or CI).

   In this question, you have a 98% confidence level and $df = n_1 + n_2 - 2 = 20 + 20 - 2 = 38$. Because the degrees of freedom is more than 30, use the number in row z, so $t^* \approx 2.32635$. Two decimal places are sufficient because you're rounding to whole numbers, so use 2.33.

2. Substitute all the values into the formula and solve:

$$CI = (170 - 140) \pm 2.33\left(\sqrt{\frac{(20-1)18^2 + (20-1)12^2}{20+20-2}}\right)\left(\sqrt{\frac{1}{20} + \frac{1}{20}}\right)$$

$$= 30 \pm 2.33\left(\sqrt{\frac{(19)(324)+(19)(144)}{38}}\right)\left(\sqrt{0.05+0.05}\right)$$

$$= 30 \pm 2.32635\left(\sqrt{\frac{6,156+2,736}{38}}\right)(0.3162)$$

$$= 30 \pm 2.33\left(\sqrt{\frac{8,892}{38}}\right)(0.3162)$$

$$= 30 \pm 2.33(\sqrt{234})(0.3162)$$

$$= 30 \pm 2.33(15.297)(0.3162)$$

$$= 30 \pm 11.2700$$

3. Subtract and add the margin of error:

$$30 - 11.27 = 18.73$$
$$30 + 11.27 = 41.27$$

4. Round to the nearest whole number to get a confidence interval of 19 to 41 pounds.

581. 43

Because both sample sizes are less than 30, you'll use a t^*-value rather than a z^*-value to calculate the confidence interval. The formula to calculate the degrees of freedom (df) for a difference in means is

$$df = n_1 + n_2 - 2$$

where n_1 is the size of the sample taken from Population 1, and n_2 is the size of the sample taken from Population 2. In this example, n_1 (men) is 20 and n_2 (women) is 25, so you get $20 + 25 - 2 = 43$.

582. 1.6

Because both sample sizes are less than 30, you'll use a t^*-value to calculate the margin of error. The formula for the margin of error (MOE) is

$$MOE = t^*_{n_1 + n_2 - 2} \left(\sqrt{\frac{(n_1 - 1)s_1^2 + (n_2 - 1)s_2^2}{n_1 + n_2 - 2}} \right) \left(\sqrt{\frac{1}{n_1} + \frac{1}{n_2}} \right)$$

Here, t^* is the critical value from the t-table (Table A-2 in the appendix) with $n_1 + n_2 - 2$ degrees of freedom, n_1 and n_2 are the two sample sizes respectively, and s_1 and s_2 are the two sample standard deviations.

First, calculate the degrees of freedom using the formula $df = n_1 + n_2 - 2 = 20 + 25 - 2 = 43$.

This value is larger than any df value in Table A-2, so use the z row. For confidence intervals, use the CI row at the bottom of the table. For a 90% level of confidence, the t^*-value (as estimated by a z-value) is 1.644854. This rounds to 1.645.

Now, plug the values into the formula and solve:

$$MOE = 1.645 \left(\sqrt{\frac{(20 - 1)3.5^2 + (25 - 1)3^2}{20 + 25 - 2}} \right) \left(\sqrt{\frac{1}{20} + \frac{1}{25}} \right)$$

$$= 1.645 \left(\sqrt{\frac{(19)(12.25) + (24)(9)}{43}} \right) \left(\sqrt{0.09} \right)$$

$$= 1.645 \left(\sqrt{\frac{232.75 + 216}{43}} \right) (0.3)$$

$$= 1.645 \left(\sqrt{\frac{448.75}{43}} \right) (0.3)$$

$$= 1.645 \left(\sqrt{10.436} \right) (0.3)$$

$$= 1.645 (3.230)(0.3)$$

$$= 1.594$$

Rounded to one decimal place, the margin of error is 1.6 thousand dollars, which is the difference in average salaries between men and women for a 90% confidence level.

583. 1.9

Because both sample sizes are less than 30, you'll use a t^*-value to calculate the margin of error. The formula for the margin of error (MOE) is

$$\text{MOE} = t^*_{n_1 + n_2 - 2} \left(\sqrt{\frac{(n_1 - 1)s_1^2 + (n_2 - 1)s_2^2}{n_1 + n_2 - 2}} \right) \left(\sqrt{\frac{1}{n_1} + \frac{1}{n_2}} \right)$$

Here, t^* is the critical value from the t-table (Table A-2 in the appendix) with $n_1 + n_2 - 2$ degrees of freedom, n_1 and n_2 are the two sample sizes respectively, and s_1 and s_2 are the two sample standard deviations.

First, calculate the degrees of freedom using the formula $df = n_1 + n_2 - 2 = 20 + 25 - 2 = 43$.

This value is larger than any df value in Table A-2, so use the z row. For confidence intervals, use the CI row at the bottom of the table. For a 95% level of confidence, the t^*-value (as estimated by a z-value) is 1.95996. This rounds to 1.96.

Now, plug the values into the formula and solve:

$$\text{MOE} = 1.96 \left(\sqrt{\frac{(20-1)3.5^2 + (25-1)3^2}{20+25-2}} \right) \left(\sqrt{\frac{1}{20} + \frac{1}{25}} \right)$$

$$= 1.96 \left(\sqrt{\frac{(19)(12.25) + (24)(9)}{43}} \right) \left((\sqrt{0.09}) \right)$$

$$= 1.96 \left(\sqrt{\frac{232.75 + 216}{43}} \right) (0.3)$$

$$= 1.96 \left(\sqrt{\frac{448.75}{43}} \right) (0.3)$$

$$= 1.96 (\sqrt{10.436})(0.3)$$

$$= 1.96 (3.230)(0.3)$$

$$= 1.899$$

Rounded to one decimal place, the margin of error is 1.9 thousand dollars, which is the difference in average salaries between men and women for a 95% confidence level.

584. 4.5 to 9.5

Use the formula for creating a confidence interval for the difference of two population means when the population standard deviation isn't known and/or the sample sizes are small (less than 30) and you can't be sure whether your data came from a normal distribution.

$$\text{CI} = (\bar{x}_1 - \bar{x}_2) \pm t^*_{n_1 + n_2 - 2} \left(\sqrt{\frac{(n_1 - 1)s_1^2 + (n_2 - 1)s_2^2}{n_1 + n_2 - 2}} \right) \left(\sqrt{\frac{1}{n_1} + \frac{1}{n_2}} \right)$$

Here, t^* is the critical value from the t-table (Table A-2 in the appendix) with $n_1 + n_2 - 2$ degrees of freedom, n_1 and n_2 are the two sample sizes respectively, $\bar{x}_1$ and $\bar{x}_2$ are the two sample means, and s_1 and s_2 are the two sample standard deviations.

Follow these steps to solve:

1. Determine the t^*-value in the t-table by finding the number in the df row that intersects with the given confidence level (or CI).

   In this question, you have a 99% confidence level and $df = n_1 + n_2 - 2 = 20 + 25 - 2 = 43$. Because the degrees of freedom is more than 30, use the number in row z, so $t^* = 2.57583$. This rounds to 2.58.

2. Substitute all the values into the formula and solve:

$$\text{CI} = (37 - 30) \pm 2.58 \left(\sqrt{\frac{(20-1)3.5^2 + (25-1)3^2}{20+25-2}} \right) \left(\sqrt{\frac{1}{20} + \frac{1}{25}} \right)$$

$$= 7 \pm 2.58 \left(\sqrt{\frac{(19)(12.25) + (24)(9)}{43}} \right) (\sqrt{0.09})$$

$$= 7 \pm 2.58 \left(\sqrt{\frac{232.75 + 216}{43}} \right) (0.3)$$

$$= 7 \pm 2.58(3.230)(0.3)$$

$$= 7 \pm 2.500$$

3. Subtract and add the margin of error:

$$7 - 2.500 = 4.500$$
$$7 + 2.500 = 9.500$$

4. Round to one decimal place to get 4.5 to 9.5 (thousands of dollars).

So, a 99% confidence interval for the true difference in average income between all North American men and women after five years of employment is 4.5 and 9.5, in thousands of dollars.

585. **E. Choices (A) and (C) (An automobile factory claims 99% of its parts meet stated specifications; an automobile factory claims that it can assemble 500 automobiles an hour when the assembly line is fully staffed.)**

A hypothesis test is a statistical procedure undertaken to test a quantifiable claim. "The best-quality cars" isn't quantifiable on its own and can't be tested in this manner, while the other two claims can.

If the "best quality" is defined as having "an average of 30 amenities per vehicle," then a hypothesis test could be devised.

586. **C. A school gives its students standardized tests to measure levels of achievement compared to prior years.**

A hypothesis test is a statistical procedure undertaken to test a quantifiable claim. No quantifiable claim is clearly defined in the preceding statement, and therefore no hypothesis test is possible without further clarification.

If you're told that last year's standardized test yielded an average math subscore of 80, you could test whether this year's students perform significantly better. But as currently stated, you're not given a clear definition of the population value (parameter) of interest.

587. $H_0: p = 0.75$

The null hypothesis is the prior claim that you want to test — in this case, that "75% of voters conclude the bond issue." The null and alternative hypotheses are always stated in terms of a population parameter (p in this case).

588. $H_a: p \neq 0.75$

The alternative hypothesis is the statement about the world that you will conclude if you have statistical evidence to reject the null hypothesis, based on the data. The null and alternative hypotheses are always stated in terms of a population parameter (in this case p).

589. impossible to tell without further information

The null and alternative hypotheses are defined by a real-world situation. Suppose that 132 is the intended mean amount of M&M's in a bag of set weight. A consumer advocacy group may want to solely test whether $\mu < 132$ to ensure that customers aren't being cheated. A quality control manager may want to test that $\mu \neq 132$ so that bags aren't being underfilled or overfilled. You can't know the alternative hypothesis without knowing the context of the situation.

590. $H_0: \mu = 10.50$

The null hypothesis is the original claim or current "best guess" at the value of interest. The null hypothesis is always written in terms of a population parameter (in this case, (μ) being equivalent to a specific value.

591. $H_0: \mu = 52$

The null hypothesis states a current claim about the condition of the world. The null hypothesis is always written in terms of a population parameter (in this case μ) being equivalent to a specific value.

592. The average number of songs on an MP3 player owned by a college student is 228.

A null hypothesis states the specific value of a population parameter, using an equal sign. In this case, you claim that the population mean μ) equals 228.

593. $H_0: p = 0.78$

A null hypothesis states the specific value of a population parameter, using an equal sign. In this case, you define that the parameter is the proportion (p) of the entire teenage population who owns cellphones.

594. $H_0: p_1 - p_2 = 0$

Both proportions are assumed to be the same, so their difference is assumed to be zero. Another way to represent this statement is $H_0: p_1 = p_2$.

Null hypotheses are always written using population parameters, in this case p_1 and p_2.

595. $H_a: p < 0.70$

The alternative hypothesis is the hypothesis that you conclude if you have sufficient evidence to reject the null hypothesis. In this case, you hope to reject the null hypothesis that 70% of Americans think Congress is doing a good job $(H_0: p = 0.70)$ and thus conclude the alternative hypothesis that the true percentage is lower, based on the data. The hypothesis test will help you decide.

596. $H_a: \mu > 2.5$

The alternative hypothesis is the hypothesis that you will conclude if you have sufficient evidence to reject the null hypothesis. In this case, you're hoping to reject the null hypothesis that the train trip takes an average of 2.5 hours $(H_0: \mu = 2.5)$ and conclude the alternative that it takes longer than 2.5 hours, based on the data.

597. $H_a: p < 0.92$

The alternative hypothesis is the hypothesis that you will conclude if you have sufficient evidence to reject the null hypothesis. In this case, you're hoping to reject the null hypothesis that the airline's planes arrive early 92% of the time $(H_0: p = 0.92)$ and conclude the alternative that the true proportion is lower, based on the data.

598. $H_a: \mu < 39$

The alternative hypothesis is the hypothesis that you will conclude if you have sufficient evidence to reject the null hypothesis. In this case, you're hoping to reject the null hypothesis that the car averages 39 miles per gallon $(H_0: \mu = 39)$ and conclude the alternative that the true average is lower, based on the data.

599. $H_a: p > 0.005$

The alternative hypothesis is the hypothesis that you will conclude if you have sufficient evidence to reject the null hypothesis. In this case, you're hoping to reject the null hypothesis that only 1 in 200 computers has a mechanical malfunction $(H_0: p = 0.005)$ and conclude the alternative hypothesis that the true proportion is higher, based on the data.

600. $H_a: p > 0.05$

The alternative hypothesis is the hypothesis that you will conclude if you have sufficient evidence to reject the null hypothesis. In this case, you're hoping to reject the null hypothesis that only 5% of patients are dissatisfied with their care $(H_0: p = 0.05)$ and conclude the alternative hypothesis that the true proportion is higher, based on the data.

601.

H_a: $\mu \neq 3{,}300$

The alternative hypothesis is the hypothesis that you will conclude if you have sufficient evidence to reject the null hypothesis, based on the data. In this case, you're hoping to reject the null hypothesis that American adults consume an average of 3,300 calories per day $\left(H_0\colon \mu = 3{,}300\right)$ and conclude the alternative hypothesis that the true average is different, based on the data.

You're not given specific direction as to whether you believe the statistic underestimates or overestimates the truth, so you need the two-sided alternative.

602.

H_a: $\mu > 1.8$

The alternative hypothesis is the hypothesis that you will conclude if you have sufficient evidence to reject the null hypothesis. In this case, you're hoping to reject the null hypothesis that adults watch an average of 1.8 hours of television per day $\left(H_0\colon \mu = 1.8\right)$ and conclude the alternative hypothesis that the true average is higher, based on the data.

603.

H_a: $\mu < 0.08$

The alternative hypothesis is the hypothesis that you will conclude if you have sufficient evidence to reject the null hypothesis. In this case, you're hoping to reject the null hypothesis that the investment company's clients make an average 8% return each year $\left(H_0\colon \mu = 0.08\right)$ and conclude the alternative hypothesis that the true average return is lower, based on the data.

Note that interest rates are continuous measures. They're not binomial proportions, as if to say "8% of the time you get a return." As such, the 8% return is an average, not a proportion.

604.

H_a: $p_1 - p_2 \neq 0.25$

The alternative hypothesis is the hypothesis that you will conclude if you have sufficient evidence to reject the null hypothesis. In this case, you're hoping to reject the null hypothesis that the difference in college attendance between the two cities is 25% $\left(H_0\colon p_1 - p_2 = 0.25\right)$ and conclude the alternative that the true difference in proportions is some other value, based on the data.

You're not given specific direction as to whether you believe the statistic underestimates or overestimates the truth. The possibility exists that the true difference is greater than 25%, or perhaps that it is significantly less than that. So, you need the two-sided alternative.

605.

−2

You calculate the test statistic by subtracting the claimed value (from the null hypothesis) from the sample statistic and dividing by the standard error. In this example, the claimed value is 4, the sample statistic is 3, and the standard error is 0.5, so the test statistic is

$$\frac{3-4}{0.5} = -2$$

606. 1

You calculate the test statistic by subtracting the claimed value (from the null hypothesis) from the sample statistic and dividing by the standard error. In this example, the claimed value is 4, the sample statistic is 4.5, and the standard error is 0.5, so the test statistic is

$$\frac{4.5 - 4}{0.5} = 1$$

607. 2.4

You calculate the test statistic by subtracting the claimed value (from the null hypothesis) from the sample statistic and dividing by the standard error. In this example, the claimed value is 4, the sample statistic is 5.2, and the standard error is 0.5, so the test statistic is

$$\frac{5.2 - 4}{0.5} = 2.4$$

608. −0.8

You calculate the test statistic by subtracting the claimed value (from the null hypothesis) from the sample statistic and dividing by the standard error. In this example, the claimed value is 4, the sample statistic is 3.6, and the standard error is 0.5, so the test statistic is

$$\frac{3.6 - 4}{0.5} = -0.8$$

609. 0.1556

The p-value tells you the probability of a result being at or beyond your test statistic, if the null hypothesis is true. Because you have a two-tailed test, you will double the table probability to account for test results both above and below the claimed value. Note that the p-value is a probability and can never be negative.

In Table A-1 in the appendix, find 1.4 in column z. Then read across the 1.4 row to find the column labeled 0.02. The number where row 1.4 intersects with column 0.02 is 0.9222.

Table A-1 shows the probability of a value below a given z-score. You want the probability of a value above 1.42, so subtract the table value from 1 (because total probability always equals 1): $1 - 0.9222 = 0.0778$.

To get the p-value in this case, double this number because H_a is "not equal to" and both the upper and the lower ends of the distribution must be included: $2(0.0778) = 0.1556$.

610. 0.1188

The p-value tells you the probability of a result being at or beyond your test statistic, if the null hypothesis is true. Because you have a two-tailed test, you will double the table probability to account for test results both above and below the claimed value.

In Table A-1 in the appendix, find –1.5 in column z. Then read across the –1.5 row to find the column labeled 0.06. The number where row –1.5 intersects with column 0.06 is 0.0594.

To get the p-value in this case, double this number because H_a is "not equal to" and both the upper and the lower ends of the distribution must be included: $2(0.0594) = 0.1188$.

611. 0.4532

The p-value tells you the probability of a result being at or beyond your test statistic, if the null hypothesis is true. Because you have a two-tailed test, you will double the table probability to account for test results in the extremes both above and below the assumed mean. Note that the p-value is a probability and can never be negative.

In Table A-1 in the appendix, find 0.7 in column z. Then read across the 0.7 row to find the column labeled 0.05. The number where row 0.7 intersects with column 0.05 is 0.7734.

Table A-1 shows the probability of a value below a given z-score. You want the probability of a value above 0.75, so subtract the table value from 1 (because total probability always equals 1): $1 - 0.7734 = 0.2266$.

To get the p-value in this case, double this number because H_a is "not equal to" (two-sided) and both the upper and the lower ends of the distribution must be included: $2(0.2266) = 0.4532$.

612. 0.4180

The p-value tells you the probability of a result being at or beyond your test statistic, if the null hypothesis is true. Because you have a two-tailed test, you will double the table probability to account for test results in the extremes both above and below the assumed mean. Note that the p-value is a probability and can never be negative.

In Table A-1 in the appendix, find –0.8 in column z. Then read across the –0.8 row to find the column labeled 0.01. The number where row –0.8 intersects with column 0.01 is 0.2090.

To get the p-value in this case, double this number because H_a is "not equal to" (two-sided) and both the upper and the lower ends of the distribution must be included: $2(0.2090) = 0.4180$.

613. Fail to reject H_0 because the p-value of your result is greater than alpha.

To reject the null hypothesis at the alpha = 0.05 level, you need a test statistic with a p-value of less than 0.05.

Your alternative hypothesis is directional (>), so you'll reject the null hypothesis only if your test statistic is positive and has a p-value of less than the alpha level of 0.05.

Using Table A-1 in the appendix, find the value where the 1.5 row intersects the 0.01 column. This value of 0.9345 is the probability of a value being less than 1.51. To find the probability of a value being greater than 1.51, subtract the table value from 1 (because total probability is always 1):

$$1 - 0.9345 = 0.0655$$

This value is the p-value of your test statistic of 1.51, if the null hypothesis is true and you use a single-tailed test (your alternative hypothesis is that the population proportion is greater than 0.45).

This p-value is greater than your alpha level of 0.05, so you will fail to reject the null hypothesis when the alpha level is 0.05. You data didn't provide enough evidence to reject the null hypothesis.

614. **Reject H_0 because the p-value of your result is less than alpha.**

To reject the null hypothesis at the alpha = 0.10 level, you need a test statistic with a p-value of less than 0.10.

Your alternative hypothesis is directional (>), so you'll reject the null hypothesis only if your test statistic is positive and has a p-value of less than the alpha level of 0.10.

Using Table A-1 in the appendix, find the value where the 1.5 row intersects the 0.01 column. This value of 0.9345 is the probability of a value being less than 1.51. To find the probability of a value being greater than 1.51, subtract the table value from 1 (because total probability is always 1):

$$1 - 0.9345 = 0.0655$$

This value is the p-value of your test statistic of 1.51, if the null hypothesis is true and you use a single-tailed test (your alternative hypothesis is that the population proportion is greater than 0.45).

This p-value is less than your alpha level of 0.10, which means that your observed test statistic was unlikely, assuming that the original claim was true. So, you will reject the null hypothesis at the 0.10 alpha level.

615. **Fail to reject H_0 because the p-value of your result is greater than alpha.**

To reject the null hypothesis at the alpha = 0.01 level, you need a test statistic with a p-value of less than 0.01.

Your alternative hypothesis is directional (>), so you'll reject the null hypothesis only if your test statistic is positive and has a p-value of less than the alpha level of 0.01.

Using Table A-1 in the appendix, find the value where the 1.9 row intersects the 0.08 column. This value of 0.9761 is the probability of a value being less than 1.98. To find the probability of a value being greater than 1.98, subtract the table value from 1 (because total probability is always 1):

$$1 - 0.9761 = 0.0239$$

This value is the p-value of your test statistic of 1.98, if the null hypothesis is true and you use a single-tailed test (your alternative hypothesis is that the population proportion is greater than 0.45).

This p-value is greater than your alpha level of 0.01, so you will fail to reject the null hypothesis at the alpha level of 0.01. Your data didn't provide enough evidence to reject the null hypothesis.

616. **Reject H_0 because the p-value of your result is less than alpha.**

To reject the null hypothesis at the alpha = 0.05 level, you need a test statistic with a p-value of less than 0.05.

Your alternative hypothesis is directional (>), so you'll reject the null hypothesis only if your test statistic is positive and has a p-value of less than the alpha level of 0.05.

Using Table A-1 in the appendix, find the value where the *1.9* row intersects the *0.08* column. This value of 0.9761 is the probability of a value being less than 1.98. To find the probability of a value being greater than 1.98, subtract the table value from 1 (because total probability is always 1):

$$1 - 0.9761 = 0.0239$$

This value is the p-value of your test statistic of 1.98, if the null hypothesis is true and you use a single-tailed test (your alternative hypothesis is that the population proportion is greater than 0.45).

This p-value is less than your alpha level of 0.05, which means that your observed test statistic was unlikely, assuming the original claim was true. So, you will reject the null hypothesis at the alpha level of 0.05.

617.

Fail to reject H$_0$ because your test statistic is negative, while the alternative hypothesis is that the population probability is greater than 0.45.

You don't have to do any calculations for this problem. Your alternative hypothesis ($p > 0.45$) indicates that your test statistic must also be positive to reject the null hypothesis. That is, you can reject the claim only if your observed value is significantly greater than 0.45. However, in this example, the test statistic is negative, so you won't reject the null hypothesis.

You could still calculate the p-value to verify. To reject the null hypothesis at the alpha $= 0.05$ level, you need a test statistic with a p-value of less than 0.05. Using Table A-1 in the appendix, find the value where the *-1.9* row intersects the *0.08* column. This value of 0.0239 is the probability of a value being less than -1.98. To find the probability of a value being greater than -1.98, subtract the table value from 1 (because total probability is always 1):

$$1 - 0.0239 = 0.9761$$

This value is the p-value of your test statistic of -1.98, if the null hypothesis is true and you use a single-tailed test (your alternative hypothesis is that the population proportion is greater than 0.45).

This p-value is much greater than your alpha level of 0.05, so you will fail to reject the null hypothesis at the 0.05 alpha level. Your data didn't provide enough evidence to reject the null hypothesis.

618.

Fail to reject H$_0$ because your test statistic is negative, while the alternative hypothesis is that the population probability is greater than 0.45.

You don't have to do any calculations for this problem. Your alternative hypothesis ($p > 0.45$) indicates that your test statistic must also be positive to reject the null hypothesis. That is, you can reject the claim only if your observed value is significantly greater than 0.45. However, in this example, the test statistic is negative, so you won't reject the null hypothesis.

You could still calculate the p-value to verify. To reject the null hypothesis at the alpha $= 0.01$ level, you need a test statistic with a p-value of less than 0.01. Using Table A-1 in the appendix, find the value where the *-3.0* row intersects the *0.00* column. This value of 0.0013 is the probability of a value being less than –3.0. To find the probability of a value being greater than –3.0, subtract the table value from 1 (because total probability is always 1):

$$1 - 0.0013 = 0.9987$$

This value is the p-value of your test statistic of –3.0, if the null hypothesis is true and you use a single-tailed test (your alternative hypothesis is that the population proportion is greater than 0.45).

This p-value is much greater than your alpha level of 0.01, so you will fail to reject the null hypothesis at the 0.01 alpha level. Your data didn't provide enough evidence to reject the null hypothesis.

619. **There is a 1% chance of getting a value at least that extreme, if the null hypothesis is true.**

The p-value of a test statistic tells you the probability of getting a specific, observed value or a value more extreme (farther from the claimed value), assuming the null hypothesis is true.

620. **Choices (A) and (C) (There is a 10% chance that you will reject the null hypothesis when it is true; you should reject the null hypothesis if your test statistic has a p-value of 0.10 or less.)**

An alpha level of 0.10 means that you have a 10% chance that you'll randomly reject the null hypothesis when it's actually true and also that you'll reject the null hypothesis if your test statistic has a p-value of 0.10 or less.

621. **Fail to reject the null hypothesis.**

You reject the null hypothesis if the p-value for your test statistic is less than the alpha level. It's never correct to say that you accept the null hypothesis (it implies that you know the null hypothesis is factually true) or that you reject the alternative hypothesis (because it's the null, not the alternative hypothesis, that you're putting to the test).

622. **highly statistically significant**

A test result with a p-value of 0.001, when the alpha level is 0.05, is typically stated as being "highly statistically significant," because it's so much smaller than the required value of 0.05.

623. **Reject H_0 because the results are statistically significant.**

The p-value is well below the significance level $\alpha = 0.05$. The results support rejecting H_0. Alpha values and hypotheses should always be set before results are calculated. It's incorrect to adjust either after the data has been gathered, in the hope of getting a significant result.

624. **B. 0.01**

H_0 is rejected if the p-value is less than the significance level ($\alpha = 0.02$)

625. **E. 0.04**

H_0 is rejected if the p-value is less than the significance level ($\alpha = 0.05$)

626. **Reject H_0.**

You reject H_0 because the p-value is less than the significance level (α level) of 0.01.

627. **Reject H_0.**

You reject H_0 because the p-value is less than the significance level (α level) of 0.03.

628. **Fail to reject H_0.**

The p-value of 0.06 is greater than the significance level of $\alpha = 0.05$ You must fail to reject H_0.

629. **Fail to reject H_0.**

The p-value of 0.42 is much greater than the significance level of $\alpha = 0.05$, so you fail to reject H_0.

630. **Fail to reject H_0.**

The p-value of 0.2 is much greater than the significance level of $\alpha = 0.02$, so you fail to reject H_0.

631. **Fail to reject H_0.**

The p-value of 0.018 is greater than the significance level of $\alpha = 0.01$, so you fail to reject H_0.

632. **impossible to determine from the given information**

Even though the p-value is less than the significance level of $\alpha = 0.05$, notice that the test statistic is negative. This means that the sample mean is *less* than the claimed population mean of 9.65. However, the alternative hypothesis states that the sample mean is expected to be *greater* than the population mean. This is a contradiction that tells you either the test statistic or the p-value was calculated incorrectly. You should return to the original data and double-check the calculations.

633. **You should fail to reject H_0.**

If the alpha level is 0.05 and the p-value of the test statistic is 0.07, you should fail to reject H_0. *Note:* It's never correct to say that you *accept* H_0.

634. **C. rejecting the null hypothesis when it is true**

You make a Type I error when the null hypothesis is true but you reject it. This error is just by random chance, because if you knew for a fact that the null was true, you certainly wouldn't reject it. But there's a slim chance (alpha level) that it could happen.

A Type I error is sometimes referred to as a "false alarm," because rejecting the null hypothesis is like sounding an alarm to change an established value. If the null is true, then there's no need for such a change.

635.   **D. failing to reject the null hypothesis when it is false**

You make a Type II error when the null hypothesis is false but you fail to reject it because your data couldn't detect it, just by chance.

This error is sometimes referred to as "missing out on a detection." The claim really was wrong, but you didn't get a random sample that would provide enough evidence (small enough p-value) to reject it.

636.   **0.01**

The alpha level (or significance level) of 0.01 indicates the probability of a Type I error — that is, the error of rejecting the null hypothesis when it's actually true.

637.   **impossible to tell without further information**

The probability of a Type II error isn't directly related to the alpha level in terms of its formula or calculations. Instead, it's determined by a combination of several factors, including sample size. However, the behavior of alpha is related to the behavior of beta (probability of Type II error). In general, they have an indirect relationship: The larger alpha is, the more likely you are to reject (regardless of the truth) and thus the less likely you are to make a Type II error.

638.   **D. the probability of rejecting the null hypothesis when it is false**

The power of the test describes the probability of rejecting the null hypothesis when it's false — that is, of making the correct decision when the null hypothesis is false. In that way, it's the opposite (or complement) of making a Type II error.

639.   **E. Choices (B) and (C) (Having a random sample of data; having a large sample size.)**

A random sample is necessary to try to get the most representative and unbiased data possible. Otherwise, a sample could be tampered with to have only data values that support the null hypothesis, forcing you to fail to reject H_0 no matter what.

A large sample size increases the power of the test and makes it more likely that you'll be able to correctly detect when H_0 is false.

640.   **E. Choices (A), (B), and (C) (having a low significance level; having a random sample of data; having a large sample size)**

A low significance level, such as $\alpha = 0.01$ or $\alpha = 0.05$, means that you'll reject only if the observed data provides a result that you'd consider very unlikely under the condition of the null hypothesis.

A random sample is necessary to try to get the most representative and unbiased data possible. Otherwise, a sample could be tampered with to have only extreme data values far from the claimed value in the null hypothesis, forcing you to reject H_0 regardless of the truth.

A large sample will reduce the variation by giving you a more accurate estimate of the truth than a smaller sample would.

The value of the p-value being low or high is out of your hands and up to the data. But a large, random sample will help ensure a more reliable p-value.

641. **the population standard deviation and a statement that soda consumption is normally distributed among U.S. teens**

To run a z-test to see whether a sample mean differs from a population mean, you need the population standard deviation; you also need to know either that the characteristic of interest is normally distributed in the population or that the sample size is at least $n = 30$. Because the sample is small ($n = 15$), you need to know that soda consumption is normally distributed among U.S. teens and the population standard deviation of soda consumption.

642. **E. Choices (A) and (C) (whether the characteristic of interest is normally distributed in the population; the population mean and standard deviation)**

In addition to needing the sample mean and sample size, you can run a z-test if you have information about the population of interest. You need the population mean and standard deviation and some knowledge of the behavior of the population. In this scenario, because the sample size is small ($n = 20$), you need to know that the characteristic of interest is normally distributed in the population.

643. $H_0: \mu = 25$

The null hypothesis is always that the population parameter is *equal to* some specific value.

For a test of one population mean, the null hypothesis is that the population mean of interest is equal to a certain claimed value, which is 25 in this case.

644. $H_a: \mu \neq 10$

The null hypothesis is always that the population mean is *equal to* the stated value. In this case, the researcher believes that the null hypothesis is wrong, but the researcher doesn't have any theory about whether the true population mean is higher or lower than that, so the alternative hypothesis is a *not equal to* hypothesis.

645. $H_a: \mu > 5$

The computer store owner believes that her customers buy more than five flash drives per year on average. In other words, she believes that her customers' population average, μ, is greater than the claimed value of five. Therefore, the alternative hypothesis is a *greater than* hypothesis.

646. $H_a: \mu < 3$

The man believes that the cost of dry-cleaning a shirt in his town is lower than the average amount of 3 dollars. Therefore, the alternative hypothesis is a *less than* hypothesis.

647. 0.0668

Using Table A-1, find –1.5 in the left-hand column, and then go across the row to the column for 0.00, where the value is 0.0668. This is the proportion of the curve area that's to the left of (less than) the test statistic value of z that you're looking up. In this case, the alternative hypothesis is a *less than* hypothesis, so you can read the p-value from the table without doing further calculations.

648. 0.1336

Using Table A-1, find –1.5 in the left-hand column, and then go across the row to the column for 0.00, where the value is 0.0668. This is the proportion of the curve area that's to the left of (less than) the value of z you're looking up. In this case, the alternative hypothesis is a *not equal to* hypothesis, so you double the outlying tail quantity (area below the z-value of –1.5) to get the p-value.

649. 0.0456

Using Table A-1, find –2.0 in the left-hand column, and then go across the row to the column for 0.0, where the value is 0.0228. This is the proportion of the curve area that's to the left of (less than) the value of z you're looking up. In this case, the alternative hypothesis is a *not equal to* hypothesis, so you double the outlying tail quantity (area below the z-value of –2.0) to get the p-value.

650. 0.2714

Using Table A-1, find 1.1 in the left-hand column, then go across the row to the column for 0.0, where the value is 0.8643. This is the area under the curve to the left of the z value of 1.1. Because total area under the curve equals 1, the area above z in this case is $1 - 0.8643 = 0.1357$.

For a *not equal to* alternative hypothesis, you double the value of the outlying tail area: $p = 2(0.1357) = 0.2714$.

651. 3.54

Start by identifying the sample mean $\bar{x}$. In this case, you're told that the same mean is 115 degrees.

Next, calculate the standard error. For a one-sample z-test, the standard error is the population standard deviation, σ_X, divided by the square root of the sample size, n:

$$\sigma_X = \frac{\sigma_X}{\sqrt{n}}$$
$$= \frac{10}{\sqrt{50}}$$
$$= 1.4142$$

To get the z-test statistic, find the difference between the sample mean, $\bar{x}$. and the claimed population mean, μ_0 and divide that by the standard error, $\sigma_{\bar{x}}$:

$$z = \frac{x - \mu_0}{\sigma_x}$$

$$= \frac{115 - 110}{1.4142}$$

$$= 3.5355678, \text{ or } 3.54 \text{ (rounded)}$$

652. −3.4031

First, find the standard error by dividing the population standard deviation, σ_X, by the square root of the sample size, n:

$$\sigma_x = \frac{\sigma_X}{\sqrt{n}}$$

$$= \frac{26.52}{\sqrt{40}}$$

$$= 4.19318$$

Then, calculate the z-statistic by subtracting the claimed population mean, μ_0 from the sample mean, $\bar{x}$, and dividing the result by the standard error, $\sigma_{\bar{x}}$:

$$z = \frac{x - \mu_0}{\sigma_x}$$

$$= \frac{172.12 - 186.39}{4.19318}$$

$$= \frac{-14.27}{4.19318}$$

$$= -3.403145, \text{ or } 3.4031 \text{ (rounded)}$$

653. 0.0132

First, set up the null and alternative hypotheses. The null hypothesis is always an *equal to* hypothesis:

H_0: $\mu = 40$

Because the research director believes that customers hope to use a pen for fewer than 40 days, you use a *less than* alternative hypothesis:

H_a: $\mu < 40$

Next, identify the sample mean, which is 36 in this case.

Then, compute the standard error by dividing the population standard deviation, σ_X, by the square root of the sample size, n:

$$\sigma_x = \frac{\sigma_X}{\sqrt{n}}$$

$$= \frac{9}{\sqrt{25}}$$

$$= 1.8$$

Now, find the z-statistic by subtracting the claimed population mean, μ_0 from the sample mean, $\bar{x}$ and dividing the result by the standard error, $\sigma_{\bar{x}}$:

$$z = \frac{x - \mu_0}{\sigma_x}$$

$$= \frac{36 - 40}{1.8}$$

$$= -2.222\bar{2}$$

Rounded to two decimal places (the degree of precision in Table A-1 in the appendix), you get –2.22. Using Table A-1, find –2.2 in the left-hand column, and then go across the row to the column for 0.02, where the value is 0.0132. This is the area to the left of –2.22. Because the alternative hypothesis is a *less than* hypothesis, the *p*-value is the same as the value you find in the table: 0.0132.

654. 0.1587

First, set up the null hypothesis and alternative hypotheses. The null hypothesis is that the mean of the farmer's population of bush yields will be the same as the claimed mean:

$$H_0: \mu = 3$$

Because the farmer expects to get less fruit than she read about, you use a *less than* alternative hypothesis:

$$H_2: \mu < 3$$

Next, identify the sample mean, which is 2.9 in this case.

Then, compute the standard error by dividing the population standard deviation, σ_X by the square root of the sample size, *n*:

$$\sigma_X = \frac{\sigma_X}{\sqrt{n}}$$
$$= \frac{1}{\sqrt{100}}$$
$$= 0.1$$

Now, find the *z*-statistic by subtracting the claimed population mean, μ_0, from the sample mean, *x*, and dividing the result by the standard error, $\sigma_{\bar{x}}$:

$$z = \frac{\bar{x} - \mu_0}{\sigma_X}$$
$$= \frac{2.9 - 3}{0.1}$$
$$= -1$$

Use Table A-1 in the appendix, find the value –1.0 in the left-hand column, and then go across the row to the column for 0.0. This value is the area under the curve to the left (less than) of this value of *z*, 0.1587. Because you have a *less than* alternative hypothesis, this value is also the *p*-value.

655. Reject the null hypothesis.

Any time the *p*-value is less than the alpha level (α also known as the significance level), you reject the null hypothesis. In this case, you're given a *p*-value of 0.02 and a significance level of $\alpha = 0.05$, which is enough information to reject the null hypothesis for this study.

656. The null hypothesis can't be rejected.

First, set up the null and alternative hypotheses. The null hypothesis is that business travelers will have the same population mean as the average airplane passenger:

$$H_0: \mu = 45$$

The alternative hypothesis is that business travelers carry less that the claimed value of luggage, so you use a *less than* hypothesis:

$$H_a: \mu < 45$$

Next, identify the sample mean, which is 44.5 in this case.

Then, compute the standard error by dividing the population standard deviation σ_X, by the square root of the sample size, n:

$$\sigma_X = \frac{\sigma_X}{\sqrt{n}}$$
$$= \frac{10}{\sqrt{250}}$$
$$= 0.6325$$

Now, find the z-statistic by subtracting the claimed population mean, μ_0, from the sample mean, $\bar{x}$, and dividing the result by the standard error, $\sigma_{\bar{x}}$:

$$z = \frac{\bar{x} - \mu_0}{\sigma_{\bar{x}}}$$
$$= \frac{44.5 - 45}{0.6325}$$
$$= -0.790513834$$

Use a Z-table, such as Table A-1 in the appendix, to find the area under the normal curve to the left of the computed test statistic value. Round the test statistic value to two decimal places (to −0.79). Find the value of −0.7 in the left-hand column, and then go across the row to the column for 0.09. Table A-1 specifies that an area of 0.2148 lies to the left of this value. For a *less than* alternate hypothesis, the p-value is the same as the table value: 0.2148.

Finally, compare the p-value (0.2148) to the stated significance level (0.05). In this case, the p-value is bigger than the significance level, so you *can't reject* the null hypothesis on the basis of this data.

657. The shopkeeper doesn't have enough evidence to conclude that his rent is cheaper on average.

The shopkeeper can't conclude that the rent on average is lower than $2.00 per month because the mean in his sample ($3.00) is higher than that amount.

658. 0.0082

First, set up the null and alternative hypotheses. The null hypothesis is that the doctor's patients have the same average temperature as humans in general:

$$H_0: \mu = 98.6$$

Because the doctor believes that her patients have a higher temperature than humans on average, you use a *greater than* alternative hypothesis:

$$H_\alpha: \mu > 98.6$$

Next, identify the sample mean, which is 98.8 in this case.

Then, compute the standard error by dividing the population standard deviation, σ_X, by the square root of the sample size, n:

$$\sigma_X = \frac{\sigma_X}{\sqrt{n}}$$

$$= \frac{0.5}{\sqrt{36}}$$

$$= \frac{0.5}{6}$$

$$= 0.0833$$

Now, find the z-statistic by subtracting the claimed population mean, μ_0, from the sample mean, $\bar{x}$, and dividing the result by the standard error, $\sigma_{\bar{x}}$:

$$z = \frac{\bar{x} - \mu_0}{\sigma_{\bar{x}}}$$

$$= \frac{98.8 - 98.6}{0.0833}$$

$$= 2.40096$$

Use a Z-table, such as Table A-1 in the appendix, find the value of 2.4 in the left-hand column, and then go across to the column for 0.00. The value of 0.9918 is the area to the left of the z-value of 2.40. Because you have a *greater than* alternative hypothesis, you need to subtract the table value from 1 (the total area under the curve) to get the p-value: $1 - 0.9918 = 0.0082$.

659. 0.0606

First, set up the null and alternative hypotheses. The null hypothesis is that the mean of the population of interest (the Northeastern division customers) is the same as the claimed mean:

$$H_0: \mu = 5$$

In this case, the researcher suspects that the population mean for the Northeastern division is higher than 5, so you use a *greater than* alternative hypothesis:

$$H_a: \mu > 5$$

Next, identify the sample mean, which is 5.1 in this case.

Then, compute the standard error by dividing the population standard deviation, σ_X, by the square root of the sample size, n:

$$\sigma_X = \frac{\sigma_X}{\sqrt{n}}$$

$$= \frac{0.5}{\sqrt{60}}$$

$$= \frac{0.5}{7.746}$$

$$= 0.0645$$

Now, find the z-statistic by subtracting the claimed population mean, μ_0, from the sample mean, $\bar{x}$, and dividing the result by the standard error, $\sigma_{\bar{x}}$:

$$z = \frac{\bar{x} - \mu_0}{\sigma_X}$$

$$= \frac{5.1 - 5.0}{0.0645}$$

$$= \frac{0.1}{0.0645}$$

$$= 1.5504$$

Use a Z-table, such Table A-1 in the appendix, find the value of 1.5 in the left-hand column, and then go across to the column for 0.05. The value of 0.9394 is the area to the left of the z-value of 1.55. Because you have a *greater than* alternative hypothesis, you need to subtract the curve area from 1 to get the p-value: $1 - 0.9394 = 0.0606$.

660.

Fail to reject the null hypothesis that the average words per minute is equal to 20.

First, set up the null and alternative hypotheses. The null hypothesis is that the mean of the population of interest (the employees of the manager's branch) is equal to the claimed mean:

$$H_0: \mu = 20$$

In this case, you use a *greater than* alternative hypothesis because the manager believes his employees have a speed greater than the claimed average:

$$H_a: \mu > 20$$

Next, identify the sample mean, which is 20.5 words per minute.

Then, compute the standard error by dividing the population standard deviation, σ_X, by the square root of the sample size, n:

$$\sigma_X = \frac{\sigma_X}{\sqrt{n}}$$

$$= \frac{3}{\sqrt{30}}$$

$$= 0.5477$$

Now, find the z-statistic by subtracting the claimed population mean, μ_0, from the sample mean, $\bar{x}$, and dividing the result by the standard error, $\sigma_{\bar{x}}$:

$$z = \frac{\bar{x} - \mu_0}{\sigma_{\bar{x}}}$$

$$= \frac{20.5 - 20}{0.5477}$$

$$= 0.9129$$

Use a Z-table, such Table A-1 in the appendix, find the value of 0.9 in the left-hand column, and then go across to the column for 0.01. The value of 0.8186 is the area to the left of the z-value of 0.91. Because you have a *greater than* alternative hypothesis, the p-value is 1 minus the table value: $1 - 0.8186 = 0.1814$.

Compare the p-value to the significance level and reject the null hypothesis only if the p-value is less than the significance level. Here, $p\text{-}value = 0.1814$, which is greater than the significance level of 0.05. This means you *fail to reject the null hypothesis*.

661.   Fail to reject H_0.

First, set up the null and alternative hypotheses. In this case, μ_0 represents the current average amount of glaze put on a single vase by workers in the workshop. The target value of 2 ounces plays the same role as a claimed value might. The potter wants to know whether her current value is significantly above 2 ounces.

$$H_0: \mu = 2$$

Because the potter believes her workshop uses more than 2 ounces, the alternative hypothesis is a *greater than* hypothesis:

$$H_a: \mu > 2$$

Next, identify the sample mean, which is 2.3 ounces in this case.

Then, compute the standard error by dividing the population standard deviation, σ_X, by the square root of the sample size, n:

$$\sigma_X = \frac{\sigma_X}{\sqrt{n}}$$
$$= \frac{0.8}{\sqrt{30}}$$
$$= 0.14606$$

Now, find the z-statistic by subtracting the claimed population mean, μ_0, from the sample mean, $\bar{x}$, and dividing the result by the standard error, $\sigma_{\bar{x}}$:

$$z = \frac{x - \mu_0}{\sigma_X}$$
$$= \frac{2.3 - 2}{0.14606}$$
$$= 2.054$$

Use a Z-table, such as Table A-1 in the appendix, find the value of 2.0 in the left-hand column, and then go across to the column for 0.05. The value of 0.9798 is the area to the left of the z-value of 2.05.

Now, find the p-value by subtracting the curve area from 1, because you have a *greater than* alternative hypothesis: $1 - 0.9798 = 0.0202$.

Finally, compare this value to your significance level. The p-value (0.0202) is greater than the significance level (0.01), so you *fail to reject the null hypothesis.*

662.   Reject H_0.

First, set up the null and alternative hypotheses. The null hypothesis is that the mean of the population of interest (the density of the tissue cultures) is equal to the claimed mean:

$$H_0: \mu = 0.0047$$

The researcher suspects that her samples are heavier than the norm, so you have a *greater than* alternative hypothesis:

$$H_a: \mu > 0.0047$$

Next, identify the sample mean, 0.005 in this case, which is greater than the claimed value, so proceed with testing.

Then, compute the standard error by dividing the population standard deviation, σ_X, by the square root of the sample size, n:

$$\sigma_{\bar{x}} = \frac{\sigma_X}{\sqrt{n}}$$
$$= \frac{0.00047}{\sqrt{40}}$$
$$= 0.00007431$$

Now, find the z-statistic by subtracting the claimed population mean, μ_0, from the sample mean, $\bar{x}$, and dividing the result by the standard error, $\sigma_{\bar{x}}$:

$$z = \frac{\bar{x} - \mu_0}{\sigma_{\bar{x}}}$$
$$= \frac{0.005 - 0.0047}{0.00007431}$$
$$= 4.037$$

This rounds to 4.04, which is above the highest value in Table A-1 in the appendix, so the probability of a value at least this extreme is less than 0.0001 (the area under the curve above the highest table value of 3.69). A p-value of 0.0001 is less than the significance level of 0.001, so you should reject the null hypothesis.

The researcher has enough evidence to conclude that the tissue cultures in her lab are denser than average. And because the p-value is so small (0.0001), she can say these results are highly significant.

---

663.   The farmer can't reject the null hypothesis at a significance level of 0.05.

First, set up the null and alternative hypotheses. The null hypothesis is that the farmer's hens lay an average of 15 eggs per month:

$$H_0: \mu = 15$$

The alternative hypothesis is that the average farmer's hens is a different value (a *not equal to* hypothesis):

$$H_a: \mu \neq 15$$

Then, compute the standard error by dividing the population standard deviation, σ_X, by the square root of the sample size, n:

$$\sigma_X = \frac{\sigma_X}{\sqrt{n}}$$
$$= \frac{5}{\sqrt{30}}$$
$$= 0.9129$$

Now, find the z-statistic by subtracting the claimed population mean, μ_0, from the sample mean, $\bar{x}$, and dividing the result by the standard error, $\sigma_{\bar{x}}$:

$$z = \frac{\bar{x} - \mu_0}{\sigma_{\bar{x}}}$$
$$= \frac{16.5 - 15}{0.9129}$$
$$= 1.6431$$

Now, find the associated area under the normal curve to the left of the value you got for z, using a Z-table, such as Table A-1 in the appendix. Find the value of 1.6 in the left-hand column, and go across the row to the column for 0.04. The value is 0.9495.

Because the alternative hypothesis is a *not equal to* hypothesis and the test statistic value is positive, you need to subtract the value you found in the table from $1(1 - 0.9495 = 0.0505)$ and then double that result to get the p-value: $2(0.0505) = 0.101$.

Because the p-value is larger than the significance level, the farmer can't reject the null hypothesis. In other words, he can't say that his hens lay eggs any differently than the norm.

664. Reject the null hypothesis.

First, set up the null and alternative hypotheses. The null hypothesis is that the average number of boxes used by Chicago families is equal to the national average:

$$H_0: \mu = 110$$

Because the Chicago company just wants to see how it compares to the national average in terms of boxes used when moving, it's interested in being either higher or lower on average. So, you use a *not equal to* alternative hypothesis:

$$H_a: \mu \neq 110$$

Then, compute the standard error by dividing the population standard deviation, σ_X, by the square root of the sample size, n:

$$\sigma_X = \frac{\sigma_X}{\sqrt{n}}$$
$$= \frac{30}{\sqrt{80}}$$
$$= 3.3541$$

Now, find the z-statistic by subtracting the claimed population mean, μ_0, from the sample mean, $\bar{x}$, and dividing the result by the standard error, $\sigma_{\bar{x}}$:

$$z = \frac{\bar{x} - \mu_0}{\sigma_X}$$
$$= \frac{103 - 110}{3.3541}$$
$$= -2.087$$

Now find the associated area under the normal curve to the left of the value you got for z, using a Z-table, such as Table A-1 in the appendix. Find the value of −2.0 in the left-hand column, and go across the row to the column for 0.09 (rounding to two decimals). The value is 0.0183, which is the area under the curve to the left of this z-value.

Because you have a *not equal to* alternative hypothesis and the test statistic value is negative, you need to double the curve area to get the p-value: $2(0.0183) = 0.0366$.

This p-value is lower than the significance level of 0.05, so you *reject the null hypothesis*. In other words, you reject the claim that the average number of boxes used by Chicago families is equal to the national average. (Because the z-value is negative, the average box count is probably less than the national average.)

665.
Fail to reject the null hypothesis.

First, set up the null and alternative hypotheses. The null hypothesis is that the average American family who uses a car to go on vacation travels an average of 382 miles from home:

$$H_0: \mu = 382$$

Because the researcher is interested in seeing whether a difference occurs in average travel distance, the alternative hypothesis is a *not equal to* hypothesis:

$$H_a: \mu \neq 382$$

Then, compute the standard error by dividing the population standard deviation, σ_X, by the square root of the sample size, *n*:

$$\sigma_X = \frac{\sigma_X}{\sqrt{n}}$$
$$= \frac{150}{\sqrt{30}}$$
$$= 27.3861$$

Now, find the *z*-statistic by subtracting the claimed population mean, μ_0, from the sample mean, $\bar{x}$, and dividing the result by the standard error, $\sigma_{\bar{x}}$:

$$z = \frac{\bar{x} - \mu_0}{\sigma_{\bar{x}}}$$
$$= \frac{398 - 382}{27.3861}$$
$$= 0.5842$$

Now find the associated area under the normal curve to the left of the value you got for *z*, using a Z-table, such as Table A-1 in the appendix. Find the value of 0.5 in the left-hand column, and go across the row to the column for 0.08. The value is 0.7190, which is the area to the left of this value. Because you have a *not equal to* alternative hypothesis, you find the *p*-value by subtracting the area under the curve from 1 and doubling the result: *p-value* = 2(1 − 0.7190) = 0.562.

This value is greater than the significance level of 0.05, so you fail to reject the null hypothesis. You can't say families who vacation by car with dogs travel a different distance on average than other families, based on this data.

666.
$H_0: \mu = 120; H_a: \mu < 120$

The question is about the average number of minutes U.S. teenagers spend texting, which represents a population. So, the symbol for the population mean μ, is in the null and alternative hypotheses (and not the symbol for the sample mean, $\bar{x}$). The claim is that the mean equals 120, so this is the value in the null hypothesis. You believe it's less than that, so your alternative hypothesis is the *less than* alternative.

667.
$H_0: \mu = 250; H_a: \mu > 250$

The null and alternative hypotheses are always about a population value (in this case, the population mean), not about a sample value. So, the symbols in the null and alternative hypothesis should be μ (population mean), not $\bar{x}$ (sample mean). In addition, the actual numbers in the null and alternative hypotheses should reference the original claim regarding the population mean (250 calories). Finally, because the researcher believes that the pizza on the college campus where he works has more calories, the alternative hypothesis is a *greater than* hypothesis.

668. **A. The population standard deviation is unknown.**

If the population has a normal distribution but the population standard deviation is unknown, you use a t-test (otherwise, use a z-test). The sample standard deviation comes from your data, so you'll always know its value.

669. **A. The population standard deviation isn't known.**

When the population standard deviation isn't known but can be estimated from the sample standard deviation, you should use a t-test instead of using the Z-distribution to test a hypothesis about a single population mean.

670. **The student should do a single population t-test.**

Whenever the population standard deviation isn't known but the sample standard deviation allows an estimate of it, you should do a t-test for a single population mean rather than use the Z-distribution.

671. $H_0: \mu_1 = \mu_0$

The null hypothesis is that the mean of the population from which the sample was drawn is equal to the claimed mean.

672. $H_a: \mu_1 < \mu_0$

Because the student believes that his friends spend less time than is claimed, the alternative hypothesis is a *less than* hypothesis.

673. **She could make the conclusion under other conditions, but the sample from her store doesn't represent all stores in this retail chain.**

An important condition of any hypothesis test is that the data going into the calculations is reliable and valid. In this case, because the question resolves around the average price of a certain hair product from a national chain, whose stores are throughout the United States, the data collected must also represent this population. In this case, the data represent 30 bottles of this product taken from her store only. So, any conclusions she makes from a hypothesis test should be rendered invalid.

674. $H_0: \mu = 3.5; H_a: \mu < 3.5$

The null hypothesis is that the population mean from which the sample was drawn is equal to the claimed population mean. The alternative hypothesis in this case is that the dentist's patients experience less pain than average, so the alternative is a *less than* hypothesis.

675. −2

To find the value for the test statistic, t, use the t-test formula:

$$t = \frac{\bar{x} - \mu_0}{s / \sqrt{n}}$$

where $\bar{x}$ is the sample mean, μ_0 is the claimed population mean, s is the sample standard deviation, and n is the sample size. Substitute these values from the question into the formula and solve:

$$t = \frac{30 - 35}{10 / \sqrt{16}}$$

$$= \frac{-5}{2.5}$$

$$= -2$$

676. −2.733

To find the value for the test statistic, t, use the t-test formula:

$$t = \frac{\bar{x} - \mu_0}{s / \sqrt{n}}$$

where $\bar{x}$ is the sample mean, μ_0 is the claimed population mean, s is the sample standard deviation, and n is the sample size. Substitute these values from the question into the formula and solve:

$$t = \frac{5.2 - 6.3}{1.8 / \sqrt{20}}$$

$$= \frac{-1.1}{0.4025}$$

$$= -2.733$$

677. 1.761310

First, figure out the degrees of freedom, which is one less than the sample size n: $df = 15 - 1 = 14$.

Because the researcher believes that population of interest has a mean greater than 6.1, you use a *greater than* alternative hypothesis, so 0.05 of the area under the curve should be in the upper tail. Using a t-table, such as Table A-2 in the appendix, find the row for 14 degrees of freedom and the column for 0.05. The critical value is 1.761310.

678. 2.26216

First, figure out the degrees of freedom, which is one less than the sample size n: $df = 10 - 1 = 9$.

Because the research believes that the sampled population of interest has a mean that differs from the claimed value, you use a *not equal to* alternative hypothesis, so half of the confidence level will be in each tail: $0.05 / 2 = 0.025$.

Using a t-table, such as Table A-2 in the appendix, find the row for 9 degrees of freedom and the column for 0.025. The critical values are 2.26216 and –2.26216; you reject the null hypothesis if your test statistic is outside the range of –2.26216 to 2.26216.

679. She would reject H_0.

The test statistic is –2.733, which is in the correct direction for a *less than* alternative hypothesis. Compare it to the critical value for a *less than* alternative hypothesis with a confidence level of 0.05, with $n-1 = 20-1 = 19$ degrees of freedom. Using a t-table, such as Table A-2 in the appendix, find the row for 19 degrees of freedom and the column for 0.05. The value you find is 1.729133. However, because this is a left-tailed test, the critical value is –1.729133 in this case.

Because Table A-2 contains only positive values, compare the absolute value of the test statistic, 2.733, with the critical value. The absolute value of the test statistic is greater, so reject the null hypothesis. The researcher has significant evidence to suggest that mothers get less sleep than the average person.

680. The test statistic t is larger than the positive critical value.

First, figure out the degrees of freedom, which is one less than the sample size, n: $df = 17-1 = 16$.

Because this is a *not equal to* alternative hypothesis (the question specified that the researcher is interested in a difference "in either direction,") half the confidence level will be in the upper tail and half in the lower tail (0.05 in each).

Then, find the critical value from a t-table, such as Table A-2 in the appendix. Find the row for 16 degrees of freedom and the column for 0.05. The critical values are –1.745884 and 1.745884.

Next, find the test statistic t, using the basic formula for t:

$$t = \frac{\bar{x} - \mu_0}{s / \sqrt{n}}$$

where $\bar{x}$ is the sample mean, μ_0 is the population mean, s is the sample standard deviation, and n is the sample size.

$$t = \frac{\bar{x} - \mu_0}{s / \sqrt{n}}$$
$$= \frac{0.0123 - 0.0112}{0.0019 / \sqrt{17}}$$
$$= 2.387061$$

This is larger than the positive critical value of t.

681. $0.20 < p\text{-value} < 0.50$

First, find the degrees of freedom, which is one less than the sample size, n: $df = 11-1 = 10$.

Then, using Table A-2 in the appendix, find the row for 10 degrees of freedom. Read from the left to find the last value that's *smaller than* your test statistic t. In this case, that's 0.699812, which corresponds to a single-tail area of 0.25. Because the alternative is two-sided, you have to double that to get 0.50 as one bound of the p-value.

On the same row, read from the left to find the first value that's *greater than* the test statistic *t*, which is 1.372184. Now, look at the column head to find the single-tail area, 0.10. Double that to get the other bound of the *p*-value, 0.20.

You can say that the *p*-value is between 0.20 and 0.50, or that 0.20 < *p*-value < 0.50.

682. 0.10 < *p*-value < 0.25

First, find the degrees of freedom, which is one less than the sample size, *n*:
$df = 29 - 1 = 28$.

Then, find the test statistic *t*, using the basic formula for *t*:

$$t = \frac{\bar{x} - \mu_0}{s / \sqrt{n}}$$

where $\bar{x}$ is the sample mean, μ_0 is the population mean, *s* is the sample standard deviation, and *n* is the sample size.

$$t = \frac{\bar{x} - \mu_0}{s / \sqrt{n}}$$
$$= \frac{89.8 - 90}{1 / \sqrt{29}}$$
$$= \frac{-0.2}{0.1857}$$
$$= -1.077$$

You have a *less than* alternative hypothesis because the researcher "believes that her sample mean is smaller than 90." Because of this, and because the sample mean is below the hypothesized population mean, you must translate the *t*-statistic into an absolute value of 1.077. Using Table A-2 in the appendix, go to the row for 28 degrees of freedom, and find the closest values to your test statistic. This value falls between the 0.25 and 0.10 columns, so the *p*-value is between 0.10 and 0.25.

683. 0.01 < *p*-value < 0.02

First, find the degrees of freedom, which is one less than the sample size, *n*:
$df = 12 - 1 = 11$.

Because the warden has no hypothesis about the direction her inmate costs may differ, the alternative hypothesis is a *not equal to* hypothesis.

Calculate the test statistic *t*, using the formula for a one-sample *t*-test,

$$t = \frac{\bar{x} - \mu_0}{s / \sqrt{n}}$$

where $\bar{x}$ is the sample mean, μ_0 is the population mean, *s* is the sample standard deviation, and *n* is the sample size.

$$t = \frac{\bar{x} - \mu_0}{s / \sqrt{n}}$$
$$= \frac{58,660 - 50,000}{10,000 / \sqrt{12}}$$
$$= \frac{8,660}{2,886.75}$$
$$= 2.9999$$

Because you have a positive value of t, you can use the t-table without worrying about the sign on t. Using Table A-2 in the appendix, read across the row for 11 degrees of freedom. The nearest numbers in the table to the test statistic are 2.71808, which has a right-tail area of 0.01 (from the column heading), and 3.10581, which has a right-tail area of 0.005.

You have a *not equal to* hypothesis because no direction is stated in the problem. Double the right-tail areas (because there's an equal area in the left tail), so the p-value is between 2(0.005) and 2(0.01), or between 0.01 and 0.02.

684. The p-value is between 0.01 and 0.025.

Using Table A-2 in the appendix, find the row for 14 degrees of freedom. The test statistic, 2.5, falls between 2.14479 (corresponding to a right-tail area of 0.025 as given in the column heading) and 2.62449 (with a right-tail area of 0.01).

Because you have a *greater than* alternative hypothesis, the p-values are the same as right-tail areas. So, the p-value is between 0.01 and 0.025.

685. 0.02 < p-value < 0.05

Because the test statistic is negative but the t-table deals with positive values, use the absolute value of the test statistic, 2.5.

On Table A-2 in the appendix, read across the row for 20 degrees of freedom. The nearest numbers in the table to the test statistic are 2.08596, which has a right-tail area of 0.025, and 2.52798, which has a right-tail area of 0.01. Because you have a *not equal to* alternative hypothesis, double both values, giving you a range of 0.02 to 0.05.

686. There is enough evidence to conclude that the average exam time is more than 45 minutes.

The claim is that the average test-taking time is 45 minutes (the null hypothesis). You believe that the average test-taking time is more than 45 minutes (the alternative hypothesis). Because the p-value from your sample is smaller than your significance level, you reject the null hypothesis, but you can go further and say that you can conclude that the average time is more than 45 minutes, based on your data. *Note:* You could be wrong, so your results don't "prove" anything; however, they do provide strong evidence against the claim.

687. Fail to reject H_0.

Because the sample size is small and you're testing the population mean with unknown population standard deviation (only the sample standard deviation is given), a t-test is in order.

The null hypothesis is that the music students have average verbal ability. The scientist believes that it's lower, so the alternative hypothesis is a *less than* alternative.

First, note the degrees of freedom, which is one less than the sample size n:
$df = 8 - 1 = 7$.

Then, to find the critical value, use Table A-2 in the appendix to get the value of 1.894579. However, because this is a left-tailed test, the critical value is –1.894579.

Next, calculate the test statistic t, using this formula:

$$t = \frac{\bar{x} - \mu_0}{s / \sqrt{n}}$$

where $\bar{x}$ is the sample mean, μ_0 is the population mean, s is the sample standard deviation, and n is the sample size.

$$t = \frac{\bar{x} - \mu_0}{s / \sqrt{n}}$$

$$= \frac{97.5 - 100}{5 / \sqrt{8}}$$

$$= \frac{-2.5}{1.7678}$$

$$= -1.4142$$

The value of this test statistic (-1.4142) is closer to zero than the critical value (-1.894579), which was found earlier. Therefore, you *can't reject* the null hypothesis. There isn't enough evidence to say that people who play musical instruments have below average verbal ability.

688.   **Reject H$_0$.**

Because the sample size is small and you're testing the population mean with unknown population standard deviation (only the sample standard deviation is given), a t-test is in order.

The null hypothesis is that the employees donate the target amount of $50 per year on average. The president believes it's lower than that, so the alternative hypothesis is a *less than* alternative.

First, note the degrees of freedom, which is one less than the sample size n: $df = 10 - 1 = 9$.

Then, find the critical value in Table A-2 in the appendix, which is -1.833113 (because this is a left-tailed test).

Next, calculate the test statistic t, using this formula:

$$t = \frac{\bar{x} - \mu_0}{s / \sqrt{n}}$$

where $\bar{x}$ is the sample mean, μ_0 is the population mean, s is the sample standard deviation, and n is the sample size.

$$t = \frac{\bar{x} - \mu_0}{s / \sqrt{n}}$$

$$= \frac{43.40 - 50}{5.2 / \sqrt{10}}$$

$$= \frac{-6.6}{1.6444}$$

$$= -4.0136$$

Because you have a negative t-test statistic and a *less than* alternative hypothesis, you reject the null hypothesis because the value of the t-test statistic is less than the critical value ($-4.0136 < -1.833113$).

So, the president has a point and can conclude that the donations of her employees to charity are lower than the targeted value on average, based on her data.

689. **Reject H₀.**

Because the sample size is small and you're testing the population mean with unknown population standard deviation (only the sample standard deviation is given), a *t*-test is in order.

The null hypothesis is that the coats are protected to an average of −5 degrees Centigrade, on average. But the coat maker believes that the average temperature is lower than that; so the alternative hypothesis is a *less than* alternative.

First, note the degrees of freedom, which is one less than the sample size n: $df = 15 - 1 = 14$.

Then, to find the critical value, use Table A-2 in the appendix, going to the row for 14 degrees of freedom and the column for 0.10; the value is 1.345030. Because this is a left-tailed test, the critical value is −1.345030.

Next, calculate the test statistic t, using this formula:

$$t = \frac{\bar{x} - \mu_0}{s / \sqrt{n}}$$

where $\bar{x}$ is the sample mean, μ_0 is the population mean, s is the sample standard deviation, and n is the sample size.

$$t = \frac{\bar{x} - \mu_0}{s / \sqrt{n}}$$
$$= \frac{-6.5 - (-5)}{1 / \sqrt{15}}$$
$$= \frac{-1.5}{0.2582}$$
$$= -5.8095$$

Because you have a *less than* alternative hypothesis and the test result is negative, you reject the null hypothesis because the test statistic is less than the critical value (−5.8095 < −1.345030).

Based on his data, the coat maker has enough evidence to reject the advertisement's claim and conclude that the average protection temperature is lower than that.

690. **Fail to reject H₀.**

Because the sample size is small and you're testing the population mean with unknown population standard deviation (only the sample standard deviation is given), a *t*-test is in order.

The null hypothesis is that the average teacher evaluations at the teacher's school are the same as those in the district. But the teacher believes that the average is lower than that; so the alternative hypothesis is a *less than* alternative.

First, note the degrees of freedom, which is one less than the sample size n: $df = 6 - 1 = 5$.

The significance level is $\alpha = 0.05$. Because you have a *less than* alternative hypothesis and the sample mean is less than the hypothesized value, you can just use the column heading of the *t*-table (Table A-2 in the appendix) to find the significance level (0.05) and read down to find the row corresponding to the degrees of freedom (5). You get a value of 2.015048. Because this is a left–tailed test, the critical value is –2.015048.

Now, calculate the test statistic t, using this formula:

$$t = \frac{\bar{x} - \mu_0}{s / \sqrt{n}}$$

where $\bar{x}$ is the sample mean, μ_0 is the population mean, s is the sample standard deviation, and n is the sample size.

$$t = \frac{\bar{x} - \mu_0}{s / \sqrt{n}}$$
$$= \frac{6.667 - 7.2}{2 / \sqrt{6}}$$
$$= \frac{-0.533}{0.8165}$$
$$= -0.6528$$

The fact that the alternative hypothesis is *less than* (because the teacher "believes that other teachers in her school get lower evaluations compared to teachers in other schools in the district") and the test statistic is not less than (is closer to zero than) the critical value (–0.6528 > –2.015048), you *fail to reject* the null hypothesis. There isn't enough evidence to conclude that the average of the teacher evaluations in the teacher's school is lower than the district.

691. Fail to reject H₀.

Because the sample size is small and you're testing the population mean with unknown population standard deviation (only the sample standard deviation is provided), a t-test is in order.

The null hypothesis is that the average popsicle temperature is 1.92 degrees Centigrade. The president believes that the average is lower than that; so the alternative hypothesis is a *less than* alternative.

First, note the degrees of freedom, which is one less than the sample size n: $df = 5 - 1 = 4$.

Then, to find the critical value, use Table A-2 in the appendix. You get a value of 3.74695 by going to the row for 4 degrees of freedom and the column for 0.01. Because this is a left–tailed test, the value is –3.74695.

Next, calculate the test statistic t, using this formula:

$$t = \frac{\bar{x} - \mu_0}{s / \sqrt{n}}$$

where $\bar{x}$ is the sample mean, μ_0 is the population mean, s is the sample standard deviation, and n is the sample size.

$$t = \frac{\bar{x} - \mu_0}{s / \sqrt{n}}$$
$$= \frac{-2.25 - (-1.92)}{1.62 / \sqrt{5}}$$
$$= \frac{-0.33}{0.7245}$$
$$= -0.4555$$

Because you have a *less than* alternative hypothesis, you fail to reject the null hypothesis because the test statistic is not less than (is closer to zero than) the critical value ($-0.4555 > -3.74695$). Based on this data, president doesn't have enough evidence to conclude that the temperature for the popsicles is set too low on average.

692. **Reject H$_0$.**

Because the sample size is small and you're testing the population mean with unknown population standard deviation, a *t*-test is in order.

The null hypothesis is that the average weight is 50 grams per object. The alternative hypothesis is that the average weight is more than 50 grams, so here you have a *greater than* alternative hypothesis.

First, note the degrees of freedom, which is one less than the sample size *n*: $df = 16 - 1 = 15$.

Then, find the critical value in Table A-2 in the appendix, which is 1.753050, by going to the row for 15 degrees of freedom and the column for 0.05.

Next, calculate the test statistic *t*, using this formula:

$$t = \frac{\bar{x} - \mu_0}{s / \sqrt{n}}$$

where $\bar{x}$ is the sample mean, μ_0 is the population mean, *s* is the sample standard deviation, and *n* is the sample size.

$$t = \frac{\bar{x} - \mu_0}{s / \sqrt{n}}$$
$$= \frac{54 - 50}{8 / \sqrt{16}}$$
$$= \frac{4}{2} = 2$$

Because the test statistic (2) is larger than the critical value (1.753050), you reject the null hypothesis. In other words, you reject the null hypothesis that the average weight is 50 grams per object in favor of the alternative hypothesis that the average weight is more than 50 grams.

693. **Fail to reject H$_0$.**

Because you're testing the population mean with unknown population standard deviation, a *t*-test is in order.

The null hypothesis is that the average number of beads per 1-pound bag is 1,200. The alternative hypothesis is that the average weight is more than that, so you have a *greater than* alternative hypothesis.

First, note the degrees of freedom, which is one less than the sample size *n*: $df = 30 - 1 = 29$.

Then, find the critical value in Table A-2 in the appendix by going to the row for 29 degrees of freedom and the column for 0.01; the critical value is 2.46202.

Next, calculate the test statistic *t*, using this formula:

$$t = \frac{\bar{x} - \mu_0}{s / \sqrt{n}}$$

where $\bar{x}$ is the sample mean, μ_0 is the population mean, s is the sample standard deviation, and n is the sample size.

$$t = \frac{\bar{x} - \mu_0}{s / \sqrt{n}}$$
$$= \frac{1,350 - 1,200}{500 / \sqrt{30}}$$
$$= \frac{150}{91.2871}$$
$$= 1.6432$$

The critical value of 2.46202 is larger than the test statistic value of 1.6432, so the conclusion is that you fail to reject the null hypothesis. There isn't enough evidence for the retailer to say that the average number of beads in a 1-pound bag is more than 1,200.

694. **Fail to reject H₀.**

Because you're testing the population mean with unknown population standard deviation, a t-test is in order.

The null hypothesis is that the average number of beads per 1-pound bag is 1,200. The alternative hypothesis is that the average weight is more than that, so you have a *greater than* alternative.

First, note the degrees of freedom, which is one less than the sample size n: $df = 30 - 1 = 29$.

Then, find the critical value in Table A-2 in the appendix by going to the row for 29 degrees of freedom and the column for 0.05; the critical value is 1.699127.

Next, calculate the test statistic t, using this formula:

$$t = \frac{\bar{x} - \mu_0}{s / \sqrt{n}}$$

where $\bar{x}$ is the sample mean, μ_0 is the population mean, s is the sample standard deviation, and n is the sample size.

$$t = \frac{\bar{x} - \mu_0}{s / \sqrt{n}}$$
$$= \frac{1,350 - 1,200}{500 / \sqrt{30}}$$
$$= \frac{150}{91.2871}$$
$$= 1.6432$$

The test statistic of 1.6432 is less than the critical value of 1.699127 so you fail to reject the null hypothesis. There isn't enough evidence to say that the average number of beads in a 1-pound bag is more than 1,200.

695. **Reject H₀.**

Because you're testing the population mean with unknown population standard deviation (only the sample standard deviation is given), a t-test is in order.

The null hypothesis is that the average amount of money spent on entertainment is $100 (maximum). The alternative hypothesis is that the average money spent is more than that, so you have a *greater than* alternative hypothesis.

First, note the degrees of freedom, which is one less than the sample size n: $df = 25 - 1 = 24$.

Then, find the critical value in Table A-2 in the appendix by going to the row for 24 degrees of freedom and the column for 0.01; the critical value is 2.49216.

Next, calculate the test statistic t, using this formula:

$$t = \frac{\bar{x} - \mu_0}{s / \sqrt{n}}$$

where $\bar{x}$ is the sample mean, μ_0 is the population mean, s is the sample standard deviation, and n is the sample size.

$$t = \frac{\bar{x} - \mu_0}{s / \sqrt{n}}$$
$$= \frac{118.44 - 100}{35 / \sqrt{25}}$$
$$= \frac{18.44}{7}$$
$$= 2.6343$$

The test statistic is greater than the critical value, so you reject H_0. There's enough evidence based on this data that the average money spent on entertainment is more than $100 per month.

696. **Reject H_0.**

Because you're testing the population mean with unknown population standard deviation (only the sample standard deviation is given), a t-test is in order.

The null hypothesis is that the average amount of cheese in the United States matches that of Europe, which is 25.83 kilograms per person per year. The alternative hypothesis is that the average amount of cheese eaten in the United States is more than that, so you have a greater than alternative hypothesis.

First, note the degrees of freedom, which is one less than the sample size n: $df = 30 - 1 = 29$.

Then, find the critical value in Table A-2 in the appendix by going to the row for 29 degrees of freedom and the column for 0.05; the critical value is 1.699127.

Next, calculate the test statistic t, using this formula:

$$t = \frac{\bar{x} - \mu_0}{s / \sqrt{n}}$$

where $\bar{x}$ is the sample mean, μ_0 is the population mean, s is the sample standard deviation, and n is the sample size.

$$t = \frac{\bar{x} - \mu_0}{s / \sqrt{n}}$$
$$= \frac{27.86 - 25.83}{6.46 / \sqrt{30}}$$
$$= \frac{2.03}{1.1794}$$
$$= 1.7212$$

The test statistic is larger than the critical value, so you reject the null hypothesis. There's enough evidence based on this data to reject the claim that the average cheese consumption in the United States is 25.83 kilograms per person and conclude that it's actually higher than that.

697. **Reject H_0.**

Because you're testing the population mean with unknown population standard deviation (only the sample standard deviation is given), a t-test is in order.

The null hypothesis is that his students match the ideal meditation time of 20 minutes. The alternative hypothesis is that the average amount differs from that, so you have a *not equal to* alternative hypothesis.

First, note the degrees of freedom, which is one less than the sample size n:
$df = 9 - 1 = 8$.

Using Table A-2 in the appendix, find the critical values, with 8 degrees of freedom and a right-tail area of 0.05 (half of 0.10 because the alternative is two-sided); the value is 1.859548. Because this is a two-tailed test, the critical values are 1.859548 and −1.859548.

Next, calculate the test statistic t, using this formula:

$$t = \frac{\bar{x} - \mu_0}{s / \sqrt{n}}$$

where $\bar{x}$ is the sample mean, μ_0 is the population mean, s is the sample standard deviation, and n is the sample size.

$$t = \frac{\bar{x} - \mu_0}{s / \sqrt{n}}$$
$$= \frac{24 - 20}{5 / \sqrt{9}}$$
$$= \frac{4}{1.6667}$$
$$= 2.39995$$

The test statistic of 2.4 (rounded up) is greater than the positive critical value of 1.859548, so you reject the null hypothesis. There's sufficient evidence to conclude that this instructor's students don't meditate the ideal amount on average.

698. **Reject H_0.**

Because you're testing the population mean with unknown population standard deviation (only the sample standard deviation is given), a t-test is in order.

The null hypothesis is that the average output of this type of laser printer before servicing is 20,000 pages. The alternative hypothesis is that the average output differs from that, so you have a *not equal to* alternative hypothesis.

First, note the degrees of freedom, which is one less than the sample size n:
$df = 16 - 1 = 15$.

Next, find the critical values. Table A-2 in the appendix gives right-tail probabilities in the column headings. The right-tail probability is half of the significance level when the alternative hypothesis is *not equal to*. This works out to half of $\alpha = 0.05$, that is,

0.025; find the value by going to the row for 15 degrees of freedom and the column for 0.025. The value is 2.13145. Because this is a two-tailed test, the critical values are 2.13145 and −2.13145.

You'll reject the null hypothesis if the test statistic value of t falls outside of the range of −2.13145 to 2.13145.

Next, calculate the test statistic t, using this formula:

$$t = \frac{\bar{x} - \mu_0}{s/\sqrt{n}}$$

where $\bar{x}$ is the sample mean, μ_0 is the population mean, s is the sample standard deviation, and n is the sample size.

$$t = \frac{\bar{x} - \mu_0}{s/\sqrt{n}}$$
$$= \frac{18{,}356 - 20{,}000}{2{,}741/\sqrt{16}}$$
$$= \frac{-1{,}644}{685.25}$$
$$= -2.39912$$

The test statistic value of −2.39912 is outside the range defined by the critical value of t (−2.13145 to +2.13145). This means that you reject the null hypothesis. There's significant evidence that this type of laser printer doesn't require servicing at 20,000 pages on average. It appears (because the test statistic is negative) that it's likely to be less than that.

699. **Fail to reject H₀.**

Because you're testing the population mean with unknown population standard deviation (only the sample standard deviation is given), a t-test is in order.

The null hypothesis is that the students' dissertations in the doctoral program match the norm of 90 pages. The alternative hypothesis is that the average amount differs from that, so you have a *not equal to* alternative hypothesis.

First, note the degrees of freedom, which is one less than the sample size n: $df = 10 - 1 = 9$.

Then, find the critical value. Because this is a *not equal to* test, the 0.05 significance level means that 5% of curve area is split between two tails, leaving 2.5% (or 0.025) in each tail. Use Table A-2 in the appendix to find the critical value of t by going to the row for 9 degrees of freedom and the column for 0.025; the value is 2.26216. Because this is a *not equal to* test, you'll fail to reject the null hypothesis if the test statistic is in the range of −2.26216 to +2.26216.

Next, calculate the test statistic t, using this formula:

$$t = \frac{\bar{x} - \mu_0}{s/\sqrt{n}}$$

where $\bar{x}$ is the sample mean, μ_0 is the population mean, s is the sample standard deviation, and n is the sample size.

$$t = \frac{\bar{x} - \mu_0}{s / \sqrt{n}}$$

$$= \frac{85.2 - 90}{7.59 / \sqrt{10}}$$

$$= \frac{-4.8}{2.40}$$

$$= -2.0$$

The test value is within the range defined by the critical values -2.26216 and $+2.26216$, so you fail to reject the null hypothesis. The student doesn't have enough support for the idea that students in her doctoral program write dissertations that differ in length from the supposed mean of 90 pages.

---

700. Reject H_0.

Because you're testing the population mean with unknown population standard deviation (only the sample standard deviation is given), a t-test is in order.

The null hypothesis is that the average age of the trees in the forest is 30 years. The alternative hypothesis is that the average age differs from that, so here you have a *not equal to* alternative hypothesis.

First, note the degrees of freedom, which is one less than the sample size n:
$df = 5 - 1 = 4$.

Then, find the critical value. In Table A-2 in the appendix, find the critical value for 4 degrees of freedom that will leave 0.25 area in each tail (because this is a *not equal to* alternative hypothesis) by going to the row for 4 degrees of freedom and the column for 0.25; the value is 0.740697. You'll reject the null hypothesis if the test statistic is outside the range of -0.740697 to $+0.740697$.

Next, calculate the test statistic t, using this formula:

$$t = \frac{\bar{x} - \mu_0}{s / \sqrt{n}}$$

where $\bar{x}$ is the sample mean, μ_0 is the population mean, s is the sample standard deviation, and n is the sample size.

$$t = \frac{\bar{x} - \mu_0}{s / \sqrt{n}}$$

$$= \frac{33 - 30}{5.6 / \sqrt{5}}$$

$$= \frac{3}{2.5044}$$

$$= 1.19789$$

The computed test statistic t-value of 1.19789 is clearly outside the range of -0.740697 to $+0.740697$, so you reject the null hypothesis. The logger has evidence to conclude that the average age of the trees in this forest isn't equal to 30 years. (And because the test statistic is positive, it is likely that the average age is higher than that.)

701. The null hypothesis should be rejected; the bank should open the new branch.

First, set up the null and alternative hypotheses:

$$H_0: p_0 = 0.10$$
$$H_a: p_0 > 0.10$$

Note: Because the bank wants to ensure that *at least* 10% of the residents will use the new branch, it's implicitly interested in a *greater than* alternative hypothesis.

Next, determine whether the sample is large enough to run a z-test by checking that both np_0 and $n(1-p_0)$ equal at least 10. In this case, $n = 100$ and $p_0 = 0.10$, so $np_0 = (100)(0.10) = 10$, and $n(1-np_0) = 100(1-0.10) = 100(0.90) = 90$.

Then, compute the standard error with this formula:

$$SE = \sqrt{\frac{p_0(1-p_0)}{n}}$$

where p_0 is the population proportion and n is the sample size. Substitute the known values into the formula to get

$$SE = \sqrt{\frac{0.10(1-0.10)}{100}}$$
$$= \sqrt{\frac{0.10(0.90)}{100}}$$
$$= \sqrt{\frac{0.09}{100}}$$
$$= \sqrt{0.0009}$$
$$= 0.03$$

Next, find the observed proportion by dividing the number who said they would consider banking at the new branch by the sample size of 100: $19/100 = 0.19$.

Then, calculate the z-test statistic, using this formula:

$$z = \frac{\hat{p} - p_0}{SE}$$

where $\hat{p}$ is the observed proportion, p_0 is the hypothesized proportion, and SE is the standard error.

$$z = \frac{0.19 - 0.1}{0.03}$$
$$= 3$$

Now, use a Z-table, such as Table A-1 in the appendix, to determine the probability of observing a z-score this high or higher. Unfortunately, the table shows the probability of observing a score of $z = 3.0$ or lower, so you have to subtract the table probability from 1 to get the probability: $1 - 0.9987 = 0.0013$.

Finally, compare the probability (that is, the p-value) with the alpha (α) level. The bank wanted to use a significance level of 0.05, so $\alpha = 0.05$, and the p-value of 0.0013 is much lower than that. So, you reject the null hypothesis.

702.
The null hypothesis should be rejected; the standard has been met and exceeded.

First, set up the null and alternative hypotheses:

$$H_0: p_0 = 0.75$$

$$H_a: p_0 > 0.75$$

Note: The alternative is a *greater than* hypothesis because the call center sets a minimum threshold for performance and is asking only whether the system meets or exceeds the threshold with a high degree of confidence.

Then, determine whether the sample is large enough to run a z-test by checking that both np_0 and $n(1-p_0)$ equal at least 10. In this case, $n = 50$ and $p_0 = 0.75$, so $np_0 = 50(0.75) = 37.5$, and $n(1-p_0) = 50(1-0.75) = 50(0.25) = 12.5$.

Compute the standard error with this formula:

$$SE = \sqrt{\frac{p_0(1-p_0)}{n}}$$

where p_0 is the population proportion and n is the sample size. Substitute the known values into the formula to get

$$SE = \sqrt{\frac{0.75(1-0.75)}{50}}$$

$$= \sqrt{\frac{0.75(0.25)}{50}}$$

$$= \sqrt{\frac{0.1875}{50}}$$

$$= \sqrt{0.00375}$$

$$= 0.061237$$

Next, find the observed proportion by dividing the number of "successes" by the sample size: $45/50 = 0.90$.

Then, calculate the z-test statistic, using this formula:

$$z = \frac{\hat{p} - p_0}{SE}$$

where $\hat{p}$ is the observed proportion, p_0 is the hypothesized proportion, and SE is the standard error.

$$z = \frac{0.90 - 0.75}{0.061237}$$

$$= 2.4495, \text{ or } 2.45 \text{ (rounded)}$$

Now, use a Z-table, such as Table A-1 in the appendix, to figure out the probability of observing a value this high or higher (because the alternative is a *greater than* hypothesis). The table gives you the proportion of the area under the curve that's less than a given value of z, which is 0.9929. To get the desired area (proportion above this z-value), you have to subtract that number from 1: $1 - 0.9929 = 0.0071$.

Finally, identify the desired alpha (α) level (0.05) and compare the probability you found from the Z-table with that. Because 0.0071 is less than 0.05, you reject the null hypothesis.

703. The null hypothesis can't be rejected; the clerk shouldn't buy the books.

First, set up the null and alternative hypotheses:

$$H_0: p_0 = 0.50$$

$$H_a: p_0 > 0.50$$

Note: The store sets a minimum threshold (50%) and is interested in finding whether a collection of books is at least that marketable or more, so the alternative is a *greater than* hypothesis.

Next, determine whether the sample is large enough to run a z-test by checking that both np_0 and $n(1-p_0)$ equal at least 10. In this case, $n = 30$ and $p_0 = 0.5$, so $np_0 = 50(0.5) = 15$, and $n(1-p_0) = 30(1-0.5) = 15$.

Calculate the observed proportion by dividing the number of books likely to sell by the number of books offered: $17/30 = 0.5667$.

Then, compute the standard error with this formula:

$$SE = \sqrt{\frac{p_0(1-p_0)}{n}}$$

where p_0 is the population proportion and n is the sample size. Substitute the known values into the formula to get

$$SE = \sqrt{\frac{0.5(1-0.5)}{30}}$$
$$= \sqrt{\frac{0.50(0.50)}{30}}$$
$$= \sqrt{\frac{0.25}{30}}$$
$$= \sqrt{0.008\overline{3}}$$
$$= 0.091287$$

Next, calculate the z-test statistic, using this formula:

$$z = \frac{\hat{p} - p_0}{SE}$$

where $\hat{p}$ is the observed proportion, p_0 is the hypothesized proportion, and SE is the standard error.

$$z = \frac{0.5667 - 0.5}{0.091287}$$
$$= 0.7307, \text{ or } 0.73 \text{ (rounded)}$$

Now, find this z-score in a Z-table, such as Table A-1 in the appendix. The table value is 0.7673, which is the area under the curve to the left, the z-test statistic you observed in the sample. However, you want to know the probability of getting a z-score this high or higher (because the alternative is a *greater than* hypothesis), so you have to subtract the table value from 1 to get the p-value for the z-statistic: $1 - 0.7673 = 0.2327$.

Finally, compare this value to the α level (0.05), and you find that the p-value is greater than α. Thus, your conclusion is that you can't reject the null hypothesis, and the clerk shouldn't offer to buy the collection of books.

704. **The null hypothesis shouldn't be rejected; the owner can't conclude that the defect rate is less than 1%.**

First, set up the null and alternative hypotheses:

$$H_0: p_0 = 0.01$$

$$H_a: p_0 > 0.01$$

Note: The alternative hypothesis is a *less than* hypothesis because the factory owner has a maximum acceptable error rate, and she wants to ensure that the process makes an error 1% or less of the time.

Then, determine whether the sample is large enough to run a z-test by checking that both np_0 and $n(1-p_0)$ equal at least 10. In this case, $n = 1,000$ and $p_0 = 0.01$, so $np_0 = 1,000(0.01) = 10$, and $n(1-p_0) = 1,000(1-0.01) = 1,000(0.99) = 999$.

Calculate the observed proportion by dividing the number of defective ball bearing by the sample size. If 6 out of 1,000 ball bearings are found to be defective, the proportion is $6/1,000 = 0.006$.

Next, compute the standard error with this formula:

$$SE = \sqrt{\frac{p_0(1-p_0)}{n}}$$

where p_0 is the population proportion and n is the sample size. Substitute the known values into the formula to get

$$SE = \sqrt{\frac{0.01(0.99)}{1,000}}$$
$$= \sqrt{0.0000099}$$
$$= 0.0031464$$

Then, calculate the z-test statistic, using this formula:

$$z = \frac{\hat{p} - p_0}{SE}$$

where $\hat{p}$ is the observed proportion, p_0 is the hypothesized proportion, and SE is the standard error.

$$z = \frac{0.006 - 0.01}{0.0031464}$$
$$= -1.2712, \text{ or } -1.27 \text{ (rounded)}$$

Now, find the probability of getting a test statistic value of z at least this far from the claimed proportion under the null hypothesis. In this case, the Z-table (Table A-1 in the appendix) gives you the proportion under the curve to the left of the z-value you look up, and the alternative hypothesis is *less than*, so you get the probability you need directly from the table. That probability is 0.1020 and represents the p-value.

Compare the p-value with the desired level of $\alpha = 0.05$. Because the p-value is greater than α, you fail to reject the null hypothesis.

In practical terms, the factory owner doesn't have significant evidence that the process is working correctly and is keeping the error rate to 1% or less.

705. $\hat{p}$ is the sample proportion, and p_0 is the claimed value for the population proportion.

When you do a hypothesis test for a proportion, the null hypothesis (H_0) makes a claim about what the population proportion is; this claimed value for the population proportion is denoted by p_0. For example, a newspaper article may say that 30% (or 0.30) of Americans wear glasses; so $p_0 = 0.30$ is a claimed value for the proportion of all Americans in the population who wear glasses. You test the claim by taking a sample of Americans and finding the proportion of people in the sample who wear glasses. This result is called the sample proportion and is denoted $\hat{p}$. So, $\hat{p}$ is a value that comes from your sample (the sample proportion), while p_0 is a value that someone claims to be the population proportion.

706.   The null hypothesis can't be rejected; the manufacturer shouldn't accept the shipment.

First, set up the null and alternative hypotheses:

$$H_0: p_0 = 0.01$$

$$H_a: p_0 > 0.01$$

Note: The alternative hypothesis is *less than* because the company wants to work only with suppliers whose parts are less than 1% defective.

Then, determine whether the sample is large enough to run a z-test by checking that both np_0 and $n(1-p_0)$ equal at least 10. In this case, $n = 10,000$ and $p_0 = 0.01$, so $np_0 = 10,000(0.01) = 100$, and $n(1-p_0) = 10,000(1-0.01) = 9,900$.

Calculate the observed proportion by dividing the number of defective items by the sample size: $90 / 10,000 = 0.009$.

Next, compute the standard error with this formula:

$$SE = \sqrt{\frac{p_0(1-p_0)}{n}}$$

where p_0 is the population proportion and n is the sample size. Substitute the known values into the formula to get

$$SE = \sqrt{\frac{0.01(1-0.01)}{10,000}}$$

$$= \sqrt{\frac{0.01(0.99)}{10,000}}$$

$$= \sqrt{\frac{0.0099}{10,000}}$$

$$= 0.000994987$$

Then, calculate the z-test statistic, using this formula:

$$z = \frac{\hat{p} - p_0}{SE}$$

where $\hat{p}$ is the observed proportion, p_0 is the hypothesized proportion, and SE is the standard error.

$$z = \frac{0.009 - 0.01}{0.000994987}$$
$$= -1.005038, \text{ or } -1.01 \text{ (rounded)}$$

Now, find the area under the curve to the left of the test statistic value of z and convert this to a p-value. According to Table A-1, the nearest value for $z = -1.01$, which gives an area of 0.1562. Because you had a *less than* alternative hypothesis and the value of z is negative, the table value is the same as the p-value.

Finally, compare the p-value with the $\alpha = 0.01$ level of significance. In this case, the p-value is greater than α, so you can't reject the null hypothesis. The company can't be reasonably sure that the shipment keeps the defect rate low enough.

707. E. None of the above.

The true value for the population proportion is denoted by p. Because p is typically unknown, it's often given a claimed value (denoted by p_0). The claimed value for p is challenged by comparing it to the results of a sample, using a hypothesis test. In the end, a probability is reported, which is called the p-value. The p-value is the probability that the results in the sample happened by random chance, while the claimed value for p is true. A small p-value indicates the sample results were unlikely to be due to chance. It gives evidence against the claimed value of p, resulting in possibly rejecting H_0, depending on how small it is.

708. There is insufficient evidence to draw a conclusion.

In this case, the dealer didn't draw a sufficient sample size. To determine whether the sample is large enough to run a z-test, you check that both np_0 and $n(1 - p_0)$ equal at least 10. In this case, $n = 10$ and $p_0 = 0.05$, so $np_0 = 10(0.05) = 0.5$. This is less than 10, so you can't use the normal approximation to the binomial.

Ultimately, there's insufficient evidence to draw a conclusion in this case when sampling only ten items.

709. 100% confidence

It's important to remember basic concepts before starting to run statistical tests. A sample is a part of a population. If impurities exist in a sample of blood specimens, then impurities certainly exist in the population of specimens from which it's drawn. The blood bank can say with 100% confidence that the blood donation population of specimens has disease in it. The z-test is irrelevant here.

710. The null hypothesis can't be rejected; the city council should work to control the pigeon population.

In this case, a little thought will save a lot of statistical testing. If the city council wants the infection rate to be 3% or less and it observes an infection rate of 3% in a sample (6/200 is 3%), then the null hypothesis can't be rejected, regardless of the significance level. Or simply consider that the numerator of the test statistic z would be 0 because the sample proportion and assumed population proportion are the same. The p-value would be 0.50, far from significant. You fail to reject the null hypothesis, and the pigeons need to be controlled better, given the policy.

711. **The null hypothesis can be rejected; the designer shouldn't accept the shipment.**

First, set up the null and alternative hypotheses:

$$H_0: p_0 = 0.25$$

$$H_a: p_0 \neq 0.25$$

Note: The alternative hypothesis is *not equal to* because the designer aims for a defect rate of 0.25 and neither significantly more or less.

Next, determine whether the sample is large enough to run a z-test by checking that both np_0 and $n(1-p_0)$ equal at least 10. In this case, $n = 50$ and $p_0 = 0.25$, so $np_0 = 50(0.25) = 12.5$ and $n(1-p_0) = 50(1-0.25) = 50(0.75) = 37.5$.

The sample proportion is 0.12, as stated in the problem.

Then, compute the standard error with this formula:

$$SE = \sqrt{\frac{p_0(1-p_0)}{n}}$$

where p_0 is the population proportion and n is the sample size. Substitute the known values into the formula to get

$$SE = \sqrt{\frac{0.25(1-0.25)}{50}}$$
$$= \sqrt{\frac{0.25(0.75)}{50}}$$
$$= \sqrt{\frac{0.1875}{50}}$$
$$= \sqrt{0.00375}$$
$$= 0.061237$$

Next, calculate the z-test statistic, using this formula:

$$z = \frac{\hat{p} - p_0}{SE}$$

where $\hat{p}$ is the observed proportion, p_0 is the hypothesized proportion, and SE is the standard error.

$$z = \frac{0.12 - 0.25}{0.061237}$$
$$= -2.1229, \text{ or } -2.12 \text{ (rounded)}$$

Find the area under the curve to the left of the test statistic value of z, and convert this to a p-value. Using Table A-1 in the appendix (or another Z-table), you find that 0.0170 of the curve lies to the left of $z = -2.12$. The alternative hypothesis in this case is *not equal to*, so that curve area must be doubled: $2(0.0170) = 0.0340$.

Now, compare the p-value with the desired α level of 0.05. Because the p-value is less than α, the null hypothesis can be rejected. In this case, the designer should reject the shipment because it has too few defects for his taste.

712. H_0: $p = 0.50$; H_a: $p \neq 0.50$

You're testing to see whether a coin is fair. If the coin is fair, the probability of getting a heads (or the proportion of heads in an infinite number of coin flips) is $p = 0.50$. If the coin isn't fair, there could be either a significantly smaller proportion of heads than 0.50 or a significantly larger proportion of heads than 0.50. In other words, if the coin isn't fair, $p \neq 0.50$. In a hypothesis test, you start by claiming that the coin is fair unless evidence shows otherwise. That means that H_0 is $p = 0.50$ and H_a is $p \neq 0.50$.

713. H_0: $p = 0.45$; H_a: $p > 0.45$

In this case, p is the proportion of correct answers Joe would get if you played this game an infinite number of times. If Joe were just guessing at the suit of each card, he'd be expected to guess 25% of the card suits correctly (because there are four possible suits, and each suit occurs equally often among each deck). In other words, if Joe is guessing, p would equal 0.25 in the long run. You decided that he has to be at least 20 percentage points above the accuracy you would expect by chance: $0.25 + 0.20 = 0.45$. So, the null and alternative hypotheses in this case are H_0: $p = 0.45$ and H_a: $p > 0.45$.

714. **The null hypothesis can't be rejected, and the sample shouldn't be rejected either.**

First, set up the null and alternative hypotheses:

$$H_0: p_0 = 0.25$$

$$H_a: p_0 \neq 0.25$$

Next, determine whether the sample is large enough to run a z-test by checking that both np_0 and $n(1 - p_0)$ equal at least 10. In this case, $n = 1,000,000$ and $p_0 = 0.25$, so $np_0 = 1,000,000(0.25) = 250,000$, and $n(1 - p_0) = 1,000,000(1 - 0.25) = 1,000,000(0.75)$.

Calculate the observed proportion by dividing the cells with the phenotype by sample size: $250,060 / 1,000,000 = 0.25006$.

Then, compute the standard error with this formula:

$$SE = \sqrt{\frac{p_0(1 - p_0)}{n}}$$

where p_0 is the population proportion and n is the sample size. Substitute the known values into the formula to get

$$SE = \sqrt{\frac{p_0(1 - p_0)}{n}}$$

$$= \sqrt{\frac{0.25(1 - 0.25)}{1,000,000}}$$

$$= \sqrt{\frac{0.25(0.75)}{1,000,000}}$$

$$= \sqrt{\frac{0.1875}{1,000,000}}$$

$$= \sqrt{0.0000001875}$$

$$= 0.000433$$

Next, calculate the z-test statistic, using this formula:

$$z = \frac{\hat{p} - p_0}{\text{SE}}$$

where $\hat{p}$ is the observed proportion, p_0 is the hypothesized proportion, and SE is the standard error.

$$z = \frac{0.25006 - 0.25}{0.000433}$$
$$= 0.1386, \text{ or } 0.14 \text{ (rounded)}$$

Find the area under the curve to the right of the test statistic value of z (that is, away from the claimed proportion), and convert this to a p-value. The closest z-value is 0.14, which corresponds to an area of 0.5557. This, however, is the area to the left of $z = 0.14$, so you find $1 - 0.5557 = 0.4443$ to be the area to the right of z. The *not equal to* alternative hypothesis means that you need the total area in two tails. To get that, multiply the result by 2:

$$p\text{-value} = 2(0.4443) = 0.8886$$

Now, compare this with the α level. The p-value says that there's an 88.86% probability of seeing a sample as or more extreme that what the biologist observed. This is evidence that a reasonably common proportion was seen. This far exceeds the threshold of $\alpha = 0.10$. Because the p-value is greater than α, the biologist should fail to reject the null hypothesis and accept this sample for further study.

715. You know you can't reject H_0 because 0.20 isn't one of the values in H_a.

Bob's theory is that 30% of the customers in the checkout line buy something (so H_0 is $p = 0.30$). You believe it could even be higher than that (so H_a is $p > 0.30$; in this case, any values less than 0.30 don't matter. Given this situation, the only hope you have of rejecting H_0 is if your sample results are higher than 30%; then it would be a matter of calculating out the hypothesis test to determine whether your sample results are high enough above 30% to reject Bob's claim. However, because your sample results (20%) are lower than 30% right out of the box, you don't have to go any further; you know you won't be able to reject H_0.

716. $H_0: \mu_1 - \mu_2 = 0; H_a: \mu_1 - \mu_2 > 0$

The null hypothesis is always a statement of equality. The alternative in this case is that the population mean of Group 1 will be larger than that of Group 2.

717. 1.645

First, determine whether a z-test for independent groups is appropriate. Because the population of scores is normally distributed and you have population standard deviation information for both groups, you can proceed to do the test.

This is a *greater than* alternative hypothesis with an alpha level of 0.05. Using Table A-1 in the appendix, find the critical value of z such that 0.05 of the probability lies above it. This value falls between 1.64 and 1.65 and just happens to round to 1.645.

718. 4.472

To calculate the standard error, use this formula:

$$SE = \sqrt{\frac{\sigma_1^2}{n_1} + \frac{\sigma_2^2}{n_2}}$$

where σ_1^2 and σ_2^2 are the variances of the two populations, and n_1 and n_2 are the two sample sizes. So, for this test, the standard error is

$$SE = \sqrt{\frac{17.32^2}{30} + \frac{17.32^2}{30}}$$
$$= \sqrt{\frac{299.98}{30} + \frac{299.98}{30}}$$
$$= 4.472$$

719. 4.2263

First, find the standard error, using this formula:

$$SE = \sqrt{\frac{\sigma_1^2}{n_1} + \frac{\sigma_2^2}{n_2}}$$

where σ_1^2 and σ_2^2 are the variances of the two populations, and n_1 and n_2 are the two sample sizes. So, for this test, the standard error is

$$SE = \sqrt{\frac{17.32^2}{30} + \frac{17.32^2}{30}}$$
$$= \sqrt{\frac{299.98}{30} + \frac{299.98}{30}}$$
$$= 4.472$$

Then, use the formula to find the test statistic:

$$z = \frac{(\bar{x}_1 - \bar{x}_2) - (\mu_1 - \mu_2)}{SE}$$

where $\bar{x}_1$ and $\bar{x}_2$ are the sample means, and μ_1 and μ_2 are the population means. The null hypothesis is that the two populations have the same mean (in other words, that the difference in population means is 0).

$$z = \frac{(33.3 - 14.4) - (0)}{4.472}$$
$$= 4.2263$$

720. **The null hypothesis can be rejected; it appears that workers who smile more are more productive.**

Set up the null and alternative hypotheses. Because the manager believes that happy workers are more productive, it makes sense to set up a *greater than* alternative hypothesis.

$$H_0: \mu_1 - \mu_2 = 0$$

$$H_a: \mu_1 - \mu_2 > 0$$

Next, identify the alpha level. The question says to use $\alpha = 0.05$.

Use a Z-table, such as Table A-1 in the appendix, to find a critical value for the z-test. Because the alternative hypothesis is *greater than*, the critical value for z will be positive and occur at the point where 0.05 of the curve area lies to the right of the z-score, putting $1 - 0.05 = 0.95$ of the area to the left of that z-score. As a result, you look for a table value of 0.95 and identify the z-value corresponding to that. It turns out to be between 1.64 and 1.65, so you can call it 1.645.

Now, calculate the standard error, using this formula:

$$SE = \sqrt{\frac{\sigma_1^2}{n_1} + \frac{\sigma_2^2}{n_2}}$$

where σ_1^2 and σ_2^2 are the variances of the two populations, and n_1 and n_2 are the two sample sizes. So, for this test, the standard error is

$$SE = \sqrt{\frac{17.32^2}{30} + \frac{17.32^2}{30}}$$
$$= \sqrt{\frac{299.98}{30} + \frac{299.98}{30}}$$
$$= 4.472$$

Then use the formula to find the test statistic:

$$z = \frac{(\bar{x}_1 - \bar{x}_2) - (\mu_1 - \mu_2)}{SE}$$

where $\bar{x}_1$ and $\bar{x}_2$ are the sample means, and μ_1 and μ_2 are the population means. The null hypothesis is that the two populations have the same mean (in other words, that the difference in population means is 0).

$$z = \frac{(33.3 - 14.4) - (0)}{4.472}$$
$$= 4.2263$$

Finally, determine whether the test statistic value for z is greater than the critical value for z, and reject the null hypothesis if so. In this case, the test statistic value (4.2263) is greater than the critical value (1.645), so you reject the null hypothesis. It appears that checkout clerks who smile more are more productive.

721. $H_0: \mu_1 = \mu_2; H_a: \mu_1 \neq \mu_2$

The null hypothesis is that the two population means don't differ, while the alternative hypothesis is that they do differ (a *not equal to* alternative hypothesis).

722. 1.96

Because this is a *not equal to* alternative hypothesis, half the alpha value of 0.05 will be in the upper tail and 0.05 in the lower tail. Using Table A-1, you can see that 1.96 is the z-value that has 0.025 of the total area above it. Doubling this probability (to include the lower tail) gives a significance level of 0.05.

723. **−1.2123**

To find the test statistic, use this formula:

$$z = \frac{\bar{x}_1 - \bar{x}_2}{\sqrt{\left(\sigma_1^2 / n_1\right) + \left(\sigma_2^2 / n_2\right)}}$$

where $\bar{x}_1$ and $\bar{x}_2$ are the sample means, σ_1^2 and σ_2^2 are the population variances, and n_1 and n_2 are the sample sizes.

Then, substitute the known values into the formula and solve:

$$z = \frac{51.9 - 52.6}{\sqrt{(5 / 30) + (5 / 30)}}$$

$$= \frac{-0.7}{\sqrt{0.1667 + 0.1667}}$$

$$= \frac{-0.7}{\sqrt{0.3334}}$$

$$= -1.2123$$

724. **The null hypothesis can't be rejected. This study doesn't support the idea that smokers and nonsmokers differ in IQ.**

First, find the test statistic, using this formula:

$$z = \frac{\bar{x}_1 - \bar{x}_2}{\sqrt{\left(\sigma_1^2 / n_1\right) + \left(\sigma_2^2 / n_2\right)}}$$

where $\bar{x}_1$ and $\bar{x}_2$ are the sample means, σ_1^2 and σ_2^2 are the population variances, and n_1 and n_2 are the sample sizes.

Then, substitute the known values into the formula and solve:

$$z = \frac{51.9 - 52.6}{\sqrt{(5 / 30) + (5 / 30)}}$$

$$= \frac{-0.7}{\sqrt{0.1667 + 0.1667}}$$

$$= \frac{-0.7}{\sqrt{0.3334}}$$

$$= -1.2123$$

As you can see in Table A-1 in the appendix, the test statistic for a *not equal to* hypothesis with an alpha level of 0.10 is 1.645. The z-statistic is only 1.2123 standard deviations away from the claimed mean of 0. In other words, because the absolute value of z is less than the critical value (1.2123 < 1.645), you don't have sufficient evidence to reject the null hypothesis.

725. **The null hypothesis can't be rejected. This study doesn't support the idea that smokers and nonsmokers differ in IQ.**

First, find the test statistic, using this formula:

$$z = \frac{\bar{x}_1 - \bar{x}_2}{\sqrt{\left(\sigma_1^2 / n_1\right) + \left(\sigma_2^2 / n_2\right)}}$$

where $\bar{x}_1$ and $\bar{x}_2$ are the sample means, σ_1^2 and σ_2^2 are the population variances, and n_1 and n_2 are the sample sizes.

Then, substitute the known values into the formula and solve:

$$z = \frac{51.9 - 52.6}{\sqrt{(5/30) + (5/30)}}$$

$$= \frac{-0.7}{\sqrt{0.1667 + 0.1667}}$$

$$= \frac{-0.7}{\sqrt{0.3334}}$$

$$= -1.2123$$

As you can see from Table A-1 in the appendix, the test statistic for a *not equal to* hypothesis with an alpha level of 0.5 is 1.96. The z-statistic is only 1.2123 standard deviations away from the claimed mean of 0. In other words, because the absolute value of z is less than the critical value (1.2123 < 1.96), you don't have sufficient evidence to reject the null hypothesis.

726. $H_0: \mu_1 = \mu_2; H_a: \mu_1 \neq \mu_2$

You may think that more sleep would lead to a better score on the memory test. But because this isn't explicitly stated, you must assume that the researcher is interested in whether the two groups differ at all. Thus, the alternative hypothesis must be *not equal to*, while the null hypothesis is that the population mean scores are equal.

727. $H_0: \mu_1 = \mu_2; H_a: \mu_1 > \mu_2$

If the researcher is interested only in whether Group 1 (the group allowed to sleep five hours) performs better than Group 2 (the group allowed to sleep only three hours), this is a *less than* alternative hypothesis. The null hypothesis is always a statement of equality.

728. 2.8803

To calculate the test statistic, use this formula:

$$z = \frac{\bar{x}_1 - \bar{x}_2}{\sqrt{\left(\sigma_1^2 / n_1\right) + \left(\sigma_2^2 / n_2\right)}}$$

where $\bar{x}_1$ and $\bar{x}_2$ are the sample means, σ_1^2 and σ_2^2 are the population variances, and n_1 and n_2 are the sample sizes.

Then, substitute the known values into the formula and solve:

$$z = \frac{62 - 58}{\sqrt{\left(6^2/40\right) + \left(6^2/35\right)}}$$

$$= \frac{4}{\sqrt{(36/40) + (36/35)}}$$

$$= \frac{4}{\sqrt{0.9 + 1.0286}}$$

$$= \frac{4}{\sqrt{1.3887}}$$

$$= 2.8803$$

729. **Reject the null hypothesis and conclude that a difference in sleep is associated with a difference in performance.**

First, find the test statistic, using this formula:

$$z = \frac{\bar{x}_1 - \bar{x}_2}{\sqrt{\left(\sigma_1^2 / n_1\right) + \left(\sigma_2^2 / n_2\right)}}$$

where $\bar{x}_1$ and $\bar{x}_2$ are the sample means, σ_1^2 and σ_2^2 are the population variances, and n_1 and n_2 are the sample sizes.

Then, substitute the known values into the formula and solve:

$$z = \frac{62 - 58}{\sqrt{\left(6^2 / 40\right) + \left(6^2 / 35\right)}}$$

$$= \frac{4}{\sqrt{(36 / 40) + (36 / 35)}}$$

$$= \frac{4}{\sqrt{0.9 + 1.0286}}$$

$$= \frac{4}{\sqrt{1.3887}}$$

$$= 2.8803$$

For a two-tailed z-test with a significance level of $\alpha = 0.01$, alpha must be split between the two tails. Using Table A-1 and looking for a probability of $0.01 / 2 = 0.005$, you find the critical value to be roughly 2.58. The test statistic is greater than this so the researcher will reject the null hypothesis. Based on the two-sided alternative hypothesis, you can conclude only that a difference in sleep is associated with a difference in performance.

730. **Reject the null hypothesis and conclude that more sleep is associated with better performance.**

First, find the test statistic, using this formula:

$$z = \frac{\bar{x}_1 - \bar{x}_2}{\sqrt{\left(\sigma_1^2 / n_1\right) + \left(\sigma_2^2 / n_2\right)}}$$

where $\bar{x}_1$ and $\bar{x}_2$ are the sample means, σ_1^2 and σ_2^2 are the population variances, and n_1 and n_2 are the sample sizes.

Then, substitute the known values into the formula and solve:

$$z = \frac{62 - 58}{\sqrt{\left(6^2 / 40\right) + \left(6^2 / 35\right)}}$$

$$= \frac{4}{\sqrt{(36 / 40) + (36 / 35)}}$$

$$= \frac{4}{\sqrt{0.9 + 1.0286}}$$

$$= \frac{4}{\sqrt{1.3887}}$$

$$= 2.8803$$

Using Table A-1 in the appendix, find the critical value for a one-tailed z-test at $\alpha = 0.05$, which is roughly 1.64. The test statistic is greater than this so the researcher will reject the null hypothesis and conclude that more sleep is associated with better performance.

731. a t-test of paired population means

The households are observed twice, once for each appeal, so the two measurements on each household are paired rather than independent. The sample size of ten is too small for a z-test, so the t-test must be used.

732. A. The wallet appeal was more successful.

For the paired t-test, you calculate differences by subtracting the second group from the first. Because the mean difference score, $\bar{d}$, is negative, on average, energy consumption was lower in the wallet condition as compared to the moral condition. However, to test whether the two conditions differed significantly, you need to conduct a paired t-test.

733. 9

The appropriate test is the paired t-test. The degrees of freedom is one less than the number of pairs: $n_{\text{pairs}} - 1 = 10 - 1 = 9$.

734. 14.67

To calculate the standard error for a paired t-test, use this formula:

$$\text{SE} = \frac{S_{\bar{d}}}{\sqrt{n_{\text{pairs}}}}$$

where $S_{\bar{d}}$ is the standard deviation of the differences from the sample, and n_{pairs} is the number of pairs.

So, the standard error is

$$\text{SE} = \frac{46.39}{\sqrt{10}}$$
$$= 14.67$$

735. −5.69

First, find the standard error with this formula:

$$\text{SE} = \frac{S_{\bar{d}}}{\sqrt{n_{\text{pairs}}}}$$

where $S_{\bar{d}}$ is the standard deviation of the differences from the sample, and n_{pairs} is the number of pairs.

So, the standard error is

$$SE = \frac{46.39}{\sqrt{10}}$$
$$= 14.67$$

Then, find the test statistic, using this formula:

$$t = \frac{\bar{d}}{SE}$$

where $\bar{d}$ is the average of the paired difference from the sample. So, substituting the numbers makes the test statistic.

$$t = \frac{-83.5}{14.67}$$
$$= -5.69189$$

---

736. **Reject the null hypothesis and conclude that moral appeal is less successful than the wallet appeal.**

You will conduct a paired t-test. This is a *not equal to* alternative hypothesis, so the null and alternative hypotheses are

$$H_0: \mu_1 - \mu_2 = 0$$

$$H_a: \mu_1 - \mu_2 \neq 0$$

To calculate the standard error for a paired t-test, use this formula:

$$SE = \frac{s_{\bar{d}}}{\sqrt{n_{\text{pairs}}}}$$

where $s_{\bar{d}}$ is the standard deviation of the differences from the sample, and n_{pairs} is the number of pairs.

$$SE = \frac{46.39}{\sqrt{10}}$$
$$= 14.67$$

Then, find the test statistic:

$$t = \frac{\bar{d}}{SE}$$
$$= \frac{-83.5}{14.67}$$
$$= -5.69189$$

where $\bar{d}$ is the average of the paired differences from the sample. This result rounds to -5.69.

The degrees of freedom are $n - 1 = 10 - 1 = 9$. You can find the critical value for an alpha level of 0.10 by first splitting the level of significance in half for each of the two tails, because this test has a two-sided alternative. Examining the column for $0.10 / 2 = 0.05$ and the row for 9 *df* in Table A-2 in the appendix, you find the critical value is 1.833113.

The test statistic is greater than the critical value, so you reject the null hypothesis. The direction of the difference in means indicates that less power was used following the wallet appeal, making the moral appeal less successful.

737. **a t-test of paired population means**

The data is paired because both brands are tested on each toy, and the lengths of time each brand of battery operated a given toy are compared.

738. $H_0: \mu_d = 0; H_a: \mu_d \neq 0$

In a paired t-test, the null hypothesis is that the mean of the difference scores is 0. As a *not equal to* hypothesis, the alternative hypothesis is that the mean of the difference scores isn't 0.

739. **−0.9286**

You calculate the mean of the difference scores, $\bar{d}$, by averaging the individual difference scores, with this formula:

$$\bar{d} = \frac{\sum\limits_{i=1}^{n} d_i}{n}$$

where d_i represents the individual difference scores, and n is the number of pairs. So, substituting the numbers into the formula, you get

$$\bar{d} = \frac{-1.7 + (-2.2) + (-0.5) + (-0.8) + (-1.1) + 0.7 + (-0.9)}{7}$$

$$= \frac{-6.5}{7}$$

$$= -0.9286$$

740. **0.3483**

To find the standard error, use this formula:

$$SE = \frac{s_d}{\sqrt{n}}$$

where s_d is the standard deviation of the difference scores in the sample, and n is the sample size.

Substitute the known values into the formula to get

$$SE = \frac{0.9214}{\sqrt{7}}$$

$$= 0.3483$$

741. **2.44691**

To find the critical value, you need the degrees of freedom and the alpha level. The degrees of freedom are one less than the number of pairs: $7 - 1 = 6$. The alpha level is given.

For a *not equal to* alternative hypothesis, you want half the alpha value of 0.05 in the lower tail and half in the upper tail. Looking in the column for 0.025 and the row for 6 *df* in the *t*-table (Table A-2 in the appendix), you find the critical value 2.44691.

742. 3.70743

To find the critical value, you need the degrees of freedom and the alpha level. The degrees of freedom are one less than the number of pairs: $7-1=6$. The alpha level is given.

For a *not equal to* alternative hypothesis, you want half the alpha value of 0.01 in the lower tail and half in the upper tail. Looking in the column for 0.005 and the row for 6 *df* in the *t*-table (Table A-2 in the appendix), you find the critical value 3.70743.

743. −2.6661

Calculate the test statistic for a paired *t*-test using this formula:

$$t = \frac{\bar{d}}{SE}$$

You calculate the mean of the difference scores, $\bar{d}$, by averaging the individual difference scores with this formula.

$$\bar{d} = \frac{\sum_{i=1}^{n} d_i}{n}$$

where the d_i are the individual difference scores, and n is the number of pairs. So, substituting the numbers into the formula, you get

$$\bar{d} = \frac{-1.7 + (-2.2) + (-0.5) + (-0.8) + (-1.1) + 0.7 + (-0.9)}{7}$$

$$= \frac{-6.5}{7}$$

$$= -0.9286$$

To find the standard error, use this formula:

$$SE = \frac{s_d}{\sqrt{n}}$$

where s_d is the standard deviation of the difference scores in the sample, and n is the sample size.

Substitute the known values into the formula to get

$$SE = \frac{0.9214}{\sqrt{7}}$$

$$= 0.3483$$

Plugging in the values, the test statistic is therefore

$$t = \frac{\bar{d}}{SE}$$

$$= \frac{-0.9286}{0.3483}$$

$$= -2.6661$$

744. **Fail to reject the null hypothesis.**

You calculate the mean of the difference scores, $\bar{d}$, by averaging the individual difference scores, with this formula:

$$\bar{d} = \frac{\sum\limits_{i=1}^{n} d_i}{n}$$

where d_i represents the individual difference scores, and n is the number of pairs. So, substituting the numbers into the formula, you get

$$\bar{d} = \frac{-1.7 + (-2.2) + (-0.5) + (-0.8) + (-1.1) + 0.7 + (-0.9)}{7}$$

$$= \frac{-6.5}{7}$$

$$= -0.9286$$

To find the standard error, use this formula:

$$SE = \frac{s_d}{\sqrt{n}}$$

where s_d is the standard deviation of the difference scores in the sample, and n is the sample size.

Substitute the known values into the formula to get

$$SE = \frac{0.9214}{\sqrt{7}}$$

$$= 0.3483$$

Then calculate the test statistic:

$$t = \frac{\bar{d}}{SE}$$

$$= \frac{-0.9286}{0.3483}$$

$$= -2.6661$$

To find the critical value, you need the degrees of freedom and the alpha level. The degrees of freedom are one less than the number of pairs: $7 - 1 = 6$, and the alpha level is given as 0.01.

For a *not equal to* alternative hypothesis, you want half the alpha value of 0.01 in the lower tail and half in the upper tail. Looking in the column for 0.005 and the row for 6 *df* in the *t*-table (Table A-2 in the appendix), you find the critical value 3.70743.

Because you have a *not equal to* alternative hypothesis, you'll reject the null hypothesis if the absolute value of your test statistic exceeds the critical value. However, the absolute value of your test statistic, 2.6661, is less than the critical value, so you don't reject the null hypothesis.

Your test statistic is closer to the mean than the critical value is, so you fail to reject the null hypothesis.

745.

Reject the null hypothesis and conclude that there's a significant difference in battery life between the two brands.

You need to calculate a t-statistic and compare it to the critical value to decide whether to accept or reject the null hypothesis. The degrees of freedom for the test statistic are one less than the number of pairs: $7 - 1 = 6$.

For a *not equal to* alternative hypothesis, you want half the alpha value of 0.01 in the lower tail and half in the upper tail. Looking in the column for 0.005 and the row for 6 *df* in the t-table (Table A-2 in the appendix), you find the critical value 2.44691.

You calculate the mean of the difference scores, $\bar{d}$, by averaging the individual difference scores with this formula:

$$\bar{d} = \frac{\sum_{i=1}^{n} d_i}{n}$$

where the d_i are the individual difference scores, and n is the number of pairs. So, substituting the numbers into the formula, you get

$$\bar{d} = \frac{-1.7 + (-2.2) + (-0.5) + (-0.8) + (-1.1) + 0.7 + (-0.9)}{7}$$
$$= \frac{-6.5}{7}$$
$$= -0.9286$$

To find the standard error, use this formula:

$$SE = \frac{s_d}{\sqrt{n}}$$

where s_d is the standard deviation of the difference scores in the sample, and n is the sample size.

Substitute the known values into the formula to get

$$SE = \frac{0.9214}{\sqrt{7}}$$
$$= 0.3483$$

Then calculate the test statistic:

$$t = \frac{\bar{d}}{SE}$$
$$= \frac{-0.9286}{0.3483}$$
$$= -2.6661$$

Your test statistic is farther from the mean than the critical value is, so you will reject the null hypothesis. Your sample data provides you with sufficient evidence to reject the null hypothesis and conclude that there is a difference between the two brands.

746.

a z-test of two population proportions

The research question is whether the proportion of security breaches differs between two independent populations, and the sample size is large enough to support a z-test.

747. H_0: $p_1 = p_2$; H_a: $p_1 \neq p_2$

The null and alternative hypotheses are always stated in terms of population parameters — in this case, population proportions p_1 and p_2. The null hypothesis is always a statement of equality; when the researcher has no initial hunch about the direction of population differences, the alternative hypothesis is written as *not equal to*, using the $\neq$ symbol.

748. **0.0211 and 0.05144**

You calculate the sample proportions, $\hat{p}_1$ and $\hat{p}_2$, for each group by dividing the number of security breaches by the number of accounts observed.

Group 1 had 1,055 breaches in 50,000 cases, so the observed proportion is

$$\hat{p}_1 = \frac{1,055}{50,000} = 0.02110$$

Group 2 had 2,572 breaches in 50,000 cases, so the observed proportion is

$$\hat{p}_2 = \frac{2,572}{50,000} = 0.05144$$

749. **0.03627**

You calculate the overall sample proportion, $\hat{p}$, by dividing the total number of security breaches by the total number of accounts observed. In this example, Group 1 had 1,055 breaches, and Group 2 had 2,572 breaches. Each group had 50,000 accounts.

$$\hat{p} = \frac{1,055 + 2,572}{50,000 + 50,000} = 0.03627$$

750. **0.0012**

Calculate the standard error using the following formula, where $\hat{p}$ is the population proportion and n_1 and n_2 are the sample sizes:

$$SE = \sqrt{\hat{p}(1-\hat{p})\left(\frac{1}{n_1} + \frac{1}{n_2}\right)}$$

$$= \sqrt{0.03627(1-0.03627)\left(\frac{1}{50,000} + \frac{1}{50,000}\right)}$$

$$= \sqrt{0.034954(0.00004)}$$

$$= 0.0011824$$

751. **−25.66**

Calculate the z-statistic using the following formula, where $\hat{p}$ is the population proportion and n_1 and n_2 are the sample sizes:

$$z = \frac{\hat{p}_1 - \hat{p}_2}{\sqrt{\hat{p}(1-\hat{p})\left(\dfrac{1}{n_1} + \dfrac{1}{n_2}\right)}}$$

$$= \frac{0.02110 - 0.05144}{\sqrt{0.03627(1 - 0.03627)\left(\dfrac{1}{50,000} + \dfrac{1}{50,000}\right)}}$$

$$= \frac{-0.03034}{\sqrt{0.034954(0.00004)}}$$

$$= \frac{-0.03034}{0.0011824}$$

$$= -25.6597, \text{ or } -25.66 \text{ (rounded)}$$

752. **Reject the null hypothesis and conclude that there's a significant difference in the security of the two types of passwords.**

You will conduct a z-test for two population proportions, for a *not equal to* question, using the null and alternative hypotheses:

H_0: $p_1 = p_2$

H_a: $p_1 \neq p_2$

To calculate the test statistic, you use this formula:

$$z = \frac{\hat{p}_1 - \hat{p}_2}{\sqrt{\hat{p}(1-\hat{p})\left(\dfrac{1}{n_1} + \dfrac{1}{n_2}\right)}}$$

where $\hat{p}_1$ and $\hat{p}_2$ are the two sample proportions, and n_1 and n_2 are the sample sizes.

You calculate the overall sample proportion, $\hat{p}$, by dividing the total number of security breaches by the total number of accounts observed. In this example, Group 1 had 1,055 breaches, and Group 2 had 2,572 breaches. Each group had 50,000 accounts.

$$\hat{p} = \frac{1,055 + 2,572}{50,000 + 50,000}$$

$$= \frac{3,627}{100,000}$$

$$= 0.03627$$

Then you find the sample proportions, $\hat{p}_1$ and $\hat{p}_2$, for each group by dividing the number of security breaches by the number of accounts observed.

Group 1 had 1,055 breaches in 50,000 cases, so the observed proportion is

$$\hat{p}_1 = \frac{1,055}{50,000}$$

$$= 0.02110$$

Group 2 had 2,572 breaches in 50,000 cases, so the observed proportion is

$$\hat{p}_2 = \frac{2,572}{50,000}$$

$$= 0.05144$$

Substitute the values in the formula and solve:

$$z = \frac{\hat{p}_1 - \hat{p}_2}{\sqrt{\hat{p}(1-\hat{p})\left(\frac{1}{n_1} + \frac{1}{n_2}\right)}}$$

$$= \frac{0.02110 - 0.05144}{\sqrt{0.03627(1-0.03627)\left(\frac{1}{50,000} + \frac{1}{50,000}\right)}}$$

$$= \frac{-0.03034}{\sqrt{0.034954(0.00004)}}$$

$$= \frac{-0.03034}{0.0011824}$$

$$= -25.6597, \text{ or } -25.66 \text{ (rounded)}$$

The critical value, using Table A-1, a significance level of 0.05, and a *not equal to* alternative hypothesis, means that the critical value for z is ± 1.96 (the value that leaves 2.5%, or 0.025, of the curve area in each tail). You'll reject the null hypothesis if the test statistic value of z is outside the range of –1.96 to +1.96.

Your test statistic of –25.66 is outside the range of –1.96 to 1.96, so you reject the null hypothesis. Thus, the more plausible explanation is the two-sided alternative hypothesis that there's a difference between the two sets of security rules. The alternative hypothesis didn't favor one set of security rules over the other, but because the test statistic is negative, you know that the first group had fewer security breaches than the second, so the extra rule about passwords seems to increase security.

753. 0.5 and 0.7

$\hat{p}_1$ and $\hat{p}_2$ are the sample proportions for Group 1 and Group 2. You calculate them by dividing the number of cells with cases of interest (behavioral disturbance) in each group by the sample size for each group.

$$\hat{p}_1 = \frac{50}{100} = 0.5$$

$$\hat{p}_2 = \frac{70}{100} = 0.7$$

754. 0.6

You find the overall sample proportion, $\hat{p}$, by dividing the total number of cells with cases of interest (those with a behavioral disturbance) by the total number of cases in the study.

$$\hat{p} = \frac{50 + 70}{100 + 100}$$

$$= \frac{120}{200}$$

$$= 0.6$$

755. 0.0693

Calculate the standard error by using this formula:

$$SE = \sqrt{\hat{p}(1-\hat{p})\left(\frac{1}{n_1} + \frac{1}{n_2}\right)}$$

where n_1 and n_2 are the two sample sizes, and the overall sample proportion, $\hat{p}$, is calculated by dividing the total number of cases of interest (those with a behavioral disturbance) by the total number of cases in the study.

$$\hat{p} = \frac{50+70}{100+100}$$
$$= \frac{120}{200}$$
$$= 0.6$$

Then, plug in the numbers for the standard error formula:

$$SE = \sqrt{\hat{p}(1-\hat{p})\left(\frac{1}{n_1} + \frac{1}{n_2}\right)}$$
$$= \sqrt{0.6(1-0.6)\left(\frac{1}{100} + \frac{1}{100}\right)}$$
$$= \sqrt{0.24(0.02)}$$
$$= 0.06928, \text{ or } 0.0693 \text{ (rounded)}$$

756. 2.58

This is a *not equal to* alternative hypothesis, so you'll split the probability of 0.01 between the upper and lower tails of the distribution, resulting in a value of $0.01/2 = 0.005$ in each tail. Using Table A-1, you find that the critical value is roughly 2.58.

757. 2.33

This is a *greater than* alternative hypothesis, so you want the entire probability of 0.01 in the upper tail of the distribution. Using Table A-1, you find that the critical value is about 2.33.

758. −2.8868

To find the test statistic, use this formula:

$$z = \frac{\hat{p}_1 - \hat{p}_2}{\sqrt{\hat{p}(1-\hat{p})\left(\frac{1}{n_1} + \frac{1}{n_2}\right)}}$$

where n_1 and n_2 are the samples sizes for the two groups and $\hat{p}$ is the sample proportion. You find the value of $\hat{p}$ by dividing the total number of cells with a behavioral disturbance by the total number of cells in the study.

$$\hat{p} = \frac{50+70}{100+100}$$
$$= \frac{120}{200}$$
$$= 0.6$$

Now, plug in the numbers and solve:

$$z = \frac{\hat{p}_1 - \hat{p}_2}{\sqrt{\hat{p}(1-\hat{p})\left(\dfrac{1}{n_1} + \dfrac{1}{n_2}\right)}}$$

$$= \frac{0.5 - 0.7}{\sqrt{0.6(1-0.6)\left(\dfrac{1}{100} + \dfrac{1}{100}\right)}}$$

$$= \frac{-0.2}{\sqrt{(0.24)(0.02)}}$$

$$= \frac{-0.2}{\sqrt{0.0048}}$$

$$= -2.8868$$

759. **Reject the null hypothesis and conclude that there's a significant difference in behavioral disturbances between the two groups.**

Compute the test statistic, using this formula:

$$z = \frac{\hat{p}_1 - \hat{p}_2}{\sqrt{\hat{p}(1-\hat{p})\left(\dfrac{1}{n_1} + \dfrac{1}{n_2}\right)}}$$

where n_1 and n_2 are the samples sizes for the two groups. You calculate the overall sample proportion, $\hat{p}$, by dividing the total number of cells with a behavioral disturbance by the total number of cells in the study.

$$\hat{p} = \frac{50+70}{100+100}$$

$$= \frac{120}{200}$$

$$= 0.6$$

Now, plug the numbers into the formula and solve:

$$z = \frac{\hat{p}_1 - \hat{p}_2}{\sqrt{\hat{p}(1-\hat{p})\left(\dfrac{1}{n_1} + \dfrac{1}{n_2}\right)}}$$

$$= \frac{0.5 - 0.7}{\sqrt{0.6(1-0.6)\left(\dfrac{1}{100} + \dfrac{1}{100}\right)}}$$

$$= \frac{-0.2}{\sqrt{(0.24)(0.02)}}$$

$$= \frac{-0.2}{\sqrt{0.0048}}$$

$$= -2.8868$$

The test statistic of –2.8868 is outside the range of –1.96 to 1.96, which are the critical values (from Table A-1) for a z-test for a *not equal to* alternative hypothesis with a significance level of 0.05. You therefore reject the null hypothesis. Thus, the more plausible explanation is the two-sided alternative hypothesis that there's a difference in behavior between allowing Internet use and not allowing it. As stated, the alternative hypothesis doesn't favor a particular course of action, but because the test statistic is negative, this shows that the second group (those without Internet access) had more behavioral disturbances than the first.

760. Reject the null hypothesis and conclude that there's a significant difference in behavioral disturbances between the two groups.

You calculate the overall sample proportion, $\hat{p}$, by dividing the total number of cells with a behavioral disturbance by the total number of cells in the study.

$$\hat{p} = \frac{50+70}{100+100}$$
$$= \frac{120}{200}$$
$$= 0.6$$

To find the test statistic, use this formula:

$$z = \frac{\hat{p}_1 - \hat{p}_2}{\sqrt{\hat{p}(1-\hat{p})\left(\frac{1}{n_1} + \frac{1}{n_2}\right)}}$$

where n_1 and n_2 are the samples sizes for the two groups.

Now, plug in the numbers and solve:

$$z = \frac{\hat{p}_1 - \hat{p}_2}{\sqrt{\hat{p}(1-\hat{p})\left(\frac{1}{n_1} + \frac{1}{n_2}\right)}}$$
$$= \frac{0.5 - 0.7}{\sqrt{0.6(1-0.6)\left(\frac{1}{100} + \frac{1}{100}\right)}}$$
$$= \frac{-0.2}{\sqrt{(0.24)(0.02)}}$$
$$= \frac{-0.2}{\sqrt{0.0048}}$$
$$= -2.8868$$

The test statistic of –2.8868 is outside the range of –2.58 to 2.58, which are the critical values (from Table A-1) for a z-test with a *not equal to* alternative hypothesis. You therefore reject the null hypothesis. Thus, the more plausible explanation is the two-sided alternative hypothesis that there's a difference in behavior between allowing Internet use and not allowing it. As stated, the alternative hypothesis doesn't favor a particular course of action, but because the test statistic is negative, this shows that the second group (those without Internet access) had more behavioral disturbances than the first.

761. all adult drivers in the metro area

The target population is the people you want your results to apply to, which, in this example, is all adult drivers in the metro area.

762. bias

Bias means that the results from your sample are unlikely to be true for the population as a whole in some systematic way.

763. Choices (A), (B), and (C) (Not everyone has a phone or a listed phone number; not everyone is at home during the day on weekdays; not everyone is willing to participate in telephone polls.)

Using published phone directories as a sampling frame and scheduling calls during only one part of the day can introduce several types of bias to a study. First, people without a phone or a published phone number have no possibility of being selected for the sample. Second, people who aren't at home during the scheduled time can't supply data. And third, phone surveys in general are subject to non-response bias because a high proportion of people contacted may refuse to participate in the survey. All these factors may bias the sample and the data, so your results don't represent the target population.

764. It indicates that one answer is preferred and may introduce bias.

In responsible polling, questions should be stated in a neutral manner.

765. non-response bias

Non-response bias occurs when those among the sample who defer to take part in a study differ in some significant way from those who do take part.

766. response bias

In this case, it's likely that some of the respondents weren't being truthful. This is when response bias occurs. For example, most people believe that voting in elections is a positive characteristic, so they're more likely to report having voted, even if they didn't.

767. because it may be impossible to detect and, thus, impossible to correct

Many potential causes of bias in survey research exist, and they're particularly problematic because the survey results alone may not give any indication of what kinds of bias, if any, affected the results. Thus, it may be impossible to compensate or correct for any biases present in the data.

768. A. calculating the mean age of all the students by using their official records

You can reasonably assume that all students in the school have official records, and the records include age. This method is the only one that guarantees gathering of information from the entire target population, which is the definition of a census.

769. **E. numbering the students by using the school's official roster and selecting the sample by using a random number generator**

This method is the only one described that will result in a simple random sample. The other methods suggested are a stratified sample (classifying the students as male or female and drawing a random sample from each), a systematic sample (using an alphabetized student roster and selecting every 15th name, starting with the first one), a cluster sample (selecting three tables at random from the cafeteria during lunch hour and asking the students at those tables for their age), and a snowball sample (selecting one student at random, asking him or her to suggest three friends to participate and continuing in this fashion until you have your sample size).

770. **undercoverage bias**

Undercoverage bias results when part of the target population is excluded from the possibility of being selected for the sample. In this case, the exclusion is due to the sampling frame not including all the firm's current employees.

771. **volunteer sample bias**

You'll receive responses only from people who happen to be watching the program and then volunteer to participate in the survey. Because you didn't select them yourself beforehand, they don't make up a statistical sample, and they probably don't represent any real population of interest.

772. **convenience sample bias**

You're sampling and interviewing a sample of people that's the most convenient for you, and there's no way to know what population they represent (if any).

773. **to reduce bias caused by always using positive or negative statements**

Some people may have a tendency to agree to positive statements rather than the equivalent question written as a negative statement, so by including both types of statements, this bias can be reduced.

774. **It asks two questions at once.**

Because this survey item asks two questions in one (should everyone go to college, and should everyone seek gainful employment), some respondents may be confused about how to answer, especially if they agree with only one part of the question. For example, they may think that everyone should seek gainful employment but not necessarily go to college.

775. **Choices (A) and (B) (because the respondent may be uninformed about the topic; because the respondent may not remember enough information to answer the question)**

Including "don't know" as a category is needed for people who don't know the answer to a question, whether because it requires knowledge that they don't possess (about a world political situation, for example) or because they don't remember the information needed to answer the question (for example, what their typical diet was 30 years ago). If you're concerned about respondents finding a question offensive, include a separate category, such as "choose not to answer."

776. E. Choices (A) and (C) (strong; positive)

This scatter plot displays a strong, positive linear relationship ($r = 0.77$) between high-school and college GPA.

777. E. 4.0

The strong, positive linear relationship between high-school and college GPA suggests that, without having any further information, the student with the highest high-school GPA will also have the highest college GPA.

778. The points slope upward from left to right.

That the points slope upward from left to right means that lower values on the X variable tend to have lower values on the Y variable as well, and higher values on the X variable tend to have higher values on the Y variable.

779. The points cluster closely around a straight line.

This relationship isn't perfect (if it was, all the points would lie on a perfectly straight line), but the points do cluster closely around a line, so the relationship is fairly strong.

780. The points would all lie on a perfectly straight line sloping upward.

With a correlation of 1.0, all the points would lie perfectly on a straight line instead of just clustering around it. The line would slope upward from left to right.

781. D. Choices (A) and (C) (The points would all lie on a straight line; all the points would slope downward from left to right.)

The scatter plot of two variables with a correlation of –1.0 would have all the points lying perfectly on a line sloping downward from left to right.

782. C. weight

Scatter plots display the relationship between two quantitative variables. Of these choices, only weight is quantitative; the others are categorical. Zip codes can't be treated as a quantitative variable; for example, you can't find the average zip code for the United States.

783. E. Choices (A) and (D) (height and age; height and weight)

Scatter plots display the relationship between two quantitative variables. Of these variable pairs, only height and age and height and weight are quantitative.

784. **B. –0.8**

Among the choices, only –0.8 indicates a relationship that's both negative and strong.

785. **C. 0.2**

Among the choices, only 0.2 indicates a relationship that's both positive and weak.

786. **E. 0.9**

Among the choices, only 0.9 indicates a relationship that's both positive and very strong.

787. **A. –0.2**

Among the choices, only –0.2 indicates a relationship that's both negative and weak.

788. **nonlinear**

Linear relationships resemble a straight line. This relationship resembles a curve (in fact, it's a quadratic relationship) and is therefore nonlinear.

789. **0**

Although these variables are strongly related, the relationship isn't linear. Correlation measures only linear relationships. In this case, there's no linear relationship.

790. **Their relationship is nonlinear.**

Correlation expresses the degree of linear relationship between two variables and isn't appropriate for nonlinear relationships.

791. **A. –0.85**

The strongest linear relationship is indicated by the largest absolute value of the correlation; a correlation of –0.85 and 0.85 represents equally strong relationships between the variables. The sign just indicates whether the relationship is uphill or downhill.

792. **C. 0.1**

The weakest relationship is indicated by the smallest absolute value of the correlation; a correlation of –0.1 and 0.1 represents equally weak relationships between the variables. The sign just indicates whether the relationship is uphill or downhill.

793. **D. Choices (A) and (C) (−1; 1)**

Both 1 and −1 represent perfect correlations (one uphill and one downhill).

794. **2.5**

To calculate the mean, add together all the values and then divide by the total number of values:

$$\bar{x} = \frac{1+2+3+4}{4}$$

$$= \frac{10}{4} = 2.5$$

795. **2.75**

To calculate the mean, add together all the values and then divide by the total number of values:

$$\bar{y} = \frac{2+2+4+3}{4}$$

$$= \frac{11}{4} = 2.75$$

796. **3**

n is the number of cases (or pairs of data in this case). In this example, $n = 4$, so $n-1 = 4-1 = 3$.

797. **1.29**

To calculate the standard deviation of the X values, use this formula:

$$s_x = \sqrt{\frac{\sum (x-\bar{x})^2}{n-1}}$$

where x is a single value, $\bar{x}$ is the mean of all the values, $\sum$ sig represents the sum of the squared differences from the mean, and n is the sample size.

$$s_x = \sqrt{\frac{(1-2.5)^2 + (2-2.5)^2 + (3-2.5)^2 + (4-2.5)^2}{4-1}}$$

$$= \sqrt{\frac{2.25 + 0.25 + 0.25 + 2.25}{3}}$$

$$= \sqrt{\frac{5}{3}}$$

$$= 1.29099$$

798. 0.96

To calculate the standard deviation of the Y values, use this formula:

$$s_y = \sqrt{\frac{\sum (y - \bar{y})^2}{n-1}}$$

where y is a single value, $\bar{y}$ is the mean of all the values, $\sum$ sig represents the sum of the squared differences from the mean, and n is the sample size.

$$s_y = \sqrt{\frac{(2-2.75)^2 + (2-2.75)^2 + (4-2.75)^2 + (3-2.75)^2}{4-1}}$$

$$= \sqrt{\frac{0.5625 + 0.5625 + 1.5625 + 0.0625}{3}}$$

$$= \sqrt{\frac{2.75}{3}}$$

$$= 0.95743$$

799. 0.67

To calculate the correlation between X and Y, divide the sum of cross products by the standard deviations of x and y, and then divide the result by $n-1$.

For this example, the sum of cross products is 2.5, n is 4, the standard deviation of X is 1.29, and the standard deviation of Y is 0.96.

$$r = \frac{1}{n-1} \left(\frac{\sum_x \sum_y (x-\bar{x})(y-\bar{y})}{s_x s_y} \right)$$

$$= \frac{1}{3} \left[\frac{2.5}{(1.29)(0.96)} \right]$$

$$= 0.6729$$

800. 0.82

To calculate the correlation between X and Y, divide the sum of cross products by the standard deviations of X and Y, and then divide the result by $n-1$. In this case, the sum of cross products is 274, the standard deviation of X is 4.47, the standard deviation of Y is 5.36, and n is 15:

$$r = \frac{1}{n-1} \left(\frac{\sum_x \sum_y (x-\bar{x})(y-\bar{y})}{s_x s_y} \right)$$

$$= \frac{1}{14} \left[\frac{274}{(4.47)(5.36)} \right]$$

$$= 0.81686$$

801. It will decrease.

The correlation will decrease because you'd divide by a larger number ($n-1=19$) rather than $n-1=14$.

802. It will increase.

The correlation will increase because you'd divide by a smaller number (4.82 rather than 5.36).

803. It will increase.

The correlation will increase because the numerator would be larger (349 rather than 274).

804. It will stay the same.

The correlation is a unitless measure, so change in the units in which variables are measured won't change their correlation.

805. D. Choices (A) and (C) (−2.64; 1.5)

Correlations are always between −1 and 1.

806. The two correlations will be the same.

In a correlation, it doesn't matter which variable is designated as X and which as Y; the correlation will be the same either way.

807. E. Choices (B) and (C) (the Y variable; the response variable)

This study examines whether text font size may influence reading comprehension, so it's logical to designate reading comprehension as the Y, or response, variable.

808. A. the X variable

This study examines whether text font size may influence reading comprehension, so it's logical to designate font size as the X, or independent, variable.

809. It won't change.

In a correlation, it doesn't matter which variable is designated as X and which as Y; the correlation will be the same either way.

810. It stays the same.

The correlation is a unitless measure, so a change in the units won't change the correlation.

811. **You made a mistake in your calculations.**

Correlations are always between −1 and 1, so if you get a value outside this range, you made a mistake in your calculations.

812. **They have a strong, negative linear relationship.**

A correlation of −0.86 indicates a strong, negative linear relationship between two variables.

813. **They have a weak, positive linear relationship.**

A correlation of 0.27 indicates a weak, positive linear relationship between two variables.

814. **the X variable**

The logical assumption in this study is that time spent studying influences grades (GPA), so it makes sense to designate "time spent studying" as the X, or explanatory, variable.

815. **E. Choices (B) and (C) (the Y variable; the response variable)**

The logical assumption in this study is that time spent studying influences GPA, so it's logical to designate "GPA" as the Y, or response, variable.

816. **It doesn't change.**

The correlation is a unitless measure, so changing the units in which variables are measured won't change their correlation.

817. **It doesn't change.**

In a correlation, it doesn't matter which variable is designated as X and which as Y; the correlation will be the same either way.

818. **They have a weak, negative linear relationship.**

A correlation of −0.23 indicates a weak, negative linear relationship between two variables.

819. **You made a mistake in your calculations.**

A correlation is always between −1 and 1, so a value outside that range indicates an error in calculation.

820. **They have a strong, negative linear relationship.**

A correlation of −0.87 indicates a strong, negative relationship.

821. **E. Choices (A), (B), and (C) (Both variables are numeric; the scatter plot indicates a linear relationship; the correlation is at least moderate.)**

Before computing a regression between two variables, you should determine that both variables are quantitative (numeric), that they're at least moderately correlated, and that the scatter plot indicates a linear relationship.

822. **C. Their relationship isn't linear.**

The variables are numeric, and a correlation of 0.75 is sufficient to perform linear regression. However, the relationship of the points in the scatter plot isn't linear. The points initially have a positive relationship but then curve downward into a negative relationship.

823. m

The slope of the equation is m. If two variables have a negative relationship, they will have a negative slope.

824. m

In the equation for the least-squares regression line, m designates the slope of the regression line.

825. b

In the equation for the least-squares regression line, b designates the y-intercept for the regression line.

826. 3

The equation for calculating the regression line is $y = mx + b$, and m represents the slope. So, in the regression line $y = 3x + 1$, the slope is 3.

827. 1

The equation for calculating the regression line is $y = mx + b$, and b represents the y-intercept. So, in the regression line $y = 3x + 1$, the y-intercept is 1.

828. 11.5

The equation for the regression line is $y = 3x + 1$. To find the value for y when $x = 3.5$, substitute 3.5 for x in the equation:

$$y = 3x + 1 = (3)(3.5) + 1 = 11.5$$

829. 2.2

The equation for the regression line is $y = 3x + 1$. To find the value for y when $x = 0.4$, substitute 0.4 for x in the equation:

$$y = 3x + 1 = (3)(0.4) + 1 = 2.2$$

830. **It decreases by 1.8.**

The equation for calculating the regression line is $y = mx + b$, and m represents the slope. In this case, the slope is -1.2. If x increases by 1.5, y changes by $(-1.2)(1.5) = -1.8$. This is a decrease of 1.8.

831. **It increases by 2.76.**

The equation for calculating the regression line is $y = mx + b$, and m represents the slope. In this case, the slope is -1.2. If x decreases by 2.3, y changes by $(-1.2)(-2.3) = 2.76$. This is an increase of 2.76.

832. **(0, 0.74)**

To find the y-intercept, or the point where the line intersects the y-axis, find the value of y when $x = 0$. To do this, substitute 0 for x in the equation:

$$y = -1.2x + 0.74 = (-1.2)(0) + 0.74 = 0.74$$

833. **strong and positive**

The correlation of 0.792 and the scatter plot showing the points clustering fairly closely around a line running upward from left to right indicate a strong, positive linear relationship.

834. **200 to 800 points**

Looking at the scatter plot, you see that no values are outside the range of 200 to 800 for either variable.

835. **0.79**

Because s_x and s_y are the same, the slope will be equal to the correlation (a rare occurrence).

836. **0.792**

The equation to calculate the slope is

$$m = r\left(\frac{s_y}{s_x}\right)$$

In this case, the correlation is 0.792, the standard deviation of y is 103.2, and the standard deviation of x is 103.2. Plug these numbers into the formula and solve:

$$m = r\left(\frac{s_y}{s_x}\right)$$
$$= 0.792\left(\frac{103.2}{103.2}\right) = 0.792$$

Note: Because s_x and s_y are the same, the slope will be equal to the correlation (a rare occurrence).

837. 107.8 points

The equation to calculate the y-intercept is $b = \bar{y} - m\bar{x}$.

In this case, you know that the mean of y is 506.1 and the mean of x is 502.9. To find the slope, divide the standard deviation of y by the standard deviation of x and then multiply by the correlation. In this case, the correlation is 0.792, the standard deviation of y is 103.2, and the standard deviation of x is 103.2.

$$m = r\left(\frac{s_y}{s_x}\right)$$
$$= 0.792\left(\frac{103.2}{103.2}\right) = 0.792$$

Now, plug the values into the formula for the y-intercept:

$$b = \bar{y} - m\bar{x}$$
$$= 506.1 - (0.792)(502.9)$$
$$= 506.1 - 398.2968$$
$$= 107.8032$$

838. $y = 0.792x + 107.8$

The equation for a regression line is $y = mx + b$. To find the calculated equation of this regression line, you first need to find the slope and y-intercept.

The equation to calculate the slope is

$$m = r\left(\frac{s_y}{s_x}\right)$$

In this case, the correlation (r) is 0.792, the standard deviation of y is 103.2, and the standard deviation of x is 103.2.

$$m = r\left(\frac{s_y}{s_x}\right)$$
$$= 0.792\left(\frac{103.2}{103.2}\right) = 0.792$$

The equation to calculate the y-intercept is $b = \bar{y} - m\bar{x}$. In this case, the mean of y is 506.1, the mean of x is 502.9, and the slope is 0.792:

$$b = \bar{y} - m\bar{x}$$
$$= 506.1 - (0.792)(502.9)$$
$$= 506.1 - 398.2968$$
$$= 107.8032$$

Now, having the values of m and b, you simply plug them into the equation of the regression line to get $y = 0.792x + 107.8$.

839. 289.96 points

To find the expected value of y (verbal score) when x (math score) is 230 points, substitute 230 for x in the equation and solve for y:

$$y = 0.792(230) + 107.8 = 289.96 \text{ points}$$

840. 166.32 points

First, you need to find the slope of this equation by dividing the standard deviation of y by the standard deviation of x and then multiplying by the correlation. In this case, the correlation is 0.792, the standard deviation of y is 103.2, and the standard deviation of x is 103.2.

$$m = r\left(\frac{s_y}{s_x}\right)$$
$$= 0.792\left(\frac{103.2}{103.2}\right) = 0.792$$

So, for every one unit increase in x, you expect to see a 0.792 unit increase in y. In other words, if x (math score) is higher by 1 point, then y (verbal score) is expected to be 0.792 points higher. Here, Student A's math score (x) is 210 points higher, so you expect Student A's verbal score (y) to be $(0.792)(210) = 166.32$ points higher (on average).

841. Student C's verbal score will be 39.6 points lower than Student D's verbal score.

First, you need to find the slope of this equation by dividing the standard deviation of y by the standard deviation of x and then multiplying by the correlation. In this case, the correlation is 0.792, the standard deviation of y is 103.2, and the standard deviation of x is 103.2.

$$m = r\left(\frac{s_y}{s_x}\right)$$
$$= 0.792\left(\frac{103.2}{103.2}\right) = 0.792$$

So, for every one point increase in x (math score), you expect to see a 0.792 point increase in y (verbal score). The opposite also applies: A one unit decrease in x results in a 0.792 decrease in y. Here, Student C's math score (x) decreases by 50 points compared to Student D, so you expect Student C's verbal score (y) to decrease by $(0.792)(50) = 39.6$ points compared to Student D. (*Note:* This value isn't the actual verbal score; it's the amount of drop in Student C's verbal score.)

842. A third variable could be causing the observed relationship between GRA_V and GRA_M.

Finding a correlation between two variables doesn't automatically establish a causal relationship between them. For example, an increase in drug dosage may cause a change in blood pressure, but an increase in shoe size doesn't cause an increase in height. A third variable could be related to the relationship. For example, some research shows that students who are good in music are more likely to be good in both math and in verbal abilities.

843. moderate and positive

The scatter plot and the computed correlation of 0.527 both indicate a moderate, positive linear relationship between home size in square feet and selling price.

844. E. 910 square feet

Although the linear relationship between home size and selling price is moderate ($r = 0.527$), it's positive. Therefore, unless you have further information to contradict the pattern, you can expect that larger houses will generally sell for higher prices.

845. 0.106

To calculate the slope of a regression line for x and y, divide the standard deviation of y by the standard deviation of x, and then multiply by the correlation. In this case, the standard deviation of y is 11.8, the standard deviation of x is 58.5, and the correlation is 0.527.

$$m = r\left(\frac{s_y}{s_x}\right)$$
$$= 0.527\left(\frac{11.8}{58.5}\right)$$
$$= 0.10630$$

846. 24.1

To calculate the intercept of the regression line for x and y, multiply the mean of x by the slope, and then subtract that product from the mean of y. In this case, you know that the mean of x is 915.1 and the mean of y is 121.1. But you need to find the slope. To calculate the slope of a regression line for x and y, divide the standard deviation of y by the standard deviation of x, and then multiply by the correlation. In this case, the standard deviation of y is 11.8, the standard deviation of x is 58.5, and the correlation is 0.527.

$$m = r\left(\frac{s_y}{s_x}\right)$$
$$= 0.527\left(\frac{11.8}{58.5}\right)$$
$$= 0.10630$$

Now, plug the values into the equation for the intercept:

$$b = \bar{y} - m\bar{x}$$
$$= 121.1 - (0.106)(915.1)$$
$$= 24.0994$$

847. $y = -0.106x + 24.1$

The equation for a regression line is $y = mx + b$. You can find the slope by dividing the standard deviation of y (11.8) by the standard deviation of x (58.5) and then multiplying by the correlation (0.527).

$$m = r\left(\frac{s_y}{s_x}\right)$$

$$= 0.527\left(\frac{11.8}{58.5}\right)$$

$$= 0.10630$$

To find the intercept, you multiply the mean of x (915.1) by the slope (0.106), and then subtract that product from the mean of y (121.1).

$$b = \bar{y} - m\bar{x}$$

$$= 121.1 - (0.106)(915.1)$$

$$= 24.0994$$

Now, plug these numbers into the equation for the regression line: $y = -0.106x + 24.1$.

848. $130,100

First, you have to find the regression equation for relating square feet (x) to selling price (y). You do that by calculating the slope and the intercept, using the information provided:

$$m = r\left(\frac{s_y}{s_x}\right)$$

$$= 0.527\left(\frac{11.8}{58.5}\right)$$

$$= 0.10630$$

$$b = \bar{y} - m\bar{x}$$

$$= 121.1 - (0.106)(915.1)$$

$$= 24.0994$$

Then you plug these numbers into the regression line equation: $y = -0.106x + 24.1$.

To find the expected selling price measured in thousands of dollars for a house of 1,000 square feet, substitute 1,000 for x in the equation:

$$y = 0.106(1,000) + 24.1 = 130.1$$

Finally, you convert that to whole dollars by multiplying by $1,000:130.1($1,000) = $130,100.

849. Cannot make an appropriate price prediction for a house of 1,500 square feet.

The sizes of the houses in this data set range from 700 to 1,000 square feet on the scatter plot. Therefore, making predictions for price is only appropriate for houses whose square footage lies within this range or is close to it. (This house has 1,500 square feet, so you can't make an appropriate price prediction.) If you did make a prediction for a house outside of the range of the data, you'd be committing an error called extrapolation.

850. $118,440

First, you have to find the regression equation for relating square feet (x) to selling price (y). You do that by calculating the slope and the intercept, using the information provided:

$$m = r\left(\frac{s_y}{s_x}\right)$$
$$= 0.527\left(\frac{11.8}{58.5}\right)$$
$$= 0.10630$$

$$b = \bar{y} - m\bar{x}$$
$$= 121.1 - (0.106)(915.1)$$
$$= 24.0994$$

Then, you plug these numbers into the regression line equation: $y = -0.106x + 24.1$.

To find the expected selling price measured in thousands of dollars for a house of 890 square feet, substitute 890 for x in the equation:

$$y = 0.106(890) + 24.1 = 118.44$$

Finally, convert to whole dollars by multiplying by 1,000: 118.44($ 1,000) = $ 118,440.

851. $9,540 more

First, you have to find the slope by dividing the standard deviation of y by the standard deviation of x and then multiplying by the correlation. In this case, the standard deviation of y is 11.8, the standard deviation of x is 58.5, and the correlation is 0.527:

$$m = r\left(\frac{s_y}{s_x}\right)$$
$$= 0.527\left(\frac{11.8}{58.5}\right)$$
$$= 0.10630$$

Due to the slope, for every unit change in square footage, price increases by 0.106 thousand dollars: $90(0.106) = 9.54$. So, a house that's 90 square feet larger is expected to cost $9,540 more.

852. House C should cost about $5,720 less than House D.

First, you have to find the slope by dividing the standard deviation of y by the standard deviation of x and then multiplying by the correlation. In this case, the standard deviation of y is 11.8, the standard deviation of x is 58.5, and the correlation is 0.527:

$$m = r\left(\frac{s_y}{s_x}\right)$$
$$= 0.527\left(\frac{11.8}{58.5}\right)$$
$$= 0.10630$$

Due to the slope, for every unit change in square footage, price increases by 0.106 thousand dollars, and for every unit decrease in square footage, price decreases by 0.106 thousand dollars. In this case, the difference in size is a decrease in 54 square feet; the difference in price is a decrease of $(54)(0.106) = 5.724$. Convert to whole dollars by multiplying by $1,000: $5.724($1,000$) = \$5,724$, or about $5,720.

853. **moderate negative**

These points are mostly clustered around a line running from the upper left to lower right, indicating a moderate negative linear relationship. They're not tightly packed enough around a line to consider the linear relationship to be "strong."

854. **−0.5**

These variables have a moderate negative correlation, as evidenced by their loose clustering around a line running from the upper left to the lower right. In fact, their correlation is −0.54.

855. **It wouldn't change.**

Correlation measures the strength of the pattern around a line as well as the direction of the line (uphill or downhill). When you switch X and Y, you don't change the strength of their relationship or the direction of the relationship. For example, if the correlation between height and weight is −0.5, the correlation between weight and height is still −0.5.

856. **Yes, because they're moderately correlated and the points suggest a linear trend.**

The scatter plot indicates a possible linear relationship between the variables, and the correlation coefficient of −0.5 typically has an absolute value high enough to justify beginning a linear regression analysis.

857. **impossible to tell without looking at the scatter plot**

A moderate correlation alone isn't sufficient to indicate that two variables are good candidates for linear regression. You also need to make a scatter plot to confirm whether their relationship is at least approximately linear. In some cases, the scatter plot shows a bit of a curve, but the correlation is still moderately high; or the correlation is weak, but the scatter plot is made to look like a strong relationship exists.

858. **no, because the correlation isn't high enough to justify a linear regression analysis**

A correlation of 0.05 is barely better than chance and, by itself, doesn't indicate that two variables are good candidates for linear regression. You don't even have to look at the scatter plot to know that this isn't a good linear relationship.

859. **E. Choices (A) and (D) (−0.9; 0.9)**

The numerical part (absolute value) of the correlation is the part that measures the strength of the relationship; the sign on the correlation determines only the direction (uphill or downhill). The correlations 0.9 and −0.9 have the same strength and are the strongest of the choices on the list (because their absolute values are closest to 1).

860. **0.65**

Correlation measures the strength and direction of the linear relationship between two variables and nothing more. Switching the X and Y variables has no influence on the value of a correlation, so the correlation of weight and height is the same as the correlation of height and weight.

861. **It won't change.**

Correlation has a universal interpretation — in other words, it doesn't depend on the units of the variables. By design, changing units has no effect on the size or direction of a correlation. Correlation is a "unit-free" statistic.

862. **The linear relationship is stronger for males than for females.**

The linear relationship between height and weight is stronger for males, as indicated by the higher correlation, and is weaker for females, as indicated by the lower correlation. (Notice that you can't add correlations together for two groups; the correlation must be between −1 and 1.)

863. **A. −1.5**

Correlations are always between −1 and +1, and −1.5 is outside of this range.

864. **no, because correlation alone doesn't establish a causal linear relationship between two variables**

An observed correlation, however strong, isn't sufficient to establish a causal linear relationship. The observed correlation could be due to numerous reasons other than a causal linear relationship, including confounding variables (variables not included in the study that could affect the results), such as age or gender. For example, if you want to lower your weight, you can't lower your height.

865. **height**

Height is the X variable because you're using height to predict weight. Height is the independent (explanatory) variable in the equation, and weight is the dependent (response) variable.

866. **strong positive linear**

Because a linear trend exists in the scatter plot of two variables, a correlation of 0.74 indicates a strong positive linear relationship between the two variables. In this case, larger numbers of minutes studying are associated with higher GPAs, and smaller numbers of minutes studying correspond with lower GPAs.

867. **E. 450 minutes**

Given a strong positive correlation between minutes studying and GPA, you can predict that students who spend the most minutes studying would have the highest GPAs on average. This isn't necessarily true for each student; it's a prediction based on the information given.

868. **weak negative**

A correlation of –0.38 indicates a weak negative linear relationship between two variables. In this case, high numbers of minutes watching TV are generally associated with low GPA, and low numbers of minutes watching TV generally correspond with high GPA, but the linear relationship is fairly weak.

869. **A. 30**

The linear relationship between minutes watching TV and GPA is moderately strong and negative; because the scatter plot looks good and the correlation is moderately strong, you'd predict that the student who spends the least amount of time watching TV would have the highest GPA.

870. **GPA**

Although this study can't establish a causal linear relationship merely from the linear relationship between these two variables, it's interesting to see whether time spent studying signifies a higher GPA instead of the other way around.

871. **B. The linear relationship is stronger for engineering majors.**

The closer the C the stronger the linear relationship becomes. The correlation of 0.48 for English majors is lower than that of 0.78 for engineering majors, indicating a stronger linear relationship for engineering majors and a weaker linear relationship for English majors.

872. **You made a mistake in your calculations.**

The value of a correlation must be between –1 and +1. The value of –2.56 is outside this range and thus indicates that you made a mistake in your calculations.

873. **income**

Income is the X variable because you're using income to predict life satisfaction. Income is the independent (explanatory) variable in the equation, and life satisfaction is the dependent (response) variable.

874. **0.58**

To find the slope of a regression line, use this formula:

$$m = \left(r \frac{s_y}{s_x} \right)$$

where s_y is the standard deviation of y, s_x is the standard deviation of x, and r is the correlation. In this case, the X variable is income because it's being used to predict life satisfaction (Y). Substitute the values into the formula to solve:

$$m = \left(r \frac{s_y}{s_x} \right)$$

$$= \left[0.77 \left(\frac{12.5}{16.7} \right) \right]$$

$$\approx 0.58$$

875. **13.7**

To find the y-intercept of a regression line, use this formula:

$$b = \bar{y} - m\bar{x}$$

where $\bar{x}$ is the mean of x, m is the slope, and $\bar{y}$ is the mean of y. In this case, the X variable is income because it's being used to predict life satisfaction (Y). Substitute the values into the formula to solve:

$$b = \bar{y} - m\bar{x}$$

$$= 60.4 - 0.58(80.5)$$

$$= 60.4 - 46.69$$

$$= 13.7$$

876. $y = 0.58x + 13.7$

This equation predicts life satisfaction (Y) from income (X), with a slope of 0.58 and a y-intercept of 13.7.

877. **E. Choices (B) and (C) (predicting satisfaction for someone with an income of $45,000; predicting satisfaction for someone with an income of $200,000)**

In regression, extrapolation means using an equation to make predictions outside of the range of data used to make the equation. In this case, incomes of $45,000 and $200,000 fall outside the range of data used to find the regression equation. Because you don't have data that extends that far, you can't be sure that the same linear trend exists for those values.

878. **$190.00**

To figure out the predicted cost, use the equation $y = 50x + 65$, replacing x with the given number of hours to complete the job. In this case, $x = 2.5$, so $y = 50(2.5) + 65 = 190.00$.

879. **$302.50**

To figure out the predicted cost of a job, use the equation $y = 50x + 65$, replacing x with the given number of hours to complete the job. In this case, $x = 4.75$, so $y = 50(4.75) + 65 = 302.50$.

ANSWERS
801–900

432   PART 2 **The Answers**

880. **$12.50**

You can solve this problem in two ways.

First, the slope measures the change in cost (Y) for a given change in the number of hours (X). So, you can simply calculate the change in hours $(3.75 - 3.50 = 0.25)$, and then multiply by slope (50) to get the difference in cost, $(0.25)(50) = \$12.50$.

Second, you can calculate the costs based on both number of hours, and then take the difference. So, substitute $x = 3.75$ (hours) into the equation, and substitute $x = 3.50$ c(hours) into the equation, calculate their y values (costs), and subtract. So, you have

$$y = -50(3.75) + -65 = -252.50$$
$$y = -50(3.50) + -65 = -240.00$$

Subtract these two values to get $\$252.50 - \$240.00 = \$12.50$.

This means the job is predicted to cost $12.50 more if the hours increase from 3.50 to 3.75.

881. **$175**

If the intercept is $75 while the slope remains the same, the new equation for predicting costs will be $y = 50x + 75$.

In this case, $x = 2$, so $y = 50(2) + 75 = 175$.

882. **$10**

Because the slope is the same in the two equations, you need to consider only the difference in intercept: $\$75 - \$65 = \$10$.

883. **$203**

If the slope is $60 and the intercept is $65, the equation to predict costs is $y = 60x + 65$.

In this case, $x = 2.3$, so $y = 60(2.3) + 65 = 203$.

884. **$36**

You can solve this problem in two ways.

The equation of the regression line predicting costs in the current year is $y = 50x + 65$. The equation for the earlier year is $y = 60x + 65$.

First, the slope (50) measures the change in cost (y) for a given change in the number of hours (x). So, you can simply calculate the change in slope $(60 - 50 = 10)$, then multiply by hour (3.6) to get the difference in cost, $(10)(3.6) = 36.00$.

Second, you can calculate the costs based on both number of hours, then take the difference. So, substitute $x = 3.6$ (hours) into both equations, calculate the cost in each year, and subtract.

$$y = 50(3.6) + 65 = 245$$
$$y = 60(3.6) + 65 = 281$$

Subtract these two values to get $281 - 245 = \$36$.

Costs in the previous year would have been $36 higher than in the current year for a job requiring 3.6 hours to complete.

885. **y – intercept = 62; slope = 1.4**

This equation is written in the form $y = mx + b$, where m is the slope and b is the y-intercept. Therefore, in the regression equation $y = 1.4x + 62$, 62 is the y-intercept and 1.4 is the slope.

886. **the amount of change expected in y for a one–unit change in x**

The slope is the amount of change you can expect in the value of the y variable when the x variable changes by one unit.

887. **the point where the regression line crosses the y-axis**

The y-intercept is the value of y when $x = 0$. This is the same as the point where the regression line crosses the y-axis. If the value of $x = 0$ isn't within the range of observed values of x, then you can't interpret the y value at that point. However, you still use the y-intercept as part of the equation of the regression line.

888. **90 points**

To find the expected job satisfaction rating, use the regression equation for predicting job satisfaction (y), substitute the value for x (years of experience), and calculate y:

$$y = 1.4x + 62$$

In this case, $x = 20$, so $y = 1.4(20) + 62 = 90$.

So, the expected job satisfaction rating for an employee with 20 years of experience, on a scale of 0 to 100, is 90 points.

889. **64.8 points**

To find the expected job satisfaction rating, use the regression equation for predicting job satisfaction (y), substitute the value for x (years of experience), and calculate y:

$$y = 1.4x + 62$$

In this case, $x = 2$, so $y = 1.4(2) + 62 = 64.8$.

So, the expected job satisfaction rating for an employee with 2 years of experience, on a scale of 0 to 100, is 64.8 points.

890. **9.8 points**

To find the expected difference in job satisfaction ratings, use the regression equation for predicting job satisfaction (y), substitute the value for x (years of experience), and calculate y:

$$y = 1.4x + 62$$

In this case, you substitute both years of experience ($x = 15$ years and $x = 8$ years) into the equation, calculate their y values (satisfaction), and then find the difference, like so:

$$y = 1.4(15) + 62 = 83$$
$$y = 1.4(8) + 62 = 73.2$$
$$83 - 73.2 = 9.8$$

Or, because the slope (1.4) measures the change in satisfaction for a given change in years of experience (x), you can simply calculate the change in job satisfaction rating ($15 - 8 = 7$) and then multiply by slope (1.4) to get the difference: $1.4(7) = 9.8$.

Whatever method you choose, the answers are the same: The person with 15 years of experience is expected to rate 9.8 points higher on job satisfaction than someone with 8 years of experience.

891. **21.0 points**

To find the expected difference in job satisfaction ratings, use the regression equation for predicting job satisfaction (y), substitute the value for x (years of experience), and calculate y:

$$y = 1.4x + 62$$

In this case, you substitute both years of experience ($x = 15$ years and $x = 0$ years) into the equation, calculate their y values (satisfaction) and then find the difference, like so:

$$y = 1.4(15) + 62 = 83$$
$$y = 1.4(0) + 62 = 62$$
$$83 - 62 = 21$$

Or, because the slope (1.4) measures the change in satisfaction for a given change in years of experience (x), you can simply calculate the change in job satisfaction rating ($15 - 0 = 15$) and then multiply by slope (1.4) to get the difference: $1.4(15) = 21$.

Whatever method you choose, the answers are the same: The person with 15 years of experience is expected to rate 21 points higher on job satisfaction than someone with 0 years of experience.

892. **16.1 points**

To find the expected difference in job satisfaction ratings, use the regression equation for predicting job satisfaction (y), substitute the value for x (years of experience), and calculate y:

$$y = 1.4x + 62$$

In this case, you substitute both years of experience ($x = 11.5$ years and $x = 0$ years) into the equation, calculate their y values (satisfaction), and then find the difference, like so:

$$y = 1.4(11.5) + 62 = 78.1$$
$$y = 1.4(0) + 62 = 62$$
$$78.1 - 62 = 16.1$$

Or, because the slope (1.4) measures the change in satisfaction for a given change in years of experience (x), you can simply calculate the change in job satisfaction rating ($11.5 - 0 = 11.5$) and then multiply by slope (1.4) to get the difference: $1.4(11.5) = 16.1$.

Whatever method you choose, the answers are the same: The person with 15 years of experience is expected to rate 16.1 points higher on job satisfaction than someone with 0 years of experience.

893. 5

The regression lines have the same slope but different y-intercepts. Therefore, for employees from the two companies with the same years of experience, the only difference in their predicted job satisfaction ratings is the difference in y-intercepts: $67 - 62 = 5$.

894. $100,100

To find the expected market value, use the regression equation for Community 1 and substitute the value for x (square footage):

$$\begin{aligned} y_1 &= 77x_1 - 15,400 \\ &= 77(1,500) - 15,400 \\ &= 115,500 - 15,400 \\ &= 100,100 \end{aligned}$$

So, the expected market value for a home of 1,500 square feet in Community 1 is $100,100.

895. $126,280

To find the expected market value, use the regression equation for Community 1 and substitute the value for x (square footage):

$$\begin{aligned} y_1 &= 77x_1 - 15,400 \\ &= 77(1,840) - 15,400 \\ &= 141,680 - 15,400 \\ &= 126,280 \end{aligned}$$

So, the expected market value for a home of 1,840 square feet in Community 1 is $126,280.

896. $99,700

To find the expected market value, use the regression equation for Community 2 and substitute the value for x (square footage):

$$y_2 = 74x_2 - 11,300$$
$$= 74(1,500) - 11,300$$
$$= 111,000 - 11,300$$
$$= 99,700$$

So, the expected market value for a home of 1,500 square feet in Community 2 is $99,700.

897. $61,220

To find the expected market value, use the regression equation for Community 2 and substitute the value for x (square footage):

$$y_2 = 74x_2 - 11,300$$
$$= 74(980) - 11,300$$
$$= 72,520 - 11,300$$
$$= 61,220$$

So, the expected market value for a home of 980 square feet in Community 2 is $61,220.

898. The home in Community 2 has an expected market value of $1,100 greater than the home in Community 1.

Because the slope and y-intercept are both different for the two equations, you need to calculate the expected value for each home and then find their difference. In both cases, $x = 1,000$.

For the home in Community 1, the expected market value is

$$y_1 = 77x - 15,400$$
$$= 77(1,000) - 15,400$$
$$= 77,000 - 15,400$$
$$= 61,600$$

For the home in Community 2, the expected market value is

$$y_2 = 74x - 11,300$$
$$= 74(1,000) - 11,300$$
$$= 74,000 - 11,300$$
$$= 62,700$$

The difference in these two values is $62,700 - 61,600 = 1,100$, with the home in Community 2 having the greater value.

899. **The home in Community 2 has an expected market value of $760 less than the home in Community 1.**

Because the slope and y-intercept are both different for the two equations, you need to calculate the expected value for each home and then find their difference. In both cases, $x = 1,620$.

For the home in Community 1, the expected market value is

$$
\begin{aligned}
y_1 &= 77x - 15,400 \\
&= 77(1,620) - 15,400 \\
&= 124,740 - 15,400 \\
&= 109,340
\end{aligned}
$$

For the home in Community 2, the expected market value is

$$
\begin{aligned}
y_2 &= 74x - 11,300 \\
&= 74(1,620) - 11,300 \\
&= 119,880 - 11,300 \\
&= 108,580
\end{aligned}
$$

The difference in these two values is $108,580 - 109,340 = -760$, with the home in Community 2 having the lesser value.

900. **The home in Community 1 has an expected market value of $1,690 more than the home in Community 2.**

Because the slope and y-intercept are both different for the two equations, you need to calculate the expected value for each home and then find their difference. In both cases, $x = 1,930$.

For the home in Community 1, the expected market value is

$$
\begin{aligned}
y_1 &= 77x - 15,400 \\
&= 77(1,930) - 15,400 \\
&= 148,610 - 15,400 \\
&= 133,210
\end{aligned}
$$

For the home in Community 2, the expected market value is

$$
\begin{aligned}
y_2 &= 74x - 11,300 \\
&= 74(1,930) - 11,300 \\
&= 142,820 - 11,300 \\
&= 131,520
\end{aligned}
$$

The difference in these two values is $131,520 - 133,210 = -1,690$, with the home in Community 1 having the greater value.

901. **The home in Community 1 has an expected market value of $8,200 less than the home in Community 2.**

Because the slope and y-intercept are both different for the two equations, you need to calculate the expected value for each home and then calculate their difference. In this case, $x_1 = 1,100$ and $x_2 = 1,200$.

For the home in Community 1, the expected market value is

$$\begin{aligned} y_1 &= 77x - 15,400 \\ &= 77(1,100) - 15,400 \\ &= 84,700 - 15,400 \\ &= 69,300 \end{aligned}$$

For the home in Community 2, the expected market value is

$$\begin{aligned} y_2 &= 74x - 11,300 \\ &= 74(1,200) - 11,300 \\ &= 88,800 - 11,300 \\ &= 77,500 \end{aligned}$$

The difference in these two values is $77,500 - 69,300 = 8,200$, with the home in Community 1 having the lesser value.

902. **E. Choices (B) and (C) (estimating the market value of a home in Community 1 with 2,900 square feet; estimating the market value of a home in Community 1 with 750 square feet)**

Extrapolation means applying a regression equation for values beyond the range included in the data set used to create the regression equation. In this case, both 750 square feet and 2,900 square feet are outside the range of the data set used to create the equation. Because you don't have data that extends that far, you can't be sure that the same linear trend exists for those values.

903. **545**

To find the answer, substitute the given x-value of 360 into the equation and solve:

$$SAT = 725 - 0.5(360) = 545$$

904. **425**

To find the answer, substitute the given x-value of 600 into the equation and solve:

$$SAT = 725 - 0.5(600) = 425$$

905. **Student A, Student C, Student B**

Because the coefficient for minutes of TV watching is negative, students who watch the least amount of TV will have the highest predicted SAT scores. In this example, Student A watches the least amount of TV, followed by Student C and Student B.

906. $42,700

Use the regression equation for Company 1 and substitute the x_1 value of 6:

$$y_1 = 6.7(6) + 2.5$$
$$= 40.2 + 2.5$$
$$= 42.7$$

This answer is in thousands of dollars, so multiply by $1,000: $42.7($1,000) = $42,700$.

907. $116,400

Use the regression equation for Company 1 and substitute the x_1 value of 17:

$$y_1 = 6.7(17) + 2.5$$
$$= 113.9 + 2.5$$
$$= 116.4$$

This answer is in thousands of dollars, so multiply by $1,000: $116.4($1,000) = $116,400$.

908. $19,200

Use the regression equation for Company 2 and substitute the x_2 value of 2.5:

$$y_2 = 7.2(2.5) + 1.2$$
$$= 18 + 1.2$$
$$= 19.2$$

This answer is in thousands of dollars, so multiply by $1,000: $19.2($1,000) = $19,200$.

909. $43,200

You can find the exact difference in two ways.

First, you can calculate the salary of each employee based on his/her respective years of experience and then find the difference, like so:

$$7.2(13) + 1.2 = 94.8$$
$$7.2(7) + 1.2 = 51.6$$
$$94.8 - 51.6 = 43.2$$

Convert this difference to thousands of dollars: $43.2($1,000) = $43,200$.

Or, because the slope measures the change in salary (y) for a given change in the years of experience (x), you can simply calculate the difference in years ($13 - 7 = 6$) and then multiply by the slope to get the difference in salary:

$$y = (6)(7.2) = 43.2$$

Convert to whole dollars: $43.2($1,000) = $43,200$.

Whatever method you choose, the answers are the same: You expect a part-time employee from Company 2 with 13 years of experience to make $43,200 more than one with 7 years of experience.

910. **$38,160**

You can find the exact difference in two ways.

First, you can calculate the salary of each employee based on his/her respective years of experience and then find the difference, like so:

$$7.2(6.5) + 1.2 = 48$$
$$7.2(1.2) + 1.2 = 9.84$$
$$48 - 9.84 = 38.16$$

Convert this difference to thousands of dollars: $38.16(\$1,000) = \$38,160$.

Or, because the slope measures the change in salary (y) for a given change in the years of experience (x), you can simply calculate the difference in years ($6.5 - 1.2 = 5.3$) and then multiply by the slope to get the difference in salary:

$$y = (5.3)(7.2) = 38.16$$

Convert to thousands of dollars: $38.16(\$1,000) = \$38,160$.

Whatever method you choose, the answers are the same: You expect a part-time employee from Company 2 with 6.5 years of experience to make $38,160 more than one with 1.2 years of experience.

911. **Company 1**

The starting salary is the salary associated with 0 years of experience. This means that the x terms fall out of the equations (being multiplied by 0, they'll both equal 0), and you find the answer by comparing the y-intercepts. A larger y-intercept means a larger starting salary.

For Company 1, the y-intercept is 2.5. For Company 2, the y-intercept is 1.2. Therefore, the starting salary is higher at Company 1.

912. **Company 2**

The rate of increase is the amount that salary is expected to increase with each additional year of employment. To find which is higher, compare the slopes; the company with the larger slope will have the higher rate of increase.

For Company 1, the slope is 6.7. For Company 2, the slope is 7.2. Therefore, the rate of increase is higher at Company 2.

913. **The employee at Company 1 is expected to make $18,200 less than the other.**

Because the slopes and y-intercepts both differ in these equations, you must calculate the expected salary for each case and then compare them. For this example, $x_1 = 3$ and $x_2 = 5.5$.

For the employee at Company 1, the expected salary (in thousands of dollars) is

$$y_1 = 6.7x + 2.5$$
$$= 6.7(3) + 2.5$$
$$= 20.1 + 2.5$$
$$= 22.6$$

For the employee at Company 2, the expected salary (in thousands of dollars) is

$$y_2 = 7.2x + 1.2$$
$$= 7.2(5.5) + 1.2$$
$$= 39.6 + 1.2$$
$$= 40.8$$

The difference between the two expected salaries is $40.8 - 22.6 = 18.2$.

This answer is in thousands of dollars, so multiply by $1,000: $18.2($1,000) = $18,200$. The employee at Company 2 has the higher expected salary.

914.

The employee at Company 1 is expected to make $4,200 less than the other.

Because the slopes and y-intercepts both differ in these equations, you must calculate the expected salary for each case and then compare them. For this example, $x_1 = 3.8$ and $x_2 = 4.3$.

For the employee at Company 1, the expected salary (in thousands of dollars) is

$$y_1 = 6.7x + 2.5$$
$$= 6.7(3.8) + 2.5$$
$$= 25.46 + 2.5$$
$$= 27.96$$

For the employee at Company 2, the expected salary (in thousands of dollars) is

$$y_2 = 7.2x + 1.2$$
$$= 7.2(4.3) + 1.2$$
$$= 30.96 + 1.2$$
$$= 32.16$$

The difference between the two expected salaries is $32.16 - 27.96 = 4.2$.

This answer is in thousands of dollars, so multiply by $1,000: $4.2($1,000) = $4,200$. The employee at Company 2 has the higher expected salary.

915.

E. Choices (A), (B), and (C) (replication of this study at other companies; a longitudinal study tracking growth in individual employee salaries as years of employment increase; adding additional variables to the model to control for other influences on salary)

Although observing a strong linear relationship between salary and years of experience isn't sufficient to draw causal inferences, additional information can strengthen your ability to draw such inferences. Examples of studies that can strengthen causal inference include replication studies, longitudinal studies, and studies including additional variables to control for other influences on the outcome variable.

916. **D. Choices (A) and (B) (blood type; country of origin)**

Blood type (A, AB, B, or O) and country of origin are both categorical because the data can take on a limited set of values, and the values have no numeric meaning. Annual income is continuous rather than categorical because the data can take on an infinite number of values, and the values have numeric meaning.

917. **E. Choices (A), (B), and (C) (gender; hair color; zip code)**

Gender, hair color, and zip code are all categorical because the data can take on a limited set of values, and the values have no numeric meaning. Although zip codes are written with numeric symbols (for example, 10024), the digits are symbols rather than numbers that can be added, subtracted, and so forth.

918. **E. Choices (C) and (D) (homeownership [yes/no]; gender)**

For a two-way table, the data must be categorical. Years of education and height are numerical and can take on an undetermined number of possible values and can even be considered continuous, not categorical. Gender and homeownership are categorical and can be used in a two-way table.

919. **E. Choices (A), (B), and (C) (whether someone is a high-school graduate; whether someone is a college graduate; highest level of school completed)**

Whether someone is a high-school graduate and *whether someone is a college graduate* both clearly have only two possible values — yes or no — and can be considered appropriate for a two-way table. *Highest level of school completed* is also a categorical variable whose possible values can be listed (for example, "less than high school," "high-school graduate," "some college," "college graduate," and "graduate school"). Therefore, it's also appropriate for a two-way table.

920. 4

A 2-x-2 table is a term that means a two-way table with exactly two rows and two columns. The two rows represent the two possible categories for one of the variables, and the two columns represent the two possible categories for the other variable. To find the number of cells in a table, multiply the number of rows by the number of columns. A 2-x-2 table has two rows and two columns, so it has four cells: $(2)(2) = 4$. These represent the four possible combinations of the values of the two variables.

921. **the number of all students polled who are both male and favor an increase**

In a cross-tabulation table, the number in an individual cell represents the cases that have the characteristics described by the row and column intersecting at that cell. In this example, 72 is in the cell at the intersection of the row for *Male* and the column for *Favor Fee Increase* so 72 represents the number of students who are both male and favor a fee increase.

922. **the number of all students polled who are female and not in favor of the increase**

> In a cross-tabulation table, the number in an individual cell represents the cases that have the characteristics described by the row and column intersecting at that cell. In this example, 132 is in the cell at the intersection of the row for *Female* and the column for *Do Not Favor Fee Increase* so 132 represents the number of students who are both female and do not favor a fee increase.

923. 48

> To find the number of students polled who are both female and favor the fee increase, find the cell where the row for *Female* and the column for *Favor Fee Increase* intersect. The value in this cell is 48. The keyword in this question is *and*, which means intersection.

924. 180

> To find the total number of male students included in this poll (out of all 360 students polled), add the values in the row labeled *Male:* $72 + 108 = 180$.

> This sum represents all the students who are male and in favor of the increase plus all the students who are male and not in favor of the increase.

925. 180

> To find the total number of female students included in this poll (out of all 360 students polled), add the values in the row labeled *Female:* $48 + 132 = 180$.

> This sum represents all the students who are female and in favor of the increase plus all the students who are female and not in favor of the increase.

926. 120

> To find the total number of students who favor the fee increase (out of all 360 students polled), add the values in the column labeled *Favor Fee Increase:* $72 + 48 = 120$.

> This sum represents all the students who are in favor of the increase and male plus all the students who are in favor of the increase and female.

927. 240

> To find the total number of students who don't favor the fee increase, add the values in the column labeled *Do Not Favor Fee Increase:* $108 + 132 = 240$.

> This sum represents all the students who are not in favor of the increase and male plus all the students who are not in favor of the increase and female.

928. 360

> To find the total number of students who took part in the poll, add the values of all four cells in the 2-x-2 table: $72 + 108 + 48 + 132 = 360$.

929.   0.40

A proportion involves a fraction, including a numerator and denominator. The denominator is the key because it represents the total number of individuals in the group you're looking at, and you may be looking at different groups from problem to problem.

To find the proportion of male students who favor the fee increase, you divide the number of students who are male and in favor of the fee increase (72) by the total number of male students in the poll (72 + 108 = 180):

$$\frac{72}{180} = 0.4$$

Note that you don't divide by 360 (all students polled) because the question asks you to find the *proportion of male students*, not both male and female students, who favor the increase. So, you divide by the total number of male students polled (180).

930.   0.73

A proportion involves a fraction, including a numerator and denominator. The denominator is the key because it represents the total number of individuals in the group you're looking at, and you may be looking at different groups from problem to problem.

To find the proportion of female students who don't favor the fee increase, you divide the number of students who are female and don't favor the fee increase (132) by the total number of female students in the poll (132 + 48 = 180):

$$\frac{132}{180} \approx 0.73$$

Note that you don't divide by 360 (all students polled) because the question asks you to find the *proportion of female students*, not both female and male students, who don't favor the increase. So, you divide by the total number of female students polled (180).

931.   0.67

A proportion involves a fraction, including a numerator and denominator. The denominator is the key because it represents the total number of individuals in the group you're looking at, and you may be looking at different groups from problem to problem.

To find the proportion of all students who don't favor the fee increase, divide the total number of students who don't favor the fee increase (108 + 132 = 240) by the total number of students in the poll (108 + 132 + 72 + 48 = 360):

$$\frac{240}{360} \approx 0.67$$

Note that you divide by 360 here because you want the proportion of *all* students.

932. **the percentage of females polled who favor the fee increase**

This pie chart represents the opinions of only the 180 females who were polled, not all the students who were polled, so you'll use 180 as the denominator to calculate the percentage of females polled. According to the table, of the 180 females polled, 48 of them favor the increase, so the proportion of females who favor the fee increase is

$$\frac{48}{180} \approx 0.27, \text{ or } 27\%$$

933. **the percentage of all students who favor the fee increase**

This pie chart represents the opinions of all 360 students polled, so you divide by 360 to do your calculations.

Of the 360 students, 120 of them favor the increase, so the proportion of all students who favor the fee increase is

$$\frac{120}{360} \approx 0.33, \text{ or } 33\%$$

934. **the percentage of all students who are female and not in favor of the increase**

Looking at the survey data from the 2-x-2 table, you see that there are 360 total students. To find the number corresponding to 37% of these students, convert 37% to a proportion by dividing by 100, and then multiply that result by 360:

$$\frac{37}{100} = 0.37$$
$$(0.37)(360) = 133.2$$

The difference between 133.2 and 132 is due to rounding:

$$\frac{132}{360} = 0.3\overline{6}$$

which rounds up to 0.37. The closest value in the table is 132, so 37% represents the percentage of all students who are female and not in favor of the increase.

Note: If the results were based only on the females in the poll, you'd divide by 180, the total number of females. If the results were based only on those students not in favor of the fee increase, you'd divide by 240, the total number of students polled not in favor of the fee increase.

You can check this by finding what percentage of students in each other cell in the table represents:

$$\frac{48}{360} = 0.1\overline{3}, \text{ or } 13\%$$
$$\frac{72}{360} = 0.2, \text{ or } 20\%$$
$$\frac{108}{360} = 0.3, \text{ or } 30\%$$

935. the percentage of all students who are male and in favor of the fee increase

Looking at the survey data from the 2-x-2 table, you see that there are 360 total students. To find the number corresponding to 20% of these students, convert 20% to a proportion by dividing by 100, and then multiply that result by 360:

$$\frac{20}{100} = 0.2$$
$$(0.2)(360) = 72$$

There are 72 students who are male and favor the fee increase, so that's who the 20% represents.

936. the percentage of all students who are male and not in favor of the fee increase

Looking at the survey data from the 2-x-2 table, you see that there are 360 total students. To find the number corresponding to 30% of these students, convert 30% to a proportion by dividing by 100, and then multiply that result by 360:

$$\frac{30}{100} = 0.3$$
$$(0.3)(360) = 108$$

There are 108 students who are male and do favor the fee increase, so that's who the 30% represents.

937. percentage of all students who are female and in favor of the fee increase

Looking at the survey data from the 2-x-2 table, you see that there are 360 total students. To find the number corresponding to 13% of these students, convert 13% to a proportion by dividing by 100, and then multiply that result by 360:

$$\frac{13}{100} = 0.13$$
$$(0.13)(360) = 46.8$$

The closest value in the table is 48, so 13% represents the percentage of students who are female and in favor of the fee increase.

The difference between 46.8 and 48 is due to rounding:

$$\frac{48}{360} = 0.1\overline{3}$$

This rounds down to 0.13.

938. 72

To find the total number of smokers, add the cells in the row labeled *Smoker*: $48 + 24 = 72$.

939. 74

To find the total number of patients with a hypertension diagnosis, add the cells in the column labeled *Hypertension Diagnosis*: $48 + 26 = 74$.

940. 26

To find the number of patients who are nonsmokers and have a hypertension diagnosis, find the cell at the intersection of the row labeled *Nonsmoker* and the column labeled *Hypertension Diagnosis:* 26.

941. 48

To find the number of patients who are smokers and have a hypertension diagnosis, find the cell at the intersection of the row labeled *Smoker* and the column labeled *Hypertension Diagnosis:* 48.

942. 148

To find the total number of patients in the study, add the numbers in each cell of the table: $48 + 24 + 26 + 50 = 148$.

943. 24

To find the number of patients who don't have a hypertension diagnosis and are smokers, find the cell at the intersection of the column labeled *Hypertension Diagnosis* and the row labeled *Smoker:* 24.

944. 50

To find the number of patients who don't have a hypertension diagnosis and are non-smokers, find the cell at the intersection of the column labeled *No Hypertension Diagnosis* and the row labeled *Nonsmoker:* 50.

945. 0.65

To find the proportion of patients with a hypertension diagnosis who are smokers, you focus on the first column of the table. The total number of patients with a hypertension diagnosis is 74: 48 of them are smokers, and 26 are nonsmokers. So, the proportion of the hypertension patients who are smokers is

$$\frac{48}{74} \approx 0.65$$

946. 0.35

To find the proportion of patients with a hypertension diagnosis who are nonsmokers, you focus on the first column of the table. The total number of patients with a hypertension diagnosis is 74: 48 of them are smokers, and 26 are nonsmokers. The proportion of the hypertension patients who are nonsmokers is

$$\frac{26}{74} = 0.35$$

947. 0.34

To find the proportion of nonsmokers with a hypertension diagnosis, you focus on the second row of the table, because you're limited to the 76 nonsmokers. Of that group,

26 of them have a hypertension diagnosis. So, you divide the number of nonsmokers with a hypertension diagnosis by the total number of nonsmokers:

$$\frac{26}{76} \approx 0.34$$

948. 0.66

To find the proportion of nonsmokers with no hypertension diagnosis, you focus on the second row of the table, because you're limited to the 76 nonsmokers. Of that group, 50 of them don't have a hypertension diagnosis. So, you divide the number of nonsmokers with no hypertension diagnosis by the total number of nonsmokers:

$$\frac{50}{76} = 0.66$$

949. 0.16

To find the proportion of all patients who are smokers with no hypertension diagnosis, divide the number of patients who are smokers and have no hypertension diagnosis (24) by the total number of patients (148):

$$\frac{24}{148} \approx 0.16$$

950. 0.34

To find the proportion of all patients who are nonsmokers with no hypertension diagnosis, divide the number of patients who are nonsmokers and have no hypertension diagnosis (50) by the total number of patients (148):

$$\frac{50}{148} \approx 0.34$$

951. the conditional probability of not smoking, given a hypertension diagnosis

The 35% portion is in the bar labeled *Hypertension Diagnosis,* which sums to 100%, so this area is a conditional probability for those who have a hypertension diagnosis. You know that in the data from the table, among people with hypertension, smoking is more common than not smoking. So, the smaller area of this bar must be the conditional probability of not smoking, given a hypertension diagnosis.

You can also calculate the conditional probability of not smoking, given a hypertension diagnosis, by dividing the number of respondents who don't smoke and have a hypertension diagnosis (26) by the total number with a hypertension diagnosis (74):

$$P(\text{nonsmoker} \mid \text{hypertension}) = \frac{26}{74} \approx 0.35$$

This probability notation has *hypertension* in the back part of the parentheses because that's the subgroup you're looking at (and why you divide by 74). The *nonsmoker* goes in the front part of the parentheses because you want to know what proportion of that subgroup are nonsmokers.

952. the conditional probability of smoking, given a hypertension diagnosis

The 65% portion is in the bar labeled *Hypertension Diagnosis*, which sums to 100%, so this area is a conditional probability for those who have a hypertension diagnosis. You know that in the data from the table, among people with hypertension, smoking is more common than not smoking. So, the larger area of this bar must be the conditional probability of smoking, given a hypertension diagnosis.

You can also calculate the conditional probability of smoking, given a hypertension diagnosis, by dividing the number of respondents who do smoke and have a hypertension diagnosis (48) by the total number with a hypertension diagnosis (74):

$$P(\text{smoker} \,|\, \text{hypertension}) = \frac{48}{74} \approx 0.65$$

This probability notation has *hypertension* in the back part of the parentheses because that's the subgroup you're looking at (and why you divide by 74). The *smoker* goes in the front part of the parentheses because you want to know what proportion of that subgroup are smokers.

953. the conditional probability of not smoking, given no hypertension diagnosis

The 68% portion is in the bar labeled *No Hypertension Diagnosis*, which sums to 100%, so this area is a conditional probability for those who don't have a hypertension diagnosis. You know that in the data from the table, among people with no hypertension diagnosis, not smoking is more common than smoking. So, the larger area of this bar must be the conditional probability of not smoking, given no hypertension diagnosis.

You can also calculate the conditional probability of not smoking, given no hypertension diagnosis, by dividing the number of respondents who don't smoke and don't have a hypertension diagnosis (50) by the total number who don't have a hypertension diagnosis (74):

$$P(\text{nonsmoker} \,|\, \text{no hypertension}) = \frac{50}{74} \approx 0.68$$

This probability notation has *no hypertension* in the back part of the parentheses because that's the subgroup you're looking at (and why you divide by 74). The *nonsmoker* goes in the front part of the parentheses because you want to know what proportion of that subgroup are nonsmokers.

954. the conditional probability of smoking, given no hypertension diagnosis

The 32% portion is in the bar labeled *No Hypertension Diagnosis*, which sums to 100%, so this area is a conditional probability for those who don't have a hypertension diagnosis. You know that in the data from the table, among people with no hypertension diagnosis, smoking is less common than not smoking. So, the smaller area of this bar must be the conditional probability of smoking, given no hypertension diagnosis.

You can also calculate the conditional probability of smoking, given no hypertension diagnosis, by dividing the number of respondents who smoke and don't have a hypertension diagnosis (24) by the total number of patients who don't have a hypertension diagnosis (74):

$$P(\text{smoker} \mid \text{no hypertension}) = \frac{24}{74} \approx 0.32$$

This probability notation has *no hypertension* in the back part of the parentheses because that's the subgroup you're looking at (and why you divide by 74). The *smoker* goes in the front part of the parentheses because you want to know what proportion of that subgroup are smokers.

955. **E. Choices (A) and (D) (Patients with a hypertension diagnosis are more likely to be smokers than nonsmokers; patients without a hypertension diagnosis are more likely to be nonsmokers than smokers.)**

Although you don't want to generalize too broadly from a single sample of data, the patterns found in this data set indicate that patients with a hypertension diagnosis are more likely to be smokers (65%) rather than nonsmokers (35%), and patients without a hypertension diagnosis are more likely to be nonsmokers (68%) than smokers (32%).

You can find these percentages in the two bar graphs.

956. **C. $P(A)$ does not depend on whether or not B occurs.**

The question states that the variables A and B are independent. Two variables are independent if the probability of one event occurring doesn't depend on whether the other event occurs; therefore, their probabilities aren't affected by the occurrence of the other event.

957. **B. Gender and choice of major are not independent.**

You don't know anything about the *number* of students in either group; you're given only percentages. If gender and choice of major were independent, you'd expect to see the same proportions of men and women enrolled in each major. In engineering, 70% of students are male, but in English 20% are male. And in engineering, 30% of students are female, while in English 80% are female.

958. **B. The same proportion of males and females choose to enroll in higher education.**

Note that although the same proportion of males and females will choose to enroll, it may not be true that the same number of males and females choose to enroll, because the senior class may not have the same number of males and females.

959. **210**

Given this data, if 60% of voters voted for the bond initiative and voting was independent of gender, you'd also expect 60% of female voters to vote for the bond initiative. To find the expected number of women who voted for the bond initiative, you multiply the total number of female registered voters by 60%: $350(0.6) = 210$.

960. **120**

Given this data, if 40% of students participate in after-school activities, then 60% don't participate, because $1.0 - 0.4 = 0.6$, or 60%.

If 60% of students don't participate in after-school activities and participation is independent of gender, you'd expect that 60% of boys wouldn't participate in after-school activities. To find out how many boys 60% is, you multiply the total number of male students by 60%: $200(0.6) = 120$.

961. 0.33

Marginal probability is the probability of having a certain characteristic of one variable, without regard to the other variable(s).

In this case, the marginal probability of a person being a vegetarian is the probability of being a vegetarian, without regard to whether that person has high cholesterol. In this data, 100 adults are vegetarian out of a total of 300 adults. To find the marginal probability, divide the number of vegetarians by the total number of adults:

$$\frac{100}{300} \approx 0.33$$

962. 0.67

Marginal probability is the probability of having a certain characteristic of one variable, without regard to the other variable(s).

In this table, there are three dietary categories: vegetarian, vegan, and regular dieter. The marginal probability of a person not being a vegan is the probability of being either vegetarian or a regular dieter, without regard to whether that person has high cholesterol. In this data, 200 adults aren't vegans, out of a total of 300 adults. To find the marginal probability, divide the number of non-vegans by the total number of adults:

$$\frac{200}{300} = 0.67$$

963. 0.33

Marginal probability is the probability of having a certain characteristic of one variable, without regard to the other variable(s).

In this example, the marginal probability of a person having high cholesterol is the probability of having high cholesterol, without regard to that person's dietary habits. In this data, 100 adults have high cholesterol out of a total of 300 adults. To find the marginal probability, divide the number of adults with high cholesterol by the total number of adults:

$$\frac{100}{300} \approx 0.33$$

964. 0.67

Marginal probability is the probability of having a certain characteristic of one variable, without regard to the other variable(s).

In this example, the marginal probability of a person not having high cholesterol is the probability of not having high cholesterol, without regard to that person's dietary habits. In this data, 200 adults don't have high cholesterol out of a total of 300 adults. To find the marginal probability, divide the number of adults without high cholesterol by the total number of adults:

$$\frac{200}{300} \approx 0.67$$

965. E. Choices (A) and (C) (The same percentage of vegetarians, vegans, and regular dieters will have high cholesterol; among those with high cholesterol, equal numbers will be vegetarians, vegans, and regular dieters.)

This data set includes equal percentages of vegans, vegetarians, and regular dieters ($100 / 300 = 33.3\%$ each). If diet and cholesterol level are unrelated, then the probability of one variable occurring doesn't affect the probability of the other variable occurring. Therefore, the same percentage of vegans, vegetarians, and regular dieters should have high cholesterol, and vegans, vegetarians, and regular dieters should be equally represented among those with high cholesterol.

966. 67

If diet and cholesterol level are independent, you can find the joint probability by multiplying the marginal frequencies and dividing by the sample size.

In this data, there are 100 vegetarians and 200 adults without high cholesterol, out of 300 total adults:

$$\frac{100(200)}{300} \approx 67$$

967. 33

If being vegetarian and having high cholesterol are independent, you can find the joint probability by multiplying the marginal frequencies and dividing by the sample size.

In this data, there are 100 vegetarians and 100 adults with high cholesterol, out of 300 total adults:

$$\frac{100(100)}{300} \approx 33$$

968. 70

The marginal total of adults with high cholesterol is 100. If 10 of these are vegetarians and 20 are vegans, the remainder must be regular dieters: $100 - (10 + 20) = 70$.

969. 90

The marginal total of vegetarians is 100. If ten of these have high cholesterol, the remainder must not have high cholesterol: $100 - 10 = 90$.

970. 65

The marginal total of regular dieters is 100. If 35 of these don't have high cholesterol, the remainder must have high cholesterol: $100 - 35 = 65$.

971. 300

To find the total number of participants, add the numbers from each cell in the table: $60 + 30 + 10 + 40 + 40 + 20 + 20 + 30 + 50 = 300$.

972. 0.33

Marginal probability is the probability of having a certain characteristic of one variable, without regard to the other variable(s).

To find the marginal probability of an individual being 41 to 65 years old, divide the number of participants ages 41 to 65 $(40 + 40 + 20 = 100)$ by the total number of participants $(60 + 30 + 10 + 40 + 40 + 20 + 20 + 30 + 50 = 300)$:

$$\frac{100}{300} \approx 0.33$$

973. 0.73

Marginal probability is the probability of having a certain characteristic of one variable, without regard to the other variable(s).

To find the marginal probability of a respondent's most commonly used type of phone not being a landline, divide the number of participants who said they most commonly used a smartphone or other mobile phone $(60 + 40 + 20 + 30 + 40 + 30 = 220)$ by the total number of participants $(60 + 30 + 10 + 40 + 40 + 20 + 20 + 30 + 50 = 300)$:

$$\frac{220}{300} \approx 0.73$$

974. 0.20

The joint probability refers to the probability of having two or more characteristics — in this case, being in a particular age group and using a particular type of phone.

To find the joint probability of a respondent being between 18 and 40 years old and most commonly using a smartphone, divide the number of participants ages 18 to 40 who most commonly use a smartphone (60) by the total number of participants $(60 + 30 + 10 + 40 + 40 + 20 + 20 + 30 + 50 = 300)$:

$$\frac{60}{300} = 0.20$$

975. 0.17

The joint probability refers to the probability of having two or more characteristics — in this case, being in a particular age group and using a particular type of phone.

To find the joint probability of a respondent being age 66 or older and most commonly using a landline phone, divide the number of participants age 66 and older who most commonly use a landline (50) by the total number of participants $(60 + 30 + 10 + 40 + 40 + 20 + 20 + 30 + 50 = 300)$:

$$\frac{50}{300} \approx 0.17$$

976. 40

To find the expected number of people ages 18 to 40 who prefer a smartphone, if age and phone preference are independent, multiply the marginal frequencies for a respondent being between 18 and 40 years old $(60 + 30 + 10 = 100)$ and for preferring a smartphone $(60 + 40 + 20 = 120)$, and divide by the total number of participants $(60 + 30 + 10 + 40 + 40 + 20 + 20 + 30 + 50 = 300)$:

$$\frac{100(120)}{300} = \frac{12,000}{300} = 40$$

977. 33

To find the expected number of people age 66 or older who prefer a mobile phone, if age and phone preference are independent, multiply the marginal frequencies for a respondent being age 66 or older $(20 + 30 + 50 = 100)$ and for preferring a mobile phone $(30 + 40 + 30 = 100)$, and divide by the total number of participants $(60 + 30 + 10 + 40 + 40 + 20 + 20 + 30 + 50 = 300)$:

$$\frac{100(100)}{300} = \frac{10,000}{300} \approx 33$$

978. **C. People age 66 or older are less likely to prefer smartphones than would be expected if age and phone preference were independent.**

You can calculate the expected number of people in each joint category (age group and phone preference) by multiplying the marginal frequencies and dividing by the total number of participants. If age and phone preference are independent, these expected values will be the same as the observed number in each category.

In this example, 20 people age 66 or older said they most commonly used a smartphone (the observed value). You can find the expected value by multiplying the number of people age 66 or older $(20 + 30 + 50 = 100)$ by the number of people who prefer a smartphone $(60 + 40 + 20 = 120)$ and then dividing by the total number of participants $(60 + 30 + 10 + 40 + 40 + 20 + 20 + 30 + 50 = 300)$:

$$\frac{(100)(120)}{300} = \frac{12,000}{300} = 40$$

Because the observed number is lower than the expected number, it's correct to say that people age 66 or older are less likely to prefer smartphones than would be expected if age and phone preference were independent.

979. **E. Choices (C) and (D) (No, because the ownership rates differ by gender; no, because the marginal ownership rate differs from the conditional ownership rates.)**

If two variables were independent, all three percentages given would have to be equal. The conditional probability of ownership (0.75) would be the same as the marginal probabilities of ownership (0.85 for males; 0.65 for females), and the marginal probabilities of ownership would be the same for both genders.

980. **12**

To find the number of cells in a two-way table, multiply the number of categories for each variable. In this case, type of residence has three categories, and annual income has four categories: $(3)(4) = 12$.

981. **100**

To find the total sample size, add together the frequencies for each category: $15 + 40 + 10 + 35 = 100$.

982. **0.73**

A *conditional probability* represents the percentage of individuals within a given group who have a certain characteristic. The number of individuals in the given group always goes in the denominator.

To find the conditional probability of owning a car, given that the person is male, divide the number of male car owners (40) by the total number of males ($40 + 15 = 55$):

$$\frac{40}{55} \approx 0.73$$

983. **0.78**

A *conditional probability* represents the percentage of individuals within a given group who have a certain characteristic. The number of individuals in the given group always goes in the denominator.

To find the conditional probability of owning a car, given that the person is female, divide the number of female car owners (35) by the total number of females ($35 + 10 = 45$):

$$\frac{35}{45} \approx 0.78$$

984. **0.75**

Marginal probability is the probability of having a certain characteristic of one variable, without regard to the other variable(s).

To find the marginal probability of owning a car, divide the number of car owners ($40 + 35 = 75$) by the total number of survey participants ($40 + 15 + 35 + 10 = 100$):

$$\frac{75}{100} \approx 0.75$$

985. 0.53

A *conditional probability* represents the percentage of individuals within a given group who have a certain characteristic. The number of individuals in the given group always goes in the denominator.

To find the conditional probability of being male, given car ownership, divide the number of male car owners (40) by the total number of car owners ($40 + 35 = 75$):

$$\frac{40}{75} \approx 0.53$$

986. 0.47

A *conditional probability* represents the percentage of individuals within a given group who have a certain characteristic. The number of individuals in the given group always goes in the denominator.

To find the conditional probability of being female, given car ownership, divide the number of female car owners (35) by the total number of car owners ($40 + 35 = 75$):

$$\frac{35}{75} \approx 0.47$$

987. 41.25

Marginal probability is the probability of having a certain characteristic of one variable, without regard to the other variable(s).

To find the marginal probability of car ownership, divide the number of car owners ($40 + 35$) by the sample size ($40 + 15 + 35 + 10 = 100$):

$$\frac{75}{100} \approx 0.75$$

To apply this probability to males, multiply the number of males ($40 + 15 = 55$) by the marginal probability of car ownership:

$$55(0.75) = 41.25$$

988. 33.75

Marginal probability is the probability of having a certain characteristic of one variable, without regard to the other variable(s).

To find the marginal probability of car ownership, divide the number of car owners ($40 + 35$) by the sample size ($40 + 15 + 35 + 10 = 100$):

$$\frac{75}{100} \approx 0.75$$

To apply this probability to females, multiply the number of females $(40 + 15 + 35 + 10 = 100)$ by the marginal probability of car ownership:

$$45(0.75) = 33.75$$

989. 0.55

Marginal probability is the probability of having a certain characteristic of one variable, without regard to the other variable(s).

The marginal probability of being male is the probability that someone chosen at random from the sample is a male. It also represents the total percentage of females in the group.

To find the marginal probability of being male, divide the number of males in the sample $(40 + 15 = 55)$ by the total sample size $(40 + 15 + 35 + 10 = 100)$:

$$\frac{55}{100} = 0.55$$

990. 0.45

Marginal probability is the probability of having a certain characteristic of one variable, without regard to the other variable(s).

The marginal probability of being female is the probability that someone chosen at random from the sample is a female. It also represents the total percentage of females in the group.

To find the marginal probability of being female, divide the number of females in the sample $(35 + 10 = 45)$ by the total sample size $(40 + 15 + 35 + 10 = 100)$:

$$\frac{45}{100} = 0.45$$

991. E. Choices (A) and (D) (In this sample, more males than females own cars; in this sample, the conditional probability of car ownership is higher for females.)

In this data, more males (40) than females (35) own cars, but the conditional probability of car ownership is higher for females.

To find the conditional probability of owning a car, given male gender, divide the number of male car owners (40) by the total number of males $(15 + 40 = 55)$:

$$P(\text{car} \mid \text{male}) = \frac{40}{55} \approx 0.73$$

To find the conditional probability of owning a car, given female gender, divide the number of female car owners (35) by the total number of females $(35 + 10 = 45)$:

$$P(\text{car} \mid \text{female}) = \frac{35}{45} \approx 0.78$$

992. **E. Choices (A), (B), and (C) (replication of the survey in other locations; replication of the survey with a larger sample; replicating the survey with a nationally representative sample)**

Drawing cause-and-effect conclusions from surveys is difficult, but replicating the same survey with different samples, including larger samples, samples from different locations, and with a nationally representative sample, would all help to strengthen the argument that a relationship exists between car ownership and gender.

993. **B. a randomized clinical trial**

Although statistical evidence of a relationship between two variables isn't enough to establish causality, the randomized clinical trial design can strengthen your ability to draw cause-and-effect conclusions by minimizing bias, using sufficient sample sizes, and controlling for other variables that may affect the outcome.

994. **9**

To find the number of cells in a table, multiply the number of rows by the number of columns. With the addition of the "Inconclusive" category to each of the two variables, this table would have three rows and three columns: $(3)(3) = 9$.

995. **0.39**

Marginal probability is the probability of having a certain characteristic of one variable, without regard to the other variable(s).

To calculate the marginal probability for screening positive for depression, divide the total number of participants who screened positive for depression $(25 + 20 = 45)$ by the total sample size $(25 + 20 + 10 + 60 = 115)$:

$$\frac{45}{115} \approx 0.39$$

996. **0.30**

Marginal probability is the probability of having a certain characteristic of one variable, without regard to the other variable(s).

To calculate the marginal probability for evaluating positive for depression, divide the total number of participants who evaluated positive for depression $(25 + 10 = 35)$ by the total sample size $(25 + 20 + 10 + 60 = 115)$:

$$\frac{35}{115} \approx 0.30$$

997. 0.44

A *conditional probability* represents the percentage of individuals within a given group who have a certain characteristic. The number of individuals in the given group always goes in the denominator.

To calculate the conditional probability of evaluating negative for depression, given a positive screening result, divide the number of participants with a positive screening result and negative evaluation (20) by the total number of participants with a positive screening result ($25 + 20 = 45$):

$$\frac{20}{45} \approx 0.44$$

998. 0.14

A *conditional probability* represents the percentage of individuals within a given group who have a certain characteristic. The number of individuals in the given group always goes in the denominator.

To calculate the conditional probability of evaluating positive for depression, given a negative screening result, divide the number of participants with a negative screening result and positive evaluation (10) by the total number of participants with a negative screening result ($10 + 60 = 70$):

$$\frac{10}{70} \approx 0.14$$

999. screened negative and evaluated negative

The joint probability refers to the probability of having two or more characteristics — in this case, having a particular screening result and having a particular evaluation result.

To find the joint probabilities of these four possible outcomes, divide the number of participants with each particular combination (screening outcome and evaluation outcome) by the total sample size:

$$P(\text{screened positive and evaluated positive}) = \frac{25}{115} \approx 0.22$$

$$P(\text{screened positive and evaluated negative}) = \frac{20}{115} \approx 0.17$$

$$P(\text{screened negative and evaluated positive}) = \frac{10}{115} \approx 0.09$$

$$P(\text{screened negative and evaluated negative}) = \frac{60}{115} \approx 0.52$$

Screened negative and evaluated negative has the highest joint probability at 0.52.

Note: Because all these joint probability calculations share the total sample size as the denominator, the outcome with the highest frequency will also have the highest probability.

1000. **No, because the conditional probabilities for evaluating positive are different depending on the screening results.**

If screening results and positive evaluation for depression were independent, you'd expect the conditional probability for evaluating positive to be the same whether a person screened positive or negative. This is not the case. As you find out when you calculate the conditional probabilities, they're quite different.

To calculate the conditional probability of evaluating positive for depression, given a positive screening result, divide the number of participants with a positive screening result and positive evaluation (25) by the total number of participants with a positive screening result ($25 + 20 = 45$):

$$P(\text{evaluated positive} \mid \text{screened positive}) = \frac{25}{45} \approx 0.56$$

To calculate the conditional probability of evaluating positive for depression, given a negative screening result, divide the number of participants with a negative screening result and positive evaluation (10) by the total number of participants with a negative screening result ($10 + 60 = 70$):

$$P(\text{evaluated positive} \mid \text{screened negative}) = \frac{10}{70} \approx 0.14$$

These two conditional probabilities aren't equal, which means that the screening process has an effect on whether the person is diagnosed as being depressed. Here, you see that if someone is screened positive, he has a higher chance of being diagnosed (0.56) than if he's not screened (0.14). Because screening has an effect on the outcome of the diagnosis, screening and diagnosis aren't independent. Their outcomes are related.

1001. **E. Choices (A), (B), and (C) (The study sample was randomly selected from the population; the people doing the evaluation had no knowledge of the screening results; this study replicated an earlier study that produced similar results.)**

Although you must observe caution when drawing causal conclusions from statistical results, several factors could increase your confidence in doing so, including working with a study sample randomly selected from the population, blinding those doing the evaluation from the screening results, and replicating results of a previous study with similar results.

Appendix

Tables for Reference

E xcerpted from Statistics For Dummies, 2nd Edition, by Deborah J. Rumsey, PhD (2011, Wiley). This material is reproduced with permission of John Wiley & Sons, Inc.

This appendix includes tables for finding probabilities for three distributions used in this book: the Z–distribution (standard normal), the t–distribution, and the binomial distribution. It also includes a table listing z*–values for selected (percentage) confidence levels.

The Z-Table

Table A–1 shows less–than–or–equal–to probabilities for the Z–distribution; that is, $p(Z \leq z)$ for a given z–value. To use Table A–1, do the following:

1. Determine the z-value for your particular problem.

The z-value should have one leading digit before the decimal point (positive, negative, or zero) and two digits after the decimal point — for example, $z = 1.28$, –2.69, or 0.13.

2. Find the row of the table corresponding to the leading digit and the first digit after the decimal point.

For example, if your z-value is 1.28, look in the *1.2* row; if $z = -1.28$, look in the *–1.2* row.

3. Find the column corresponding to the second digit after the decimal point.

For example, if your z-value is 1.28 or –1.28, look in the *0.08* column.

4. Intersect the row and column from Steps 2 and 3.

This is the probability that Z is less than or equal to your z-value. In other words, you've found $p(Z \leq z)$. For example, if $z = 1.28$, you see $p(Z \leq 1.28) = 0.8997$. For $z = -1.28$, you see $p(Z \leq -1.28) = 0.1003$.

TABLE A-1   The Z-Table

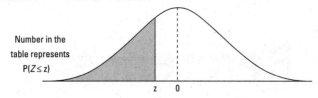

Number in the
table represents
$P(Z \le z)$

z	0.00	0.01	0.02	0.03	0.04	0.05	0.06	0.07	0.08	0.09
−3.6	.0002	.0002	.0001	.0001	.0001	.0001	.0001	.0001	.0001	.0001
−3.5	.0002	.0002	.0002	.0002	.0002	.0002	.0002	.0002	.0002	.0002
−3.4	.0003	.0003	.0003	.0003	.0003	.0003	.0003	.0003	.0003	.0002
−3.3	.0005	.0005	.0005	.0004	.0004	.0004	.0004	.0004	.0004	.0003
−3.2	.0007	.0007	.0006	.0006	.0006	.0006	.0006	.0005	.0005	.0005
−3.1	.0010	.0009	.0009	.0009	.0008	.0008	.0008	.0008	.0007	.0007
−3.0	.0013	.0013	.0013	.0012	.0012	.0011	.0011	.0011	.0010	.0010
−2.9	.0019	.0018	.0018	.0017	.0016	.0016	.0015	.0015	.0014	.0014
−2.8	.0026	.0025	.0024	.0023	.0023	.0022	.0021	.0021	.0020	.0019
−2.7	.0035	.0034	.0033	.0032	.0031	.0030	.0029	.0028	.0027	.0026
−2.6	.0047	.0045	.0044	.0043	.0041	.0040	.0039	.0038	.0037	.0036
−2.5	.0062	.0060	.0059	.0057	.0055	.0054	.0052	.0051	.0049	.0048
−2.4	.0082	.0080	.0078	.0075	.0073	.0071	.0069	.0068	.0066	.0064
−2.3	.0107	.0104	.0102	.0099	.0096	.0094	.0091	.0089	.0087	.0084
−2.2	.0139	.0136	.0132	.0129	.0125	.0122	.0119	.0116	.0113	.0110
−2.1	.0179	.0174	.0170	.0166	.0162	.0158	.0154	.0150	.0146	.0143
−2.0	.0228	.0222	.0217	.0212	.0207	.0202	.0197	.0192	.0188	.0183
−1.9	.0287	.0281	.0274	.0268	.0262	.0256	.0250	.0244	.0239	.0233
−1.8	.0359	.0351	.0344	.0336	.0329	.0322	.0314	.0307	.0301	.0294
−1.7	.0446	.0436	.0427	.0418	.0409	.0401	.0392	.0384	.0375	.0367
−1.6	.0548	.0537	.0526	.0516	.0505	.0495	.0485	.0475	.0465	.0455
−1.5	.0668	.0655	.0643	.0630	.0618	.0606	.0594	.0582	.0571	.0559
−1.4	.0808	.0793	.0778	.0764	.0749	.0735	.0721	.0708	.0694	.0681
−1.3	.0968	.0951	.0934	.0918	.0901	.0885	.0869	.0853	.0838	.0823
−1.2	.1151	.1131	.1112	.1093	.1075	.1056	.1038	.1020	.1003	.0985
−1.1	.1357	.1335	.1314	.1292	.1271	.1251	.1230	.1210	.1190	.1170
−1.0	.1587	.1562	.1539	.1515	.1492	.1469	.1446	.1423	.1401	.1379
−0.9	.1841	.1814	.1788	.1762	.1736	.1711	.1685	.1660	.1635	.1611
−0.8	.2119	.2090	.2061	.2033	.2005	.1977	.1949	.1922	.1894	.1867
−0.7	.2420	.2389	.2358	.2327	.2296	.2266	.2236	.2206	.2177	.2148
−0.6	.2743	.2709	.2676	.2643	.2611	.2578	.2546	.2514	.2483	.2451
−0.5	.3085	.3050	.3015	.2981	.2946	.2912	.2877	.2843	.2810	.2776
−0.4	.3446	.3409	.3372	.3336	.3300	.3264	.3228	.3192	.3156	.3121
−0.3	.3821	.3783	.3745	.3707	.3669	.3632	.3594	.3557	.3520	.3483
−0.2	.4207	.4168	.4129	.4090	.4052	.4013	.3974	.3936	.3897	.3859
−0.1	.4602	.4562	.4522	.4483	.4443	.4404	.4364	.4325	.4286	.4247
−0.0	.5000	.4960	.4920	.4880	.4840	.4801	.4761	.4721	.4681	.4641

© *John Wiley & Sons, Inc.*

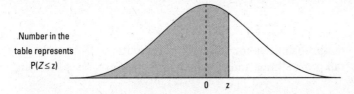

Number in the
table represents
$P(Z \le z)$

z	0.00	0.01	0.02	0.03	0.04	0.05	0.06	0.07	0.08	0.09
0.0	.5000	.5040	.5080	.5120	.5160	.5199	.5239	.5279	.5319	.5359
0.1	.5398	.5438	.5478	.5517	.5557	.5596	.5636	.5675	.5714	.5753
0.2	.5793	.5832	.5871	.5910	.5948	.5987	.6026	.6064	.6103	.6141
0.3	.6179	.6217	.6255	.6293	.6331	.6368	.6406	.6443	.6480	.6517
0.4	.6554	.6591	.6628	.6664	.6700	.6736	.6772	.6808	.6844	.6879
0.5	.6915	.6950	.6985	.7019	.7054	.7088	.7123	.7157	.7190	.7224
0.6	.7257	.7291	.7324	.7357	.7389	.7422	.7454	.7486	.7517	.7549
0.7	.7580	.7611	.7642	.7673	.7704	.7734	.7764	.7794	.7823	.7852
0.8	.7881	.7910	.7939	.7967	.7995	.8023	.8051	.8078	.8106	.8133
0.9	.8159	.8186	.8212	.8238	.8264	.8289	.8315	.8340	.8365	.8389
1.0	.8413	.8438	.8461	.8485	.8508	.8531	.8554	.8577	.8599	.8621
1.1	.8643	.8665	.8686	.8708	.8729	.8749	.8770	.8790	.8810	.8830
1.2	.8849	.8869	.8888	.8907	.8925	.8944	.8962	.8980	.8997	.9015
1.3	.9032	.9049	.9066	.9082	.9099	.9115	.9131	.9147	.9162	.9177
1.4	.9192	.9207	.9222	.9236	.9251	.9265	.9279	.9292	.9306	.9319
1.5	.9332	.9345	.9357	.9370	.9382	.9394	.9406	.9418	.9429	.9441
1.6	.9452	.9463	.9474	.9484	.9495	.9505	.9515	.9525	.9535	.9545
1.7	.9554	.9564	.9573	.9582	.9591	.9599	.9608	.9616	.9625	.9633
1.8	.9641	.9649	.9656	.9664	.9671	.9678	.9686	.9693	.9699	.9706
1.9	.9713	.9719	.9726	.9732	.9738	.9744	.9750	.9756	.9761	.9767
2.0	.9772	.9778	.9783	.9788	.9793	.9798	.9803	.9808	.9812	.9817
2.1	.9821	.9826	.9830	.9834	.9838	.9842	.9846	.9850	.9854	.9857
2.2	.9861	.9864	.9868	.9871	.9875	.9878	.9881	.9884	.9887	.9890
2.3	.9893	.9896	.9898	.9901	.9904	.9906	.9909	.9911	.9913	.9916
2.4	.9918	.9920	.9922	.9925	.9927	.9929	.9931	.9932	.9934	.9936
2.5	.9938	.9940	.9941	.9943	.9945	.9946	.9948	.9949	.9951	.9952
2.6	.9953	.9955	.9956	.9957	.9959	.9960	.9961	.9962	.9963	.9964
2.7	.9965	.9966	.9967	.9968	.9969	.9970	.9971	.9972	.9973	.9974
2.8	.9974	.9975	.9976	.9977	.9977	.9978	.9979	.9979	.9980	.9981
2.9	.9981	.9982	.9982	.9983	.9984	.9984	.9985	.9985	.9986	.9986
3.0	.9987	.9987	.9987	.9988	.9988	.9989	.9989	.9989	.9990	.9990
3.1	.9990	.9991	.9991	.9991	.9992	.9992	.9992	.9992	.9993	.9993
3.2	.9993	.9993	.9994	.9994	.9994	.9994	.9994	.9995	.9995	.9995
3.3	.9995	.9995	.9995	.9996	.9996	.9996	.9996	.9996	.9996	.9997
3.4	.9997	.9997	.9997	.9997	.9997	.9997	.9997	.9997	.9997	.9998
3.5	.9998	.9998	.9998	.9998	.9998	.9998	.9998	.9998	.9998	.9998
3.6	.9998	.9998	.9999	.9999	.9999	.9999	.9999	.9999	.9999	.9999

The t-Table

Table A-2 shows right-tail probabilities for selected t-distributions. Follow these steps to use Table A-2 to find right-tail probabilities and p-values for hypothesis tests involving t:

1. **Find the t-value for which you want the right-tail probability (call it t), and find the sample size (for example, n).**

2. **Find the row corresponding to the degrees of freedom (df) for your problem (for example, $n - 1$). Follow the row across to find the two t-values between which your t falls.**

   For example, if your t is 1.60, and your n is 7, you look in the row for $df = 7 - 1 = 6$. Across that row, you find your t lies between t-values 1.44 and 1.94.

3. **Look at the top of the columns containing the two t-values from Step 2.**

   The right-tail (greater-than) probability for your t-value is somewhere between the two values at the top of these columns. For example, your $t = 1.60$ is between t-values 1.44 and 1.94 ($df = 6$), so the right-tail probability for your t is between 0.10 (column heading for $t = 1.44$) and 0.05 (column heading for $t = 1.94$).

The row near the bottom with z in the df column gives right-tail (greater-than) probabilities from the Z-distribution.

To use Table A-2 to find $t*$-values (critical values) for a confidence interval involving t, do the following

1. **Determine the confidence level you need (as a percentage).**

2. **Determine the sample size (for example, n).**

3. **Look at the bottom row of the table where the percentages are shown. Find your % confidence level there.**

4. **Intersect this column with the row representing $n - 1$ degrees of freedom (df).**

   This is the t-value you need for your confidence interval. For example, a 95% confidence interval with $df = 6$ has $t^* = 2.45$. (Find 95% on the last line and follow it up to row 6.)

TABLE A-2 The *t*-Table

Numbers in each row of the table are values on a *t*-distribution with
(*df*) degrees of freedom for selected right-tail (greater-than) probabilities (*p*).

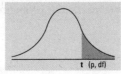

df/p	0.40	0.25	0.10	0.05	0.025	0.01	0.005	0.0005
1	0.324920	1.000000	3.077684	6.313752	12.70620	31.82052	63.65674	636.6192
2	0.288675	0.816497	1.885618	2.919986	4.30265	6.96456	9.92484	31.5991
3	0.276671	0.764892	1.637744	2.353363	3.18245	4.54070	5.84091	12.9240
4	0.270722	0.740697	1.533206	2.131847	2.77645	3.74695	4.60409	8.6103
5	0.267181	0.726687	1.475884	2.015048	2.57058	3.36493	4.03214	6.8688
6	0.264835	0.717558	1.439756	1.943180	2.44691	3.14267	3.70743	5.9588
7	0.263167	0.711142	1.414924	1.894579	2.36462	2.99795	3.49948	5.4079
8	0.261921	0.706387	1.396815	1.859548	2.30600	2.89646	3.35539	5.0413
9	0.260955	0.702722	1.383029	1.833113	2.26216	2.82144	3.24984	4.7809
10	0.260185	0.699812	1.372184	1.812461	2.22814	2.76377	3.16927	4.5869
11	0.259556	0.697445	1.363430	1.795885	2.20099	2.71808	3.10581	4.4370
12	0.259033	0.695483	1.356217	1.782288	2.17881	2.68100	3.05454	43178
13	0.258591	0.693829	1.350171	1.770933	2.16037	2.65031	3.01228	4.2208
14	0.258213	0.692417	1.345030	1.761310	2.14479	2.62449	2.97684	4.1405
15	0.257885	0.691197	1.340606	1.753050	2.13145	2.60248	2.94671	4.0728
16	0.257599	0.690132	1.336757	1.745884	2.11991	2.58349	2.92078	4.0150
17	0.257347	0.689195	1.333379	1.739607	2.10982	2.56693	2.89823	3.9651
18	0.257123	0.688364	1.330391	1.734064	2.10092	2.55238	2.87844	3.9216
19	0.256923	0.687621	1.327728	1.729133	2.09302	2.53948	2.86093	3.8834
20	0.256743	0.686954	1.325341	1.724718	2.08596	2.52798	2.84534	3.8495
21	0.256580	0.686352	1.323188	1.720743	2.07961	2.51765	2.83136	3.8193
22	0.256432	0.685805	1.321237	1.717144	2.07387	2.50832	2.81876	3.7921
23	0.256297	0.685306	1.319460	1.713872	2.06866	2.49987	2.80734	3.7676
24	0.256173	0.684850	1.317836	1.710882	2.06390	2.49216	2.79694	3.7454
25	0.256060	0.684430	1.316345	1.708141	2.05954	2.48511	2.78744	3.7251
26	0.255955	0.684043	1.314972	1.705618	2.05553	2.47863	2.77871	3.7066
27	0.255858	0.683685	1.313703	1.703288	2.05183	2.47266	2.77068	3.6896
28	0.255768	0.683353	1.312527	1.701131	2.04841	2.46714	2.76326	3.6739
29	0.255684	0.683044	1.311434	1.699127	2.04523	2.46202	2.75639	3.6594
30	0.255605	0.682756	1.310415	1.697261	2.04227	2.45726	2.75000	3.6460
z	0.253347	0.674490	1.281552	1.644854	1.95996	2.32635	2.57583	3.2905
CI	——	——	80%	90%	95%	98%	99%	99.9%

The Binomial Table

Table A–3 shows probabilities for the binomial distribution. To use Table A–3, do the following:

1. **Find these three numbers for your particular problem:**

   - The sample size, n
   - The probability of success, p
   - The x-value for which you want $p(X = x)$

2. **Find the section of Table A-3 that's devoted to your n.**

3. **Look at the row for your x-value and the column for your p.**

4. **Intersect that row and column.** You have found $p(X = x)$.

5. **To get the probability of being less than, greater than, greater than or equal to, less than or equal to, or between two values of X, you add the appropriate values of Table A-3.**

   For example, if $n = 10$, $p = 0.6$, and you want $p(X = 9)$, go to the $n = 10$ section, the $x = 9$ row, and the $p = 0.6$ column to find 0.04.

TABLE A-3 The Binomial Table

Numbers in the table represent $p(X=x)$ for a binomial distribution with n trials and probability of success p.

Binomial probabilities:

$$\binom{n}{x} p^x (1-p)^{n-x}$$

							p					
n	x	0.1	0.2	0.25	0.3	0.4	0.5	0.6	0.7	0.75	0.8	0.9
1	0	0.900	0.800	0.750	0.700	0.600	0.500	0.400	0.300	0.250	0.200	0.100
	1	0.100	0.200	0.250	0.300	0.400	0.500	0.600	0.700	0.750	0.800	0.900
2	0	0.810	0.640	0.563	0.490	0.360	0.250	0.160	0.090	0.063	0.040	0.010
	1	0.180	0.320	0.375	0.420	0.480	0.500	0.480	0.420	0.375	0.320	0.180
	2	0.010	0.040	0.063	0.090	0.160	0.250	0.360	0.490	0.563	0.640	0.810
3	0	0.729	0.512	0.422	0.343	0.216	0.125	0.064	0.027	0.016	0.008	0.001
	1	0.243	0.384	0.422	0.441	0.432	0.375	0.288	0.189	0.141	0.096	0.027
	2	0.027	0.096	0.141	0.189	0.288	0.375	0.432	0.441	0.422	0.384	0.243
	3	0.001	0.008	0.016	0.027	0.064	0.125	0.216	0.343	0.422	0.512	0.729
4	0	0.656	0.410	0.316	0.240	0.130	0.063	0.026	0.008	0.004	0.002	0.000
	1	0.292	0.410	0.422	0.412	0.346	0.250	0.154	0.076	0.047	0.026	0.004
	2	0.049	0.154	0.211	0.265	0.346	0.375	0.346	0.265	0.211	0.154	0.049
	3	0.004	0.026	0.047	0.076	0.154	0.250	0.346	0.412	0.422	0.410	0.292
	4	0.000	0.002	0.004	0.008	0.026	0.063	0.130	0.240	0.316	0.410	0.656
5	0	0.590	0.328	0.237	0.168	0.078	0.031	0.010	0.002	0.001	0.000	0.000
	1	0.328	0.410	0.396	0.360	0.259	0.156	0.077	0.028	0.015	0.006	0.000
	2	0.073	0.205	0.264	0.309	0.346	0.313	0.230	0.132	0.088	0.051	0.008
	3	0.008	0.051	0.088	0.132	0.230	0.313	0.346	0.309	0.264	0.205	0.073
	4	0.000	0.006	0.015	0.028	0.077	0.156	0.259	0.360	0.396	0.410	0.328
	5	0.000	0.000	0.001	0.002	0.010	0.031	0.078	0.168	0.237	0.328	0.590
6	0	0.531	0.262	0.178	0.118	0.047	0.016	0.004	0.001	0.000	0.000	0.000
	1	0.354	0.393	0.356	0.303	0.187	0.094	0.037	0.010	0.004	0.002	0.000
	2	0.098	0.246	0.297	0.324	0.311	0.234	0.138	0.060	0.033	0.015	0.001
	3	0.015	0.082	0.132	0.185	0.276	0.313	0.276	0.185	0.132	0.082	0.015
	4	0.001	0.015	0.033	0.060	0.138	0.234	0.311	0.324	0.297	0.246	0.098
	5	0.000	0.002	0.004	0.010	0.037	0.094	0.187	0.303	0.356	0.393	0.354
	6	0.000	0.000	0.000	0.001	0.004	0.016	0.047	0.118	0.178	0.262	0.531
7	0	0.478	0.210	0.133	0.082	0.028	0.008	0.002	0.000	0.000	0.000	0.000
	1	0.372	0.367	0.311	0.247	0.131	0.055	0.017	0.004	0.001	0.000	0.000
	2	0.124	0.275	0.311	0.318	0.261	0.164	0.077	0.025	0.012	0.004	0.000
	3	0.023	0.115	0.173	0.227	0.290	0.273	0.194	0.097	0.058	0.029	0.003
	4	0.003	0.029	0.058	0.097	0.194	0.273	0.290	0.227	0.173	0.115	0.023
	5	0.000	0.004	0.012	0.025	0.077	0.164	0.261	0.318	0.311	0.275	0.124
	6	0.000	0.000	0.001	0.004	0.017	0.055	0.131	0.247	0.311	0.367	0.372
	7	0.000	0.000	0.000	0.000	0.002	0.008	0.028	0.082	0.133	0.210	0.478

(continued)

Numbers in the table represent $p(X=x)$ for a binomial distribution with n trials and probability of success p.

Binomial probabilities:

$$\binom{n}{x} p^x (1-p)^{n-x}$$

							p					
n	x	0.1	0.2	0.25	0.3	0.4	0.5	0.6	0.7	0.75	0.8	0.9
8	0	0.430	0.168	0.100	0.058	0.017	0.004	0.001	0.000	0.000	0.000	0.000
	1	0.383	0.336	0.267	0.198	0.090	0.031	0.008	0.001	0.000	0.000	0.000
	2	0.149	0.294	0.311	0.296	0.209	0.109	0.041	0.010	0.004	0.001	0.000
	3	0.033	0.147	0.208	0.254	0.279	0.219	0.124	0.047	0.023	0.009	0.000
	4	0.005	0.046	0.087	0.136	0.232	0.273	0.232	0.136	0.087	0.046	0.005
	5	0.000	0.009	0.023	0.047	0.124	0.219	0.279	0.254	0.208	0.147	0.033
	6	0.000	0.001	0.004	0.010	0.041	0.109	0.209	0.296	0.311	0.294	0.149
	7	0.000	0.000	0.000	0.001	0.008	0.031	0.090	0.198	0.267	0.336	0.383
	8	0.000	0.000	0.000	0.000	0.001	0.004	0.017	0.058	0.100	0.168	0.430
9	0	0.387	0.134	0.075	0.040	0.010	0.002	0.000	0.000	0.000	0.000	0.000
	1	0.387	0.302	0.225	0.156	0.060	0.018	0.004	0.000	0.000	0.000	0.000
	2	0.172	0.302	0.300	0.267	0.161	0.070	0.021	0.004	0.001	0.000	0.000
	3	0.045	0.176	0.234	0.267	0.251	0.164	0.074	0.021	0.009	0.003	0.000
	4	0.007	0.066	0.117	0.172	0.251	0.246	0.167	0.074	0.039	0.017	0.001
	5	0.001	0.017	0.039	0.074	0.167	0.246	0.251	0.172	0.117	0.066	0.007
	6	0.000	0.003	0.009	0.021	0.074	0.164	0.251	0.267	0.234	0.176	0.045
	7	0.000	0.000	0.001	0.004	0.021	0.070	0.161	0.267	0.300	0.302	0.172
	8	0.000	0.000	0.000	0.000	0.004	0.018	0.060	0.156	0.225	0.302	0.387
	9	0.000	0.000	0.000	0.000	0.000	0.002	0.010	0.040	0.075	0.134	0.387
10	0	0.349	0.107	0.056	0.028	0.006	0.001	0.000	0.000	0.000	0.000	0.000
	1	0.387	0.268	0.188	0.121	0.040	0.010	0.002	0.000	0.000	0.000	0.000
	2	0.194	0.302	0.282	0.233	0.121	0.044	0.011	0.001	0.000	0.000	0.000
	3	0.057	0.201	0.250	0.267	0.215	0.117	0.042	0.009	0.003	0.001	0.000
	4	0.011	0.088	0.146	0.200	0.251	0.205	0.111	0.037	0.016	0.006	0.000
	5	0.001	0.026	0.058	0.103	0.201	0.246	0.201	0.103	0.058	0.026	0.001
	6	0.000	0.006	0.016	0.037	0.111	0.205	0.251	0.200	0.146	0.088	0.011
	7	0.000	0.001	0.003	0.009	0.042	0.117	0.215	0.267	0.250	0.201	0.057
	8	0.000	0.000	0.000	0.001	0.011	0.044	0.121	0.233	0.282	0.302	0.194
	9	0.000	0.000	0.000	0.000	0.002	0.010	0.040	0.121	0.188	0.268	0.387
	10	0.000	0.000	0.000	0.000	0.000	0.001	0.006	0.028	0.056	0.107	0.349
11	0	0.314	0.086	0.042	0.020	0.004	0.000	0.000	0.000	0.000	0.000	0.000
	1	0.384	0.236	0.155	0.093	0.027	0.005	0.001	0.000	0.000	0.000	0.000
	2	0.213	0.295	0.258	0.200	0.089	0.027	0.005	0.001	0.000	0.000	0.000
	3	0.071	0.221	0.258	0.257	0.177	0.081	0.023	0.004	0.001	0.000	0.000
	4	0.016	0.111	0.172	0.220	0.236	0.161	0.070	0.017	0.006	0.002	0.000
	5	0.002	0.039	0.080	0.132	0.221	0.226	0.147	0.057	0.027	0.010	0.000
	6	0.000	0.010	0.027	0.057	0.147	0.226	0.221	0.132	0.080	0.039	0.002
	7	0.000	0.002	0.006	0.017	0.070	0.161	0.236	0.220	0.172	0.111	0.016
	8	0.000	0.000	0.001	0.004	0.023	0.081	0.177	0.257	0.258	0.221	0.071
	9	0.000	0.000	0.000	0.001	0.005	0.027	0.089	0.200	0.258	0.295	0.213
	10	0.000	0.000	0.000	0.000	0.001	0.005	0.027	0.093	0.155	0.236	0.384
	11	0.000	0.000	0.000	0.000	0.000	0.000	0.004	0.020	0.042	0.086	0.314

© John Wiley & Sons, Inc.

Numbers in the table represent $p(X=x)$ for a binomial distribution with n trials and probability of success p.

Binomial probabilities:

$$\binom{n}{x} p^x(1-p)^{\,n-x}$$

| | | | | | | | p | | | | | |
n	x	0.1	0.2	0.25	0.3	0.4	0.5	0.6	0.7	0.75	0.8	0.9
12	0	0.282	0.069	0.032	0.014	0.002	0.000	0.000	0.000	0.000	0.000	0.000
	1	0.377	0.206	0.127	0.071	0.017	0.003	0.000	0.000	0.000	0.000	0.000
	2	0.230	0.283	0.232	0.168	0.064	0.016	0.002	0.000	0.000	0.000	0.000
	3	0.085	0.236	0.258	0.240	0.142	0.054	0.012	0.001	0.000	0.000	0.000
	4	0.021	0.133	0.194	0.231	0.213	0.121	0.042	0.008	0.002	0.001	0.000
	5	0.004	0.053	0.103	0.158	0.227	0.193	0.101	0.029	0.011	0.003	0.000
	6	0.000	0.016	0.040	0.079	0.177	0.226	0.177	0.079	0.040	0.016	0.000
	7	0.000	0.003	0.011	0.029	0.101	0.193	0.227	0.158	0.103	0.053	0.004
	8	0.000	0.001	0.002	0.008	0.042	0.121	0.213	0.231	0.194	0.133	0.021
	9	0.000	0.000	0.000	0.001	0.012	0.054	0.142	0.240	0.258	0.236	0.085
	10	0.000	0.000	0.000	0.000	0.002	0.016	0.064	0.168	0.232	0.283	0.230
	11	0.000	0.000	0.000	0.000	0.000	0.003	0.017	0.071	0.127	0.206	0.377
	12	0.000	0.000	0.000	0.000	0.000	0.000	0.002	0.014	0.032	0.069	0.282
13	0	0.254	0.055	0.024	0.010	0.001	0.000	0.000	0.000	0.000	0.000	0.000
	1	0.367	0.179	0.103	0.054	0.011	0.002	0.000	0.000	0.000	0.000	0.000
	2	0.245	0.268	0.206	0.139	0.045	0.010	0.001	0.000	0.000	0.000	0.000
	3	0.100	0.246	0.252	0.218	0.111	0.035	0.006	0.001	0.000	0.000	0.000
	4	0.028	0.154	0.210	0.234	0.184	0.087	0.024	0.003	0.001	0.000	0.000
	5	0.006	0.069	0.126	0.180	0.221	0.157	0.066	0.014	0.005	0.001	0.000
	6	0.001	0.023	0.056	0.103	0.197	0.209	0.131	0.044	0.019	0.006	0.000
	7	0.000	0.006	0.019	0.044	0.131	0.209	0.197	0.103	0.056	0.023	0.001
	8	0.000	0.001	0.005	0.014	0.066	0.157	0.221	0.180	0.126	0.069	0.006
	9	0.000	0.000	0.001	0.003	0.024	0.087	0.184	0.234	0.210	0.154	0.028
	10	0.000	0.000	0.000	0.001	0.006	0.035	0.111	0.218	0.252	0.246	0.100
	11	0.000	0.000	0.000	0.000	0.001	0.010	0.045	0.139	0.206	0.268	0.245
	12	0.000	0.000	0.000	0.000	0.000	0.002	0.011	0.054	0.103	0.179	0.367
	13	0.000	0.000	0.000	0.000	0.000	0.000	0.001	0.010	0.024	0.055	0.254
14	0	0.229	0.044	0.018	0.007	0.001	0.000	0.000	0.000	0.000	0.000	0.000
	1	0.356	0.154	0.083	0.041	0.007	0.001	0.000	0.000	0.000	0.000	0.000
	2	0.257	0.250	0.180	0.113	0.032	0.006	0.001	0.000	0.000	0.000	0.000
	3	0.114	0.250	0.240	0.194	0.085	0.022	0.003	0.000	0.000	0.000	0.000
	4	0.035	0.172	0.220	0.229	0.155	0.061	0.014	0.001	0.000	0.000	0.000
	5	0.008	0.086	0.147	0.196	0.207	0.122	0.041	0.007	0.002	0.000	0.000
	6	0.001	0.032	0.073	0.126	0.207	0.183	0.092	0.023	0.008	0.002	0.000
	7	0.000	0.009	0.028	0.062	0.157	0.209	0.157	0.062	0.028	0.009	0.000
	8	0.000	0.002	0.008	0.023	0.092	0.183	0.207	0.126	0.073	0.032	0.001
	9	0.000	0.000	0.002	0.007	0.041	0.122	0.207	0.196	0.147	0.086	0.008
	10	0.000	0.000	0.000	0.001	0.014	0.061	0.155	0.229	0.220	0.172	0.035
	11	0.000	0.000	0.000	0.000	0.003	0.022	0.085	0.194	0.240	0.250	0.114
	12	0.000	0.000	0.000	0.000	0.001	0.006	0.032	0.113	0.180	0.250	0.257
	13	0.000	0.000	0.000	0.000	0.000	0.001	0.007	0.041	0.083	0.154	0.356
	14	0.000	0.000	0.000	0.000	0.000	0.000	0.001	0.007	0.018	0.044	0.229

(continued)

Numbers in the table represent $p(X=x)$ for a binomial
distribution with *n* trials and probability of success *p*.

Binomial probabilities:

$$\binom{n}{x} p^x (1-p)^{n-x}$$

| | | | | | | | | *p* | | | | | |
n	x	0.1	0.2	0.25	0.3	0.4	0.5	0.6	0.7	0.75	0.8	0.9
15	0	0.206	0.035	0.013	0.005	0.000	0.000	0.000	0.000	0.000	0.000	0.000
	1	0.343	0.132	0.067	0.031	0.005	0.000	0.000	0.000	0.000	0.000	0.000
	2	0.267	0.231	0.156	0.092	0.022	0.003	0.000	0.000	0.000	0.000	0.000
	3	0.129	0.250	0.225	0.170	0.063	0.014	0.002	0.000	0.000	0.000	0.000
	4	0.043	0.188	0.225	0.219	0.127	0.042	0.007	0.001	0.000	0.000	0.000
	5	0.010	0.103	0.165	0.206	0.186	0.092	0.024	0.003	0.001	0.000	0.000
	6	0.002	0.043	0.092	0.147	0.207	0.153	0.061	0.012	0.003	0.001	0.000
	7	0.000	0.014	0.039	0.081	0.177	0.196	0.118	0.035	0.013	0.003	0.000
	8	0.000	0.003	0.013	0.035	0.118	0.196	0.177	0.081	0.039	0.014	0.000
	9	0.000	0.001	0.003	0.012	0.061	0.153	0.207	0.147	0.092	0.043	0.002
	10	0.000	0.000	0.001	0.003	0.024	0.092	0.186	0.206	0.165	0.103	0.010
	11	0.000	0.000	0.000	0.001	0.007	0.042	0.127	0.219	0.225	0.188	0.043
	12	0.000	0.000	0.000	0.000	0.002	0.014	0.063	0.170	0.225	0.250	0.129
	13	0.000	0.000	0.000	0.000	0.000	0.003	0.022	0.092	0.156	0.231	0.267
	14	0.000	0.000	0.000	0.000	0.000	0.000	0.005	0.031	0.067	0.132	0.343
	15	0.000	0.000	0.000	0.000	0.000	0.000	0.000	0.005	0.013	0.035	0.206
20	0	0.122	0.012	0.003	0.001	0.000	0.000	0.000	0.000	0.000	0.000	0.000
	1	0.270	0.058	0.021	0.007	0.000	0.000	0.000	0.000	0.000	0.000	0.000
	2	0.285	0.137	0.067	0.028	0.003	0.000	0.000	0.000	0.000	0.000	0.000
	3	0.190	0.205	0.134	0.072	0.012	0.001	0.000	0.000	0.000	0.000	0.000
	4	0.090	0.218	0.190	0.130	0.035	0.005	0.000	0.000	0.000	0.000	0.000
	5	0.032	0.175	0.202	0.179	0.075	0.015	0.001	0.000	0.000	0.000	0.000
	6	0.009	0.109	0.169	0.192	0.124	0.037	0.005	0.000	0.000	0.000	0.000
	7	0.002	0.055	0.112	0.164	0.166	0.074	0.015	0.001	0.000	0.000	0.000
	8	0.000	0.022	0.061	0.114	0.180	0.120	0.035	0.004	0.001	0.000	0.000
	9	0.000	0.007	0.027	0.065	0.160	0.160	0.071	0.012	0.003	0.000	0.000
	10	0.000	0.002	0.010	0.031	0.117	0.176	0.117	0.031	0.010	0.002	0.000
	11	0.000	0.000	0.003	0.012	0.071	0.160	0.160	0.065	0.027	0.007	0.000
	12	0.000	0.000	0.001	0.004	0.035	0.120	0.180	0.114	0.061	0.022	0.000
	13	0.000	0.000	0.000	0.001	0.015	0.074	0.166	0.164	0.112	0.055	0.002
	14	0.000	0.000	0.000	0.000	0.005	0.037	0.124	0.192	0.169	0.109	0.009
	15	0.000	0.000	0.000	0.000	0.001	0.015	0.075	0.179	0.202	0.175	0.032
	16	0.000	0.000	0.000	0.000	0.000	0.005	0.035	0.130	0.190	0.218	0.090
	17	0.000	0.000	0.000	0.000	0.000	0.001	0.012	0.072	0.134	0.205	0.190
	18	0.000	0.000	0.000	0.000	0.000	0.000	0.003	0.028	0.067	0.137	0.285
	19	0.000	0.000	0.000	0.000	0.000	0.000	0.000	0.007	0.021	0.058	0.270
	20	0.000	0.000	0.000	0.000	0.000	0.000	0.000	0.001	0.003	0.012	0.122

z*-Values for Selected Confidence Levels

Table A-4 gives you the particular $z*$-value that you need to get the confidence level (also known as percentage confidence) you want when you're calculating two types of confidence intervals in this book:

>> Confidence intervals for a population mean where the population standard deviation σ is known

>> Confidence intervals for a population proportion where the two conditions are met to use the normal approximation

- $n\hat{p} \geq 10$
- $n(1-\hat{p}) \geq 10$

Note: You don't use Table A-4 if you're calculating confidence intervals for a population mean when the population standard deviation, σ, is unknown. For this type of confidence interval, you use the t-table (Table A-2).

For the two appropriate scenarios in the preceding list, some of the more commonly used confidence levels, along with their corresponding $z*$-values, are in Table A-4. Here is how you find the $z*$-value you need:

1. **Determine the confidence level needed for the confidence interval you're doing (this is typically given in the problem).**

   Find the row pertaining to this confidence level. For example, you may be asked to find a 95% confidence interval for the mean. In that case, the confidence level is 95%, so look in that row.

2. **Find the corresponding $z*$-value in the second column of that same row in the table.**

   For example, for a 95% confidence level, the $z*$-value is 1.96.

3. **Take the $z*$-value from the table and plug it into the appropriate confidence interval formula you need.**

Note: To find a $z*$-value for a confidence level that isn't included in Table A-4, you use the Z-table (Table A-1) with a modification. The Z-table shows the z-value corresponding to the percentage *below* a number. For a confidence interval, you want a $z*$-value corresponding to the percentage *between* two numbers. To modify the Z-table to find what you need, take your original between percentage (confidence level) and convert it to a *less-than* percentage. Do this by taking your original percentage (confidence level) and adding half of what remains when you subtract it from 1. Look up this new percentage in the body of the table, and see what z-value it belongs to in the matching row/column of the table. That's the $z*$-value you use in your appropriate confidence interval formula.

For example, a 95% confidence level means the between probability is 95%, so the less-than probability is 95% plus 2.5% (half of what's left), or 97.5%. Look up 0.975 in the body of the Z-table and find $z^* = 1.96$ for a 95% confidence level.

TABLE A-4 z*-Values for Selected (Percentage) Confidence Levels

Percentage Confidence	z*-Value
80	1.28
90	1.645
95	1.96
98	2.33
99	2.58

Illustration by Ryan Sneed

Index

R

random sample, 127, 362–363
random variables
  binomial, 37–38, 41–42, 204
  continuous, 196, 219
  defined, 2, 256
  discrete, 34, 196
  probability distribution of, 35
  questions about, 36–37
  sampling distributions, 59
range of data set, 194
reference tables
  binomial table, 468–472
  t-table, 466–467
  z^*-values, 473–474
  Z-table, 463–465
registering for online access, 4
regression, and relationship, 139
regression equation, 440
regression line
  finding y-intercept for, 141, 426–427
  formula for, 422
  questions about, 140
  slope of, 141, 431
  variables and, 141
relationships
  correlation and, 425
  between mean and median, 186
  overview, 2
  regression and, 139
  weak, 417
relative standing, 16
research designs, 163
response bias, 128, 414
Rumsey, Deborah J., 4, 463

S

sample
  bias in, 300, 302, 414
  checking size for z-test, 388
  convenience, 127
  defined, 167
  fractional sizes, 314
  large, 263–264

from normal population, 263
  questions about, 8
  random, 127, 362–363
  replicating, 459
  selecting for surveys, 128–130
  standard deviation, 172–174
  variance, 179
  volunteer, 127
sample mean
  normal population and, 263
  population mean and, 266
  probability for, 66–68
  questions about, 59, 62–64, 66
sample proportions
  calculating, 411
  calculating standard error for, 276
  converting to z-value, 283
  formula for standard error, 327–328
  matching to z-scores, 69
  population proportion and, 327, 334
  questions about, 68
sample size
  calculating needed, 314–317
  confidence intervals and, 86–87
  margin of error and, 73
  small, 95–96
sampling distributions
  central limit theorem versus, 57
  defined, 256
  notation, 60–61
  overview, 57
  questions about, 58–59
  sample mean and, 62–64
  standard error, 59–60, 61–62
  symbols, 60–61
sampling error, 297
sampling frame, 127
scatter plots
  questions about, 132–134, 136–137
  regression line and, 140
  relationships displayed, 416
self-selected sample, 127
σ symbol, 205, 259
significance level, 362
skewed left/right, 172, 182, 187, 190, 192

Publisher's Acknowledgments

Acquisitions Editor: Lindsay Lefevere

Project Editor: Rick Kughen

Copy Editor: Rick Kughen

Production Editor: Mohammed Zafar Ali

Project Manager: Michelle Hacker

Cover Image: Monitor © spaxiax/Adobe Stock